OXFORD READINGS IN

THE PHILOSOPHY OF MATHEMATICS

Also published in this series

The Problem of Evil, edited by Marilyn McCord Adams and Robert Merrihew Adams
The Philosophy of Artificial Intelligence, edited by Margaret A. Boden
The Philosophy of Artificial Life, edited by Margaret A. Boden
Self-Knowledge, edited by Quassim Cassam
Perceptual Knowledge, edited by Jonathan Dancy
The Philosophy of Law, edited by Ronald M. Dworkin
Environmental Ethics, edited by Robert Elliot
Theories of Ethics, edited by Philippa Foot
The Philosophy of History, edited by Patrick Gardiner
The Philosophy of Mind, edited by Jonathan Glover
Scientific Revolutions, edited by Ian Hacking
The Philosophy of Mathematics, edited by W. D. Hart
Conditionals, edited by Frank Jackson
The Philosophy of Time, edited by Robin Le Poidevin and Murray MacBeath
The Philosophy of Religion, edited by Basil Mitchell
Meaning and Reference, edited by A. W. Moore
A Priori Knowledge, edited by Paul K. Moser
Political Philosophy, edited by Anthony Quinton
Explanation, edited by David-Hillel Ruben
The Philosophy of Social Explanation, edited by Alan Ryan
Propositions and Attitudes, edited by Nathan Salmon and Scott Soames
Consequentialism and its Critics, edited by Samuel Scheffler
Applied Ethics, edited by Peter Singer
Causation, edited by Ernest Sosa and Michael Tooley
Theories of Rights, edited by Jeremy Waldron
Free Will, edited by Gary Watson
Demonstratives, edited by Palle Yourgrau

Other volumes are in preparation

THE PHILOSOPHY OF MATHEMATICS

edited by

W. D. HART

OXFORD UNIVERSITY PRESS

1996

Oxford University Press, Walton Street, Oxford OX2 6DP
Oxford New York
Athens Auckland Bangkok Bombay
Calcutta Cape Town Dar es Salaam Delhi
Florence Hong Kong Istanbul Karachi
Kuala Lumpur Madras Madrid Melbourne
Mexico City Nairobi Paris Singapore
Taipei Tokyo Toronto
and associated companies in
Berlin Ibadan

Oxford is a trade mark of Oxford University Press

Published in the United States
by Oxford University Press Inc., New York

Introduction and selection ©
Oxford University Press 1996

All rights reserved. No part of this publication may be reproduced,
stored in a retrieval system, or transmitted, in any form or by any means,
without the prior permission in writing of Oxford University Press.
Within the UK, exceptions are allowed in respect of any fair dealing for the
purpose of research or private study, or criticism or review, as permitted
under the Copyright, Designs and Patents Act, 1988, or in the case of
reprographic reproduction in accordance with the terms of the licences
issued by the Copyright Licensing Agency. Enquiries concerning
reproduction outside these terms and in other countries should be
sent to the Rights Department, Oxford University Press,
at the address above

This book is sold subject to the condition that is shall not, by way
of trade or otherwise, be lent, re-sold, hired out or otherwise circulated
without the publisher's prior consent in any form of binding or cover
other than that in which it is published and without a similar condition
including this condition being imposed on the subsequent purchaser

British Library Cataloguing in Publication Data
Data available

Library of Congress Cataloguing in Publication Data
The philosophy of mathematics / edited by W. D. Hart.
p. cm.—(Oxford readings in philosophy)
Includes bibliographical references and index.
1. Mathematics—Philosophy. I. Hart, W. D. (Wilbur Dyre), 1943–
II. Series.
QA8.4.P478 1996   510'.1 –dc20   95–49208
ISBN 0–19–875119–2
ISBN 0–19–875120–6 (Pbk)

1 3 5 7 9 10 8 6 4 2

Typeset by Alliance Phototypesetters
Printed in Great Britain
on acid-free paper by
Bookcraft (Bath) Ltd., Midsomer Norton, Somerset

CONTENTS

	Introduction *W. D. Hart*	1
I.	Mathematical Truth *Paul Benacerraf*	14
II.	Two Dogmas of Empiricism *W. V. Quine*	31
III.	Access and Inference *W. D. Hart*	52
IV.	The Philosophical Basis of Intuitionistic Logic *Michael Dummett*	63
V.	Mathematical Intuition *Charles Parsons*	95
VI.	Perception and Mathematical Intuition *Penelope Maddy*	114
VII.	Truth and Proof: The Platonism of Mathematics *W. W. Tait*	142
VIII.	Mathematics without Foundations *Hilary Putnam*	168
IX.	The Consistency of Frege's *Foundations of Arithmetic* *George Boolos*	185
X.	Arithmetical Truth and Hidden Higher-Order Concepts *Daniel Isaacson*	203
XI.	Conservativeness and Incompleteness *Stewart Shapiro*	225
XII.	Is Mathematical Knowledge Just Logical Knowledge? *Hartry Field*	235
XIII.	The Structuralist View of Mathematical Objects *Charles Parsons*	272

Notes on the Contributors	310
Suggestions for Further Reading	311
Index	315

INTRODUCTION

W. D. HART

Mathematics has attracted philosophical reflection since Plato, and most of the great philosophers since then have had at least something to say about it. That is not because of the importance of mathematics in human life. Agriculture is at least as important, but there is no philosophy of farming. It is instead distinctive and articulable philosophical problems that have attracted philosophical attention to mathematics.

Paul Benacerraf's 'Mathematical Truth' (Chapter I) articulates a philosophical dilemma that has drawn considerable attention to mathematics in recent decades. The first horn of his dilemma starts from truth. It is true, for example, that someone stabbed Caesar. That it is true depends, in part, on the man Julius Caesar. So whatever it is that has the property of truth should have some sort of segmentation into bits, and in our example one of these bits should be pertinent to Caesar. Sentences wear their segmentations into words on their inscribed faces, and our name 'Caesar' for Caesar is one word in the sentence 'Someone stabbed Caesar'. It is not so easy to segment propositions, thoughts, or statements without reading back into them the segmentations of the sentences used to express them. So let us take truth in our example as a property of the sentence 'Someone stabbed Caesar'.

How shall we account for the truth of this sentence? Note a few banal facts. First, the name 'Caesar' denotes the man Caesar. The verb 'stabbed' is transitive, and requires a subject and an object. Let us use the letters 'x' and 'y' to mark the blanks flanking the verb for its subject and object thus: x stabbed y. Call these letters variables. They are *not* names at all. A name purports to denote a unique thing, but a variable has values, such as all the objects denoted by names that could be substituted for the variables flanking our verb, but also anything else in the offing, whether named or not. Our second banal fact is that 'x stabbed y' is true of a value of 'x' and a value of 'y' if and only if that value of 'x' stabbed that value of 'y'. Names may supplant variables. That is how we get the intransitive verb phrase 'x stabbed Caesar', which is true of a value of 'x' if and only if that value stabbed the denotation of 'Caesar'. Variables may also be bound, as it is put, by quantifiers, which are expressions like 'everything' and 'something' that pertain to how extensively a predicate is true of things. That is how we get the sentence 'Someone stabbed Caesar'. Our third banal fact is that this sentence is true if and only if there is a value of the variable 'x' which is a person of whom the phrase 'x stabbed Caesar' is true. Putting our three facts together, it follows that the sentence 'Someone stabbed Caesar' is true if and only if some person is a value of 'x' and stabbed Caesar. Since any person is a value of any variable, it follows that our sentence is true if and only if someone stabbed Caesar. We can derive a similarly banal, and thus agreeable, result for just about any of the infinitely many sentences generable from a suitable finite stock of simple predicates like 'stabbed' and names like 'Caesar' using quantifiers and truth-functional connectives like 'not', 'and', and 'or'. So our miniature illustrates many ideas central to Tarski's account of truth.[1] Tarski himself assimilated his account to the traditional correspondence theory of truth. But note that Tarski needs no facts to which true sentences as wholes correspond. Instead, his conception of the relations between words and things at the centre of truth focuses on things denoted by names and things, as values of variables, of which predicates are true. So we might say that in Tarski's account, the old slogan that truth is correspondence to fact reduces to the idea that truth requires reference (denotation by names, values of variables) to objects.

It is an ancient and honourable view that mathematics is a body of truths, and indeed that the known truths of mathematics are the most absolute and unqualified truths known most certainly by us. Euclid's *Elements* contains a lovely proof that there are infinitely many prime numbers. Proofs establish truth, so 'There are infinitely many prime numbers' is true. But truth requires reference to objects, so there are objects (namely numbers, infinitely many of which are prime) of which 'is prime' is true. Just as truths of history require the existence of historical figures like Caesar, so truths of mathematics require the existence of prime numbers like 13.

This result already begins to stick in many people's metaphysical craws. Some might hope to allow for reference to objects in accounting for historical truth, but somehow to avoid it in accounting for mathematical truth. That hope runs risks. For example, one banal fact which Tarski used is that the conjunction by 'and' of two sentences A and B is true if and only if A is true

---

[1] Tarski's basic paper is 'The Concept of Truth in Formalized Languages', in *Logic, Semantics, and Metamathematics*, trans. J. H. Woodger (New York: Oxford University Press, 1956), 152–278. He presents his ideas more popularly in 'The Semantic Conception of Truth and the Foundations of Semantics', *Journal of Philosophy and Phenomenological Research*, 4 (1944): 341–75; repr. in H. Feigl and W. Sellars (eds.), *Readings in Philosophical Analysis* (New York: Appleton-Century-Crofts, 1949), and L. Linsky (ed.), *Semantics and the Philosophy of Language* (Urbana: University of Illinois Press, 1952). For another exposition, see Mark Platts, *Ways of Meaning* (London: Routledge and Kegan Paul, 1979).

and B is true. Suppose A comes from mathematics and B from history. If we separate varieties of truth too sharply, we look to risk turning Tarski's banal fact into a hopeless equivocation. More generally, we seem to depend on a unitary account of truth in building up a unified system of the single world that exists from the sum of our separate pursuits of knowledge.

None the less, metaphysical anxiety about mathematical objects like numbers seems only to increase when we reflect on the objectivity we expect of mathematical truths. Objectivity seems to require independent objects. Not only are there infinitely many primes, but also, since Euclid's proof makes no reference to living creatures, there would have been infinitely many primes even if life had never evolved. So the objects required by the truth of his theorem cannot be mental. Nor does its truth depend on the existence of this or that material thing; so it would seem that the objects required by its truth are not physical either. So mathematical truth seems to require the existence of objects which are neither physical nor mental but, as they are often called, abstract. This platonism, the doctrine that there are independent abstract objects, will stick in many metaphysical craws.[2]

But why? What is worrisome about abstract objects? This brings us to the second horn of Benacerraf's dilemma. Mathematics is not just a body of truths; it is also a body of knowledge. We seem happiest in accounting for how a person knows, for example, that an object is a certain way when we can make out some sort of transaction, commerce, or connection between the person and the object that issues in him being justified in what he believes about the object that makes his belief true. This formula is vague or metaphorical or over-general; the locus of difficulty lies in the terms 'transaction', 'commerce', or 'connection'. But most of us seem to regard sense perception as the only generally acceptable mode of such transaction between people who know and the objects that make what they know true. Perception may be embedded in textures of testimony and reasoning; but it seems that perception is our preferred basic mode of contact with the objects required for the truth of our knowledge. Even Plato, in the allegory of the cave, and Descartes, in the metaphor of the light of reason, seem to adopt perceptual pictures of knowledge.

Could there be perception of the abstract objects required for mathematical truth? What is it to be abstract anyway? Perhaps physics is the place to look to learn from the laws of matter what it is to be physical; likewise, perhaps mathematics is the place to look to learn from the laws of numbers, sets, functions, and so forth what it is to be abstract. But maybe there are a few

---

[2] Crispin Wright attempts a more anodyne platonism in *Frege's Conception of Numbers as Objects* (Aberdeen: Aberdeen University Press, 1983). For more on platonism, see Penelope Maddy, *Realism in Mathematics* (Oxford: Clarendon Press, 1990).

general, negative things we can say about abstract objects. Numbers, unlike stones, do not seem to be anywhere; very abstract objects, like numbers but unlike geometrical figures, lack spatial location. More to the present point, numbers do not reflect light, nor do they bump into anything. Numbers would seem to be utterly inert, utterly immune to the causal processes of nature.

But Paul Grice argues convincingly that perception is, by its nature, causal.[3] So if knowledge requires commerce with the objects known, and if, say because the favoured form of such commerce is perceptual, such connections should be causal, then it would seem impossible that there should be knowledge of objects as utterly inert as very abstract objects like numbers would be. That is the problem: what seems necessary for mathematical truth seems to make mathematical knowledge impossible.[4]

My exposition of the second horn of Benacerraf's dilemma differs from his. Where he emphasized the causal theories of knowledge proposed to answer Gettier,[5] I have emphasized perception. So in my exposition, the dilemma is a conflict between platonism, the ontology of abstract objects required for mathematical truth, and empiricism, the view that perception is basic to knowledge of objects. Even so, the second horn of his dilemma remains epistemological in my exposition. But perhaps it has a metaphysical generalization. Traditionally, metaphysics had two branches: ontology asked after the basic kinds of things that there are, whereas cosmology, before physicists hijacked the word, asked how the elements fit into the system of the world. Our inarticulate cosmological consensus seems to be what used to be called naturalist; our presupposition seems to be that causation is, in Hume's phrase, the cement of the universe.[6] For example, a dualist about minds and bodies should, we think, paste them together causally into human beings. But if there really are utterly inert numbers, as well as the matter (and perhaps minds too) bound up in the causal nexus of nature, then there is also a conflict between platonism and the naturalism of our conventional cosmology.

---

[3] H. P. Grice, 'The Causal Theory of Perception', *Proceedings of the Aristotelian Society*, suppl. vol. 35 (1961); repr. in part in R. J. Swartz (ed.), *Perceiving, Sensing and Knowing* (Garden City, NY: Anchor, 1965). His reasoning is summarized in W. D. Hart, *The Engines of the Soul* (Cambridge: Cambridge University Press, 1988), 51–5.

[4] There is an attempt to undercut Benacerraf's dilemma in Mark Steiner's *Mathematical Knowledge* (Ithaca, NY: Cornell University Press, 1975). This is discussed in the review of Steiner's book in *Journal of Philosophy*, 74 (Feb. 1977): 118–29, to which Steiner replied in 'Mathematics, Explanation, and Scientific Knowledge', *Noûs*, 12 (Mar. 1978): 17–28.

[5] Edmund Gettier, 'Is Justified True Belief Knowledge?', *Analysis*, 23 (June 1963): 121–3; and Alvin I. Goldman, 'A Causal Theory of Knowing', *Journal of Philosophy*, 22 (June 1967): 357–72.

[6] David Hume, *A Treatise of Human Nature*, ed. L. A. Selby-Bigge, 2nd edn., ed. P. H. Nidditch (Oxford: Clarendon Press, 1978), 662.

Benacerraf's dilemma is not the only philosophical problem about mathematics, but it is certainly basic to metaphysical and epistemological concerns about mathematics. The dilemma gives us a perspective from which to organize many, especially contemporary, philosophical discussions of mathematics. For if the dilemma is as real as it seems, and if the ontology of platonism is incompatible with the epistemology of empiricism outlined above, then consistency demands that at least one horn of the dilemma yield. So one question to ask about an essay on the dilemma is which horn it seeks to blunt, and how.

In this regard, Quine's 'Two Dogmas of Empiricism' (Chapter II) is seminal. For much of our century, analytical philosophers have tried to argue that the truths of mathematics are analytic. Consider, for example, the now trite example 'All bachelors are unmarried'. To claim that this sentence is analytic means that it is not about bachelors but true by virtue of the meanings of the words like 'bachelor' and 'unmarried' in it. So viewed, analyticity seemed like a species of truth without reference, and so to offer mathematical truth without platonism. Or one might think that 'All bachelors are unmarried' is as much about worldly bachelors as 'All bachelors are sexually frustrated'; but suppose that while knowledge of the second would require experience of bachelors, the first can be known just from understanding its meaning without examining bachelors. So viewed, analyticity seemed like a species of knowledge without perception, and so to offer mathematical knowledge without empiricism. The first half of Quine's essay is a very influential critique of analyticity, and so might seem to reinforce the dilemma.

But the second half contains, among other riches, an epistemological contribution *avant la lettre* toward solving the dilemma. If knowledge always requires experience of the subject-matter of the truth known, then knowledge of scientific theory would seem no less problematic than mathematical knowledge. No one has ever seen, or will ever see, a dinosaur. What fascinated us as children in science museums were not dinosaur bones but fossils, which are bone-shaped rocks which sometimes occur in groups that can be assembled into what look like skeletons, though of no animal whose hide anyone ever patted or will pat. These rocks we explain by supposing that there were once huge animals, some of whom died and fell into mud, where ground water leached out the calcium, replacing it with other minerals, which, under aeons of heat and pressure, petrified. Fossils are evidence for dinosaurs because we need the hypothesis of dinosaurs in order best to explain (some) of the fossils we observe. The direction of confirmation is the reverse of that of explanation; when we need to suppose A in order best to explain B, observation of B is good evidence for A. C. S. Peirce, who first articulated this principle, originally called it 'abduction', and later, the 'method of hypothesis'. When Gilbert

Harman rediscovered it, he called it 'inference to the best explanation'.[7] While it may not be the whole story of non-demonstrative reasoning, it certainly seems like a hero of that tale. Note especially that it allows for justified belief in objects one needs to suppose in order best to explain what one observes; so, since it seems to articulate reasoning in empirical science well, it can be taken as an empiricist mode of justifying belief in objects that one does not, and even cannot, perceive.

In 'Two Dogmas' Quine takes something like abduction from Pierre Duhem and Rudolf Carnap. To this he adds the reflection that sophisticated natural science seems always to have its mathematical aspect. In other words, sophisticated natural science as it comes is always to be formalized as an extension, in the logician's sense, of some mathematics, often number theory and typically analysis. Equations are obvious to anyone reading serious science. So, by abduction, we are justified in believing to be true at least as much mathematics as we need for the best scientific explanations of what we observe. Since the truth of that much mathematics requires very abstract objects, Quine thereby began an empiricist justification for belief in the abstract objects required for mathematical truth. Numbers, like electrons and dinosaurs, are, on this sort of view, theoretical posits. Hilary Putnam's little book *Philosophy of Logic*[8] is an excellent treatment of Quine's construction, which is also developed and defended in my 'Access and Inference' (Chapter III).

There is a price to be paid for buying into Quine's epistemological construction, however. Tarski's account of truth is inductive in the mathematical sense, at least initially. This means, roughly, that it requires an ordering of sentences by something like grammatical complexity. Conditions for the truth of simple sentences are stated outright, and conditions for the truth of more complex sentences (like conjunctions) may be stated in terms of those for other sentences, but only less complex sentences (like the conjuncts) lower in the ordering. A foundations conception of knowledge might have expected it to be constructed along the lines of truth. Known sentences simple in the order for truth would be known by perception directly, while known sentences of greater complexity would be known by inference from known sentences of lesser complexity. But on Quine's epistemology no such homology between truth and knowledge is guaranteed. For truth retains the inductive structure which Tarski discerned, while there is no guaranteed uniform order of explanation and evidence; so knowledge, on Quine's view, is much less

---

[7] On Peirce, see Ian Hacking, *The Taming of Chance* (Cambridge: Cambridge University Press, 1990), 207. For Harman, see his 'Inference to the Best Explanation', *Philosophical Review*, 74 (1965): 88–95.

[8] New York: Harper, 1971. One response provoked by Quine's argument was Charles S. Chihara's *Ontology and the Vicious Circle Principle* (Ithaca, NY: Cornell University Press, 1973). For his later views, see *Constructibility and Mathematical Existence* (Oxford: Clarendon Press, 1990).

uniformly ordered than truth. Explanations come from relatively large, intuitively heterogeneous, and overlapping bodies of hypotheses on which nature and truth impose no uniform order. Quine's epistemology is holistic.

Can knowledge stick closer to its objects than holism allows? For our purposes, an exciting aspect of Penelope Maddy's 'Perception and Mathematical Intuition' (Chapter VI) is her suggestion that we see, know, and refer to at least some sets, even if perception, knowledge, and reference require causal interaction with their objects. In making this suggestion, Maddy drives such sets (centrally, those of nearby visible physical objects) closer to material things. One can worry about whether her drive is too fast and loose to deliver the goods. But suppose she can, after all, see some sets that reflect light into her eyes. Has Maddy then driven those sets so far from the sets of pure mathematics that she has added a peculiar chapter to familiar empirical knowledge without touching mathematical knowledge at all? Or can the wall between the concrete and the abstract be broken down?

But it would seem that for many contemporary philosophers, platonism is anathema. Such a philosopher might expect epistemological or cosmological comforts, but Benacerraf's dilemma should lead us to expect him to exert his metaphysical creativity on truth in mathematics. A fairly extreme position of this sort would involve simply rejecting truth in mathematics as conceived above. Michael Dummett's reworking of intuitionism (Chapter IV) starts from a radical critique of truth of this sort. To see how this is supposed to work, we should think for a bit about meaning. No speaker of a language is infallible; it is always possible for a speaker to misunderstand something he says or hears. It seems natural to describe such misunderstanding in terms of a difference between what his words mean and what he means by his words, and this description makes sense only if we distinguish between an expression's meaning and what a speaker means by an expression. Meaning is, then, in a sense, public, rather than private to individual psyches. Among expressions, it might also seem natural to think of sentences as made up of words, and so to ask first after the meanings of words, expecting some sort of semantic chemistry to yield from them molecules for sentences to mean. If we do, however, it seems difficult to avoid thinking of word meanings as ideas or as what speakers mean by those words at least conventionally, standardly, or as a rule. Even if we could eliminate this suspicious mention of rules or standards or conventions, we would still seem to have run roughshod over our distinction between expression and speaker meaning. At this point, Davidson and Dummett argue that Frege reversed the priority between word and sentence meaning for semantic theory.[9] Taking indicative sentences as basic, the

---

[9] On Davidson, see his 'Truth and Meaning', *Synthese*, 17 (1967): 304–23; repr. in *Inquiries into Truth and Interpretation* (Oxford, 1984). For Dummett, see his *Frege: Philosophy of Language*

idea is that Frege understood a sentence's meaning to be its truth conditions, and then a word's meaning is whatever has to be associated with it in order to account for the truth conditions of sentences in which it occurs. For example, 'Il pluit' means that it is raining because the sentence is true exactly in case it is raining, and rain should be associated with 'pluit' because rain is the stuff that figures in the conditions for sentences containing 'pluit' to be true.

But, Dummett emphasizes, language is used for communication, and just about everybody comes to be able to speak and understand his mother tongue. So, since understanding is knowledge of meaning, whatever meaning turns out to be, it should be knowable by just about everyone. Just about everyone can recognize rain, and can acquire, exhibit, and recognize mastery of a sentence like 'It is raining'. But truth as classically conceived allows for the possibility of sentences utterly immune to all our best possible efforts to settle whether they are true or false. The meaning of a sentence for which we cannot say in advance even what would settle it one way or the other eludes, Dummett thinks, that secure epistemic grasp of public meaning necessary for communication. There is, then, Dummett thinks, a tension between verification-transcendent truth and the public nature of meaning required for communication.

Dummett chooses to resolve this tension in favour of the acquisition and manifestation of meaning by substituting assertibility for truth conditions in Frege's schematic account of meaning; so Dummett revives a kind of verifiability account of meaning.[10] But, unlike the earlier logical positivists, Dummett's verificationism is utterly general, and so extends to logic and mathematics, where proof is the paradigmatic mode of verification. For example, to be entitled to assert that there is an object of a certain sort, it is not on this view enough that there be an object of which one is entitled to assert that it is of that sort; one should be able to find, or at least exhibit a method for finding, an object and a proof that it is of the specified sort. Where that object lies among an infinity of its kind, one is not in general guaranteed that a search among its kind will yield either an example or a finite proof that none is to be found. So, on this sort of view, one is not in general entitled to assert that an object of a certain sort exists just because one can deduce a contradiction by classical logic from the supposition that all are not of that sort. This illustrates how intuitionistic logic is more demanding than classical

---

(London: Duckworth, 1973); *The Interpretation of Frege's Philosophy* (London: Duckworth, 1981); and *Frege and Other Philosophers* (Oxford: Clarendon Press, 1991).

[10] Michael Dummett, 'What is a Theory of Meaning?', in W. Guttenplan (ed.), *Mind and Language* (Oxford: Clarendon Press, 1975); and 'What is a Theory of Meaning? (II)', in Evans and McDowell (eds.), *Truth and Meaning* (Oxford: Clarendon Press, 1976); both repr. in *The Seas of Language* (Oxford: Clarendon Press, 1993), 1–93.

logic; and, partly for that reason, intuitionistic mathematics can look like a fragment of classical mathematics.[11] There is a price to be paid in classical mathematics for abandoning classical truth on Dummett's reworking of intuitionism.

But earlier intuitionists, from Brouwer on, did not just pick and choose intuitionistically acceptable results from the wares of classical mathematicians. They also reinterpreted mathematics. Brouwer once said that a mathematician's mathematical assertions should be taken as reports of occurrences in his mind. Dummett, hewing to what he calls the extrusion of private thought into public language, does not make much of this aspect of earlier intuitionism. But an advantage which this sort of psychologism seems to offer is that it might allow one to extend the certainty and independence of perception of external objects from knowledge of, say, one's blinding headache to one's logical and mathematical knowledge. Brouwer got this aspect of his intuitionism from Kant, and this is probably responsible for his use of 'intuition' and its cognates. Kant's view was part of his transcendental idealism, and intuitionism can be seen as a sort of mathematical idealism. One difficulty with such an idealism is the threat it seems to pose to the independence from our opinions apparently required by the objectivity of mathematics. Charles Parsons's 'Mathematical Intuition' (Chapter V) is a first-rate contemporary examination of this positive aspect of intuitionism.

Dummett rejects proof-transcendent mathematical truth. In a different way, so does W. W. Tait in his influential 'Truth and Proof: The Platonism of Mathematics' (Chapter VII). But where Dummett seems epistemologically motivated, Tait might be seen to be more motivated by cosmology, or anticosmology. The Platonist looks to accept a single, all-embracing category of objects; the laws of absolutely all objects might be just the laws of logic. He then divides this category in two or three: physical objects, perhaps minds if these are not just brains ticking over, and abstract objects like numbers. His cosmological questions have in part to do with how any two of these sorts of objects can be unified into the system of the single world. But suppose that there simply is no single, all-embracing category of objects. Yes, there are numbers, and the known theorems of number theory give some of their laws. Yes, there are thoughtful people and good ideas, and maybe one day we will work out a systematic psychology. Yes, there are objects, like galaxies and glaciers, quartz crystals and maybe quarks, and various of the natural sciences give some of their laws. But a philosophical insistence on a unified cosmology of 'objects', whether abstract, mental, or really physical, is in vain, and the separate disciplines are answerable only to themselves except in so far

---

[11] There is a masterful survey of intuitionism in Michael Dummett's *Elements of Intuitionism* (Oxford: Clarendon Press, 1977).

as we can get more inclusive disciplines up and running, which philosophical cosmologists have signally failed to do. Tait may not always be easy to understand, but read in this light, he may be following a lead of the later Wittgenstein, as Dummett sees himself following another such lead ('meaning is use').

In 'Arithmetical Truth and Hidden Higher-Order Concepts' (Chapter X), Daniel Isaacson discusses mathematics and mathematical logic that is somewhat hairier than that described elsewhere in this volume; but the less mathematical reader can get from his paper a description of recent technical work, motives to master it, and a philosophically intriguing response to the theorems of our century, beginning with Gödel, on the limited capacity of consistent axiomatic proof to incorporate all mathematical truth. That may be why Isaacson's essay has sustained interest among specialists. Isaacson argues that in so far as number-theoretic claims can be fully understood just in terms of their intended subject-matter without recourse to extraneous 'higher-order' syntactical notions like that of a formal derivation, then the so-called incompleteness results of our time do not bear out the view that number-theoretic truth outruns provability in Peano arithmetic. If such a view could be worked out and established for all of mathematics, then perhaps proof could be made to yield a conception of mathematical truth alternative to the platonist's conception of truth as requiring reference to objects.

In 'Mathematics without Foundations' (Chapter VIII), Hilary Putnam also tries to get away from the idea that truth in mathematics requires reference to special, abstract mathematical objects. There is an old idea that actual abstract objects either are or should be traded in for possibilities of concrete objects. Instead of saying, for example, that there is an actual infinity of abstract natural numbers, we should say that there could be an infinity of concrete objects (like inscriptions of the numerals in the standard order) which, if actual, would be a model for number theory. We need possibility here, because there are not, and never will be, enough actual inscriptions of numerals to supplant the infinity of natural numbers; so, while actual inscriptions are visible and thus have a comfortable epistemology, the infinite here pushes us beyond the actual and perceptible, and thereby seems to raise an analogue for modality of Benacerraf's original dilemma for abstract objects: what commerce between us and the independent though mere possibility of an infinity of inscriptions of numerals could justify our belief in such a possibility?[12] Putnam's original purpose was not so much to replace platonism with a kind of 'modalism' as to argue from the viability of both approaches that mathematics needs no foundations. But his influential essay inspired a

---

[12] For more on the modal (and other) analogue of Benacerraf's dilemma, see W. D. Hart, 'The Price of Possibility', *Pacific Philosophical Quarterly*, Sept. 1989, 225–39.

number of philosophers to investigate modal treatments of mathematics.[13] Set-theoretically sophisticated readers of Putnam's essay should be aware that he was assuming that for each cardinal number, even infinite, there could be a space no less concrete than ours but whose dimensions have that cardinal. Here one sees a connection between modalism and geometry to which we shall return below. But for now, observe that a platonist might wonder whether, in order to account for the objectivity of mathematical truth, possibilities of concrete objects should answer to something (like possible worlds, perhaps) independent of our convictions, and so whether modalism can really avoid problems like those raised for platonism by Benacerraf's dilemma.

Much recent philosophical discussion of mathematics has centred around Hartry Field's book *Science without Numbers*.[14] There Field took the view that Quine's abductive construction is the only good reason to believe in abstract objects like numbers. So, if he can show that Quine's indispensability thesis is false and that good physical science has no need of abstract mathematics, then the case for platonism collapses, and nominalism (the view that everything is concrete) is vindicated. His argument against Quine's indispensability thesis has two aspects. At the beginning and the end of his book, the argument is logical. Suppose we could split natural science in which mathematics is applied into two parts, one in which only concrete things are mentioned, the other all the rest. Field argues that any purely nominalist conclusion that can be obtained from the sum of the two can be obtained from the first alone, so we do not need mathematics to do natural science. There is an excellent examination of the logic involved here in Stewart Shapiro's 'Conservativeness and Incompleteness' (Chapter XI).[15] In the central core of his book, Field argues that we can secure the required factorization. The crux of the argument is that numbers, functions, and sets from mathematics used in natural science may be replaced there by geometrical points, lines, and regions of space. Here one might wonder whether geometrical objects are concrete enough to vindicate a full-blooded nominalism. Field said in a footnote that in modern physics, space-time is no less involved in causal processes than massive particles, so space-time and its bits are nominalistically acceptable. If true, his reply might ameliorate some cosmological

---

[13] The philosophical term of art 'modality' arises thus. In English, possibility and necessity are often expressed using the verb fragments 'can' and 'must'. Grammarians call these words 'modal auxiliaries', partly after the subjunctive mood, which is often used in other languages to express possibility and necessity less periphrastically. It is these auxiliaries from which philosophers borrowed their names for modalities.

[14] Oxford: Blackwell, 1980.

[15] Field replies to Shapiro in 'On Conservativeness and Incompleteness', repr. in Hartry Field, *Realism, Mathematics and Modality* (Oxford: Blackwell, 1989).

doubts about platonism, but it seems inadequate to Benacerraf's original epistemological considerations: for one no more sees dimensionless points or breadthless lines, let alone their continuity, than one sees numbers. Later on, Field seems to have traded in at least some geometry for extensions of logic into a species of modality. He sets out this view in 'Is Mathematical Knowledge Just Logical Knowledge?' (Chapter XII), an essay in which he happily acknowledges his debt to Putnam.

The general idea that mathematics is logic elaborated is coeval with analytic philosophy. Kant claimed that logic is analytic, which may have been plausible for the impoverished logic of his day. But the mathematics of space and number Kant thought synthetic, yet necessary and known a priori. Einstein's abandonment of Euclid has made us less sure than Kant about geometry. But the mathematics of number includes existence claims—for example, that there are infinitely many primes—and at least since part IX of Hume's *Dialogues concerning Natural Religion*, it has seemed hard to believe that any existence claims are analytic. Consider too the perfect numbers— that is, numbers like 6 that are equal to the sums of the lesser numbers that divide them, namely, 1, 2, and 3 in the case of 6. The next perfect number is 28, and the next, 496. The Greeks knew a routine for describing all and only the even perfect numbers, of which there are infinitely many. You now understand the claim that there is an odd perfect number, for you now know what the only novel term in it, 'perfect', means. So if analytic claims are such that knowledge of their meaning suffices to establish them, you should know whether there is an odd perfect number. But nobody knows the answer to this ancient unsolved problem. So it is hard to believe that mathematics is analytic. None the less, Frege began analytic philosophy when he set out to refute Kant by showing that mathematics is reducible to logic, for then, granted that logic is analytic, it should follow that mathematics is too. (Yet it is hard not to read Frege as a platonist too; and his logic had considerable ontological commitments.) But to reduce mathematics to logic, Frege beefed up logic, and, as Russell discovered, Frege's beefed-up logic was inconsistent. It is not analytically true; it is false. But even if mathematics is not just an elaboration of ontologically non-committal logic plus definitions, Frege's achievement still seems considerable. George Boolos's 'The Consistency of Frege's *Foundations of Arithmetic*' (Chapter IX) sets out to describe and evaluate what can be saved from what Boolos calls Frege's 'profound analysis of arithmetic' in second-order logic—that is, logic in which we may quantify into predicate positions in sentences, as well as subject, or singular-term, positions.

Structuralism is another perennial way of trying to avoid commitment to mathematical objects. For example, early in his *Introduction to Mathematical Philosophy*, Russell toyed with the idea that it makes no difference which objects play the roles of 0, 1, 2, and so forth in the sequence of natural numbers; all that matters for mathematical purposes is the structure of any such sequence.[16] Charles Parsons provides a masterful survey of contemporary versions of such ideas in 'The Structuralist View of Mathematical Objects' (Chapter XIII).

The essays in this volume follow recent discussions in emphasizing Benacerraf's dilemma and metaphysical and epistemological issues close to it. But there are other issues about mathematics that have also elicited philosophical comment. For example, mathematics and logic seem special among the sciences in that, as the consensus pretty much has it, all mathematical and logical truths are necessarily true, whereas some of the laws of nature are contingent, or at least not so necessary. Then, too, mathematicians have no laboratories, and make no field trips to examine, say, unusual prime numbers. So, whereas natural science is justified by experience, pure mathematics does not seem to be, but rather is known, as Kant put it, a priori. (The claim that mathematics is known a priori can be seen as a rejection of the empiricism implicit in the epistemological horn of Benacerraf's dilemma, but that just points the horn toward the question of how mathematics can be justified nonempirically.) The issues of necessity and a priori knowledge in mathematics have not gone away.[17] But the issue of platonism has been a central focus of recent discussions.

One trusts that none of the authors represented here regards all his words as final, absolute, and definitive; philosophy is not like that. Each contribution is offered as an invitation to an area of philosophy no less lively today than it was in ancient Greece.

---

[16] For an interesting combination of modalism and structuralism, see Geoffrey Hellman's *Mathematics without Numbers: Towards a Modal Structural Interpretation* (Oxford: Clarendon Press, 1989).

[17] On the a priori, see Philip Kitcher's *The Nature of Mathematical Knowledge* (Oxford: Oxford University Press, 1984).

I

MATHEMATICAL TRUTH

PAUL BENACERRAF

Although this symposium is entitled 'Mathematical Truth', I will also discuss issues which are somewhat broader but which nevertheless have the notion of mathematical truth at their core, which themselves depend on how truth in mathematics is properly explained. The most important of these is mathematical knowledge. It is my contention that two quite distinct kinds of concerns have separately motivated accounts of the nature of mathematical truth: (1) the concern for having a homogeneous semantical theory in which semantics for the propositions of mathematics parallel the semantics for the rest of the language,[1] and (2) the concern that the account of mathematical truth mesh with a reasonable epistemology. It will be my general thesis that almost all accounts of the concept of mathematical truth can be identified with serving one or another of these masters *at the expense of the other*. Since I believe further that both concerns must be met by any adequate account, I find myself deeply dissatisfied with any package of semantics and epistemology that purports to account for truth and knowledge both within and outside of mathematics. For, as I will suggest, accounts of truth that treat mathematical and non-mathematical discourse in relevantly similar ways do so at the cost of leaving it unintelligible how we can have any mathematical knowledge whatsoever; whereas those which attribute to mathematical propositions the kinds of truth conditions we can clearly know to obtain, do so at the expense of failing to connect these conditions with any analysis of the sentences which shows how the assigned conditions are conditions of their *truth*. What this means must ultimately be spelled out in some detail if I am to make out my case, and I cannot hope to do that within this limited context. But I will try to make it sufficiently clear to permit you to judge whether or not there is likely to be anything in the claim.

First published in *Journal of Philosophy*, 70/19 (8 Nov. 1973): 661–79. Reproduced by permission of the *Journal of Philosophy* and Paul Benacerraf.

[1] I am indulging here in the fiction that we *have* semantics for 'the rest of language', or, more precisely, that the proponents of the views that take their impetus from this concern often think of themselves as having such semantics, at least for philosophically important segments of the language.

MATHEMATICAL TRUTH

I take it to be obvious that any philosophically satisfactory account of truth, reference, meaning, and knowledge must embrace them all, and must be adequate for all the propositions to which these concepts apply.[2] An account of knowledge that *seems* to work for certain empirical propositions about medium-sized physical objects but which fails to account for more theoretical knowledge is unsatisfactory—not only because it is incomplete, but because it may be incorrect as well, even as an account of the things it seems to cover quite adequately. To think otherwise would be, among other things, to ignore the interdependence of our knowledge in different areas. And similarly for accounts of truth and reference. A theory of truth for the language we speak, argue in, theorize in, mathematize in, etc., should by the same token provide similar truth conditions for similar sentences. The truth conditions assigned to two sentences containing quantifiers should reflect in relevantly similar ways the contribution made by the quantifiers. Any departure from a theory thus homogeneous would have to be strongly motivated to be worth considering. Such a departure, for example, might manifest itself in a theory that gave an account of the contribution of quantifiers in mathematical reasoning different from that in normal everyday reasoning about pencils, elephants, and vice-presidents. David Hilbert urged such an account in 'On the Infinite',[3] which is discussed briefly below. Later on, I will try to say more about what conditions I would expect a satisfactory general theory of truth for our language to meet, as well as more about how such an account is to mesh with what I take to be a reasonable account of knowledge. Suffice it to say here that, although it will often be convenient to present my discussion in terms of theories of mathematical truth, we should always bear in mind that what is really at issue is our over-all philosophical view. I will argue that, *as an over-all view*, it is unsatisfactory—not so much because we lack a seemingly satisfactory account of mathematical truth or because we lack a seemingly satisfactory account of mathematical knowledge—as because we lack any account that satisfactorily brings the two together. I hope that it is possible ultimately to produce such an account; I hope further that this paper will help to bring one about by bringing into sharper focus some of the obstacles that stand in its way.

---

[2] I shall in fact have nothing to say about meaning in this paper. I believe that the concept is in much deserved disrepute, but I don't dismiss it for all that. Recent work, most notably by Kripke, suggests that what passed for a long time for meaning—namely, the Fregean 'sense'—has less to do with truth than Frege or his immediate followers thought it had. Reference is what is presumably most closely connected with truth, and it is for *this* reason that I will limit my attention to reference. If it is granted that change of reference can take place without a corresponding change in meaning, and that truth is a matter of reference, then talk of meaning is largely beside the point of the cluster of problems that concern us in this paper. These comments are not meant as arguments, but only as explanation.

[3] Trans. and repr. in Paul Benacerraf and Hilary Putnam (eds.), *Philosophy of Mathematics*, 2nd edn. (Cambridge: Cambridge University Press, 1983), 183–201.

1. TWO KINDS OF ACCOUNT

Consider the following two sentences:

(1)   There are at least three large cities older than New York.
(2)   There are at least three perfect numbers greater than 17.

Do they have the same logico-grammatical form? More specifically, are they both of the form

(3)   There are at least three *FG*s that bear *R* to *a*,

where 'There are at least three' is a numerical quantifier eliminable in the usual way in favour of existential quantifiers, variables, and identity; '*F*' and '*G*' are to be replaced by one-place predicates, '*R*' by a two-place predicate, and '*a*' by the name of an element of the universe of discourse of the quantifiers? What are the truth conditions of (1) and (2)? Are they relevantly parallel? Let us ignore both the vagueness of 'large' and 'older than' and the peculiarities of attributive-adjective constructions in English which make a large city not something large and a city but more (although not exactly) like something large *for* a city. With those complications set aside, it seems clear that (3) accurately reflects the form of (1), and thus that (1) will be true if and only if the thing named by the expression replacing '*a*' ('New York') bears the relation designated by the expression replacing '*R*' ('① is older then ②') to at least three elements (of the domain of discourse of the quantifiers) which satisfy the predicates replacing '*F*' and '*G*' ('large' and 'city', respectively). This, I gather, is what a suitable truth definition would tell us. And I think it's right. Thus, if (1) is true, it is because certain cities stand in a certain relation to each other, etc.

But what of (2)? May we use (3) in the same way as a matrix in spelling out the conditions of *its* truth? That sounds like a silly question, to which the obvious answer is 'Of course'. Yet the history of the subject (the philosophy of mathematics) has seen many other answers. Some (including one of my past and present selves[4]), reluctant to face the consequences of combining what I shall dub such a 'standard' semantical account with a platonistic view of the nature of numbers, have shied away from supposing that numerals are names and thus, by implication, that (2) is of the form (3). David Hilbert (see n. 3) chose a different but equally divergent approach, in his case in an attempt to arrive at a satisfactory account of the use of the notion of infinity in mathematics. On one construal, Hilbert can be seen as segregating a class of statements and methods, those of 'intuitive' mathematics, as those which needed no further justification. Let us suppose that these are all 'finitely verifiable' in

---

[4] See my 'What Numbers Could Not Be', *Philosophical Review*, 74/1 (Jan. 1965): 47–73.

some sense that is not precisely specified. Statements of arithmetic that do not share this property—typically, certain statements containing quantifiers—are seen by Hilbert as instrumental devices for going from 'real' or 'finitely verifiable' statements to 'real' statements, much as an instrumentalist regards theories in natural science as a way of going from observation sentences to observation sentences. These mathematically 'theoretical' statements Hilbert called 'ideal elements', likening their introduction to the introduction of points 'at infinity' in projective geometry: they are introduced as a convenience to make simpler and more elegant the theory of the things you really care about. If their introduction does not lead to contradiction, and if they have these other uses, then it is justified: hence the search for a consistency proof for the full system of first-order arithmetic.

If this is a reasonable, if sketchy, account of Hilbert's view, it indicates that he did not regard all quantified statements semantically on a par with one another. A semantics for arithmetic as he viewed it would be very hard to give. But hard or not, it would certainly not treat the quantifier in (2) in the same way as the quantifier in (1). Hilbert's view as outlined represents a flat denial that (3) is the model according to which (2) is constructed.

On other such accounts, the truth conditions for arithmetic sentences are given as their formal derivability from specified sets of axioms. When coupled with the desire to attribute a truth value to each closed sentence of arithmetic, these views were torpedoed by the incompleteness theorems. They could be restored at least to internal consistency either by the liberalization of what counts as derivability (e.g., by including the application of an ω-rule in permissible derivations) or by abandoning the desire for completeness. For lack of a better term and because they almost invariably key on the syntactic (combinatorial) features of sentences, I will call such views 'combinatorial' views of the determinants of mathematical truth. The leading idea of combinatorial views is that of assigning truth values to arithmetic sentences on the basis of certain (usually proof-theoretic) syntactic facts about them. Often, truth is defined as (formal) derivability from certain axioms. (Frequently a more modest claim is made—the claim to truth-in-S, where S is the particular system in question.) In any event, in such cases truth is conspicuously not explained in terms of reference, denotation, or satisfaction. The 'truth' predicate is syntactically defined.

Similarly, certain views of truth in arithmetic on which the Peano axioms are claimed to be 'analytic' of the concept of number are also 'combinatorial' in my sense. And so are conventionalist accounts, since what marks them as conventionalist is the contrast between them and the 'realist' account that analyses (2) by assimilating it to (1), via (3).

Finally, to make one further distinction, a view is not automatically 'combinatorial' if it interprets mathematical propositions as being about combinatorial matters, either self-referentially or otherwise. For such a view might analyse mathematical propositions in a 'standard' way in terms of the names and quantifiers they might contain and in terms of the properties they ascribe to the objects within their domains of discourse—which is to say that the underlying concept of *truth* is essentially Tarski's. The difference is that its proponents, although realists in their analysis of mathematical language, part ways with the Platonists by construing the mathematical universe as consisting exclusively of mathematically unorthodox objects: mathematics for them is limited to metamathematics, and that to syntax.

I will defer to later sections my assessment of the relative merits of these various approaches to the truth of such sentences as (2). At this point I wish only to introduce the distinction between, on the one hand, those views which attribute the obvious syntax (and the obvious semantics) to mathematical statements, and, on the other, those which, ignoring the apparent syntax and semantics, attempt to state truth conditions (or to specify and account for the existing distribution of truth values) on the basis of what are evidently non-semantic syntactic considerations. Ultimately I will argue that each kind of account has its merits and defects: each addresses itself to an important component of a coherent over-all philosophic account of truth and knowledge.

But what are these components, and how do they relate to one another?

II. TWO CONDITIONS

A. The first component of such an over-all view is more directly concerned with the concept of truth. For present purposes we can state it as the requirement that there be an over-all theory of truth in terms of which it can be certified that the account of mathematical truth is indeed an account of mathematical *truth*. The account should imply truth conditions for mathematical propositions that are evidently conditions of their truth (and not *simply*, say, of their theoremhood in some formal system). This is not to *deny* that being a theorem of some system can be a truth condition for a given proposition or class of propositions. It is rather to require that any theory that proffers theoremhood as a condition of truth also *explain the connection between truth and theoremhood*.

Another way of putting this first requirement is to demand that any theory of mathematical truth be in conformity with a general theory of truth—a theory of truth theories, if you like—which certifies that the property of sentences that the account calls 'truth' is indeed truth. This, it seems to me, can be done only on the basis of some general theory for at least the language as a

MATHEMATICAL TRUTH

whole (I assume that we skirt paradoxes in some suitable fashion). Perhaps the applicability of this requirement to the present case amounts only to a plea that the semantical apparatus of mathematics be seen as part and parcel of that of the natural language in which it is done, and thus that whatever *semantical* account we are inclined to give of names or, more generally, of singular terms, predicates, and quantifiers in the mother tongue include those parts of the mother tongue which we classify as mathematese.

I suggest that, if we are to meet this requirement, we shouldn't be satisfied with an account that fails to treat (1) and (2) in parallel fashion, on the model of (3). There may well be *differences*, but I expect these to emerge at the level of the analysis of the reference of the singular terms and predicates. I take it that we have only one such account: Tarski's, and that its essential feature is to define truth in terms of reference (or satisfaction) on the basis of a particular kind of syntactico-semantical analysis of the language, and thus that any putative analysis of mathematical truth must be an analysis of a concept which is a truth concept at least in Tarski's sense. Suitably elaborated, I believe this requirement to be inconsistent with all the accounts that I have termed 'combinatorial'. On the other hand, the account that assimilates (2) above to (1) and (3) obviously meets this condition, as do many variants of it.

B. My second condition on an over-all view presupposes that we have mathematical knowledge and that such knowledge is no less knowledge for being mathematical. Since our knowledge is of truths, or can be so construed, an account of mathematical truth, to be acceptable, must be consistent with the possibility of having mathematical knowledge: the conditions of the truth of mathematical propositions cannot make it impossible for us to know that they are satisfied. This is not to argue that there cannot be unknowable truths—only that not all truths can be unknowable, for we know some. The minimal requirement, then, is that a satisfactory account of mathematical truth must be consistent with the possibility that some such truths be knowable. To put it more strongly, the concept of mathematical truth, as explicated, must fit into an over-all account of knowledge in a way that makes it intelligible how we have the mathematical knowledge that we have. An acceptable semantics for mathematics must fit an acceptable epistemology. For example, if I know that Cleveland is between New York and Chicago, it is because there exists a certain relation between the truth conditions for that statement and my present 'subjective' state of belief (whatever may be our accounts of truth and knowledge, they must connect with each other in this way). Similarly, in mathematics, it must be possible to link up what it is for p to be true with my belief that p. Though this is extremely vague, I think one can see how the second condition tends to rule out accounts that satisfy the first, and to admit many of those which do not. For a typical 'standard' account (at least in the case of number theory or set theory) will depict truth conditions in terms of conditions on objects whose nature, as normally conceived, places them beyond the reach of the better understood means of human cognition (e.g., sense perception and the like). The 'combinatorial' accounts, on the other hand, usually arise from a sensitivity to precisely this fact and are hence almost always motivated by epistemological concerns. Their virtue lies in providing an account of mathematical propositions based on the procedures we follow in justifying truth claims in mathematics: namely, proof. It is not surprising that *modulo* such accounts of mathematical truth, there is little mystery about how we can obtain mathematical knowledge. We need only account for our ability to produce and survey formal proofs.[5] However, squeezing the balloon at that point apparently makes it bulge on the side of truth: the more nicely we tie up the concept of proof, the more closely we link the definition of proof to combinatorial (rather than semantic) features, the more difficult it is to connect it up with the truth of what is being thus 'proved'—or so it would appear.

These, then, are the two requirements. Separately, they seem innocuous enough. In the balance of this paper I will both defend them further and flesh out the argument that jointly they seem to rule out almost every account of mathematical truth that has been proposed. I will consider in turn the two basic approaches to mathematical truth that I mentioned above, weighing their relative advantages in light of the two fundamental principles that I am advancing. I hope that the principles themselves will receive some illumination and support as I do so.

III. THE STANDARD VIEW

I call the 'platonistic' account that analyses (2) as being of the form (3) 'the standard view'. Its virtues are many, and it is worth enumerating them in some detail before passing to a consideration of its defects.

As I have already pointed out, this account assimilates the logical form of mathematical propositions to that of apparently similar empirical ones: empirical and mathematical propositions alike contain predicates, singular terms, quantifiers, etc.

But what of sentences that are not composed (or correctly analysable as being composed) of names, predicates, and quantifiers? More directly to the

---

[5] Properly done, this is of course an enormous task. Nevertheless, it sets to one side accounting for the burden that is borne by the semantics of the system and by our understanding of it, concentrating instead on our ability to determine that certain formal objects have certain syntactically defined properties.

point, what of sentences that do not belong to the kind of language for which Tarski has showed us how to define truth? I would say that we need for such languages (if there are any) an account of truth of the sort that Tarski supplied for 'referential' languages. I assume that the truth conditions for the language (e.g., English) to which mathematese appears to belong are to be elaborated much along the lines that Tarski articulated. So, to some extent, the question posed in the previous section—how are truth conditions for (2) to be explained?—may be interpreted as asking whether the sublanguage of English in which mathematics is done is to receive the same sort of analysis as I am assuming is appropriate for much of the rest of English. If so, then the qualms I shall sketch in the next section concerning how to fit mathematical knowledge into an over-all epistemology clearly apply—though they can perhaps be laid to rest by a suitable modification of theory. If, on the other hand, mathematese is not to be analysed along referential lines, then we are clearly in need not only of an account of truth (i.e., a semantics) for this new kind of language, but also of a new *theory of truth theories* that relates truth for referential (quantificational) languages to truth for these new (newly analysed) languages. Given such an account, the task of accounting for mathematical knowledge would still remain; but it would presumably be an easier task, since the new semantical picture of mathematese would in most cases have been prompted by epistemological considerations. However, I do not give this alternative serious consideration in this paper because I don't think that anyone has ever actually chosen it. For to choose it is explicitly to consider *and reject* the 'standard' interpretation of mathematical language, despite its superficial and initial plausibility, and then to provide an alternative semantics as a substitute.[6] The 'combinatorial' theorists whom I discuss or refer to have usually wanted to have their cake and eat it too: they have not realized that the truth conditions that their account supplies for mathematical language have not been connected to the referential semantics which they assume is *also* appropriate for that language. Perhaps the closest candidate for an exception is Hilbert in the view I sketched briefly in the opening pages of this paper. But to pursue this further here would take us too far afield. Let us return, therefore, to our praise of the 'standard view'.

One of its primary advantages is that the truth definitions for individual mathematical theories thus construed will have the same recursion clauses as those employed for their less lofty empirical cousins. Or, to put it another way, they can all be taken as parts of the same language, for which we provide a single account for quantifiers, regardless of the subdiscipline under

---

[6] I sometimes think this is one of the things that Hilary Putnam wants to do in his stimulating article 'Mathematics without Foundations', *Journal of Philosophy*, 64/1 (19 Jan. 1967): 5–22; reproduced here as Ch. VIII.

consideration. Mathematical and empirical disciplines will not be distinguished in point of logical grammar. I have already underscored the importance of this advantage: it means that the logico-grammatical theory we employ in less recondite and more tractable domains will serve us well here. We can do with one, uniform account, and need not invent another for mathematics. This should hold true on virtually any grammatical theory coupled with semantics adequate to account for truth. My bias for what I call a Tarskian theory stems simply from the fact that he has given us the only viable systematic general account we have of truth. So, one consequence of the economy attending the standard view is that logical relations are subject to uniform treatment: they are invariant with subject-matter. Indeed, they help define the concept of 'subject-matter'. The same rules of inference may be used and their use accounted for by the same theory which provides us with our ordinary account of inference, thus avoiding a double standard. If we reject the standard view, mathematical inference will need a new and special account. As it is, standard uses of quantifier inferences are justified by some sort of soundness proof. The formalization of theories in first-order logic requires for *its* justification the assurance (provided by the completeness theorem) that all the logical consequences of the postulates will be forthcoming as theorems. The standard account delivers these guarantees. The obvious answers seem to work. To reject the standard view is to discard these answers. New ones would have to be found.

So much for the obvious virtues of this account. What are its faults?

As I suggested above, the principal defect of the standard account is that it appears to violate the requirement that our account of mathematical truth be susceptible to integration into our over-all account of knowledge. Quite obviously, to make out a persuasive case to this effect, it would be necessary to sketch the epistemology I take to be at least roughly correct and on the basis of which mathematical truths, standardly construed, do not seem to constitute knowledge. This would require a lengthy detour through the general problems of epistemology. I will leave that to another time, and content myself here with presenting a brief summary of the salient features of that view which bear most immediately on our problem.

IV. KNOWLEDGE

I favour a causal account of knowledge on which for X to know that S is true requires some causal relation to obtain between X and the referents of the names, predicates, and quantifiers of S. I believe in addition in a causal theory of *reference*, thus making the link to my saying knowingly that S *doubly* causal. I hope that what follows will dispel some of the fog which surrounds this formulation.

For Hermione to know that the black object she is holding is a truffle is for her (or at least requires her) to be in a certain (perhaps psychological) state.[7] It also requires the co-operation of the rest of the world, at least to the extent of permitting the object she is holding to be a truffle. Further—and this is the part I would emphasize—in the normal case, that the black object she is holding is a truffle must figure in a suitable way in a causal explanation of her belief that the black object she is holding is a truffle. But what is a 'suitable way'? I will not try to say. A number of authors have published views that seem to point in this direction,[8] and, despite differences among them, there seems to be a core intuition which they share and which I think is correct although very difficult to pin down.

That some such view must be correct and underlies our conception of knowledge is indicated by what we would say under the following circumstances. It is claimed that X knows that p. We think that X could not know that p. What reasons can we offer in support of our view? If we are satisfied that X has normal inferential powers, that p is indeed true, etc., we are often thrown back on arguing that X could not have come into possession of the relevant evidence or reasons: that X's four-dimensional space-time worm does not make the necessary (causal) contact with the grounds of the truth of the proposition for X to be in possession of evidence adequate to support the inference (if an inference was relevant). The proposition p places restrictions on what the world can be like. Our knowledge of the world, combined with our understanding of the restrictions placed by p, given by the truth conditions of p, will often tell us that a given individual could not have come into possession of evidence sufficient to come to know p, and we will thus deny his claim to the knowledge.

As an account of our knowledge about medium-sized objects, in the present, this is along the right lines. It will involve, causally, some direct reference to the facts known and, through that, reference to these objects themselves.

---

[7] If possible, I would like to avoid taking any stand on the cluster of issues in the philosophy of mind or psychology concerning the nature of psychological states. Any view on which Hermione can learn that the cat is on the mat by looking at a real cat on a real mat will do for my purposes. If looking at a cat on a mat puts Hermione into a state and you wish to call that state a physical, or psychological, or even physiological state, I will not object so long as it is understood that such a state, if it is her state of knowledge, is causally related in an appropriate way to the cat's having been on the mat when she looked. If there is no such state, then so much the worse for my view.

[8] To cite but a few: Gilbert H. Harman, *Thought* (Princeton, Princeton University Press, 1973); Alvin I. Goldman, 'A Causal Theory of Knowing', *Journal of Philosophy*, 64/12 (22 June 1967): 357–72; Brian Skyrms, 'The Explication of "X knows that p"', *Journal of Philosophy*, 64/12 (22 June 1967): 373–89.

Furthermore, such knowledge (of houses, trees, truffles, dogs, and bread boxes) presents the clearest case and the easiest to deal with.

Other cases of knowledge can be explained as being based on inferences based on cases such as these, although there must evidently be interdependencies. This is meant to include our knowledge of general laws and theories and, through them, our knowledge of the future and much of the past. This account follows closely the lines that have been proposed by empiricists, but with the crucial modification introduced by the explicitly causal condition mentioned above—but often left out of modern accounts, largely because of attempts to draw a careful distinction between 'discovery' and 'justification'.

In brief, in conjunction with our other knowledge, we use p to determine the range of possible relevant evidence. We use what we know of X (the putative knower) to determine whether there could have been an appropriate kind of interaction, whether X's current belief that p is causally related in a suitable way with what is the case because p is true—whether his evidence is drawn from the range determined by p. If not, then X could not know that p. The connection between what must be the case if p is true and the causes of X's belief can vary widely. But there is always *some* connection, and the connection relates the grounds of X's belief to the subject matter of p.

It must be possible to establish an appropriate sort of connection between the truth conditions of p (as given by an adequate truth definition for the language in which p is expressed) and the grounds on which p is said to be known, at least for propositions that one must *come to know*—that are not innate. In the absence of this, no connection has been established between *having those grounds* and *believing a proposition which is true*. Having those grounds cannot be fitted into an explanation of *knowing p*. The link between p and justifying a belief in p *on those grounds* cannot be made. But for that knowledge which is properly regarded as some form of justified true belief, then the link *must* be made. (Of course, not *all* knowledge need be justified true belief for the point to be a sound one.)

It will come as no surprise that this has been a preamble to pointing out that combining *this* view of knowledge with the 'standard' view of mathematical truth makes it difficult to see how mathematical knowledge is possible. If, for example, numbers are the kinds of entities they are normally taken to be, then the connection between the truth conditions for the statements of number theory and any relevant events connected with the people who are supposed to have mathematical knowledge cannot be made out.[9] It will be impossible to account for how anyone knows any properly number-theoretic propositions. This second condition on an account of mathematical truth will not be

---

[9] For an expression of healthy scepticism concerning this and related points, see Mark Steiner, 'Platonism and the Causal Theory of Knowledge', *Journal of Philosophy*, 70/3 (8 Feb. 1973): 57–66.

satisfied, because we have no account of how we know that the truth conditions for mathematical propositions obtain. One obvious answer—that some of these propositions are true if and only if they are derivable from certain axioms via certain rules—will not help here. For, to be sure, we can ascertain that *those* conditions obtain. But in such a case, what we lack is the link between truth and proof, when truth is directly defined in the standard way. In short, although it may be a truth condition of certain number-theoretic propositions that they be derivable from certain axioms according to certain rules, *that* this is a truth condition must also follow from the account of *truth* if the condition referred to is to help connect truth and knowledge, if it is by their proofs that we know mathematical truths.

Of course, given some set-theoretical account of arithmetic, both the syntax and the semantics of *arithmetic* can be set out so as superficially to meet the conditions we have laid down. But the regress that this invites is transparent, for the same questions must then be asked about the set theory in terms of which the answers are couched.

V. TWO EXAMPLES

There are many accounts of mathematical truth and mathematical knowledge. The theses I have been defending are intended to apply to them all. Rather than try to be comprehensive, however, I will devote the remaining pages to the examination of two representative cases: one 'standard' view and one 'combinatorial' view. First the standard account, as expressed by one of its most explicit and lucid proponents, Kurt Gödel.

Gödel is thoroughly aware that on a realist (i.e., standard) account of mathematical truth our explanation of how we know the basic postulates must be suitably connected with how we interpret the referential apparatus of the theory. Thus, in discussing how we can resolve the continuum problem, once it has been shown to be undecidable by the accepted axioms, he paints the following picture:

... the objects of transfinite set theory ... clearly do not belong to the physical world and even their indirect connection with physical experience is very loose ...

But, despite their remoteness from sense experience, we do have a perception also of the objects of set theory, as is seen from the fact that the axioms force themselves upon us as being true. I don't see why we should have less confidence in this kind of perception, i.e., in mathematical intuition, than in sense perception, which induces us to build up physical theories and to expect that future sense perceptions will agree with them and, moreover, to believe that a question not decidable now has meaning and may be decided in the future.[10]

[10] Kurt Gödel, 'What Is Cantor's Continuum Problem?', revised version in Benacerraf and Putnam (eds.), *Philosophy of Mathematics*, 2nd edn., 483–4.

I find this picture both encouraging and troubling. What troubles me is that without an account of *how* the axioms 'force themselves upon us as being true', the analogy with sense perception and physical science is without much content. For what is missing is *precisely* what my second principle demands: an account of the link between our cognitive faculties and the objects known. In physical science we have at least a start on such an account, and it is causal. We accept as knowledge only those beliefs which we can appropriately relate to our cognitive faculties. Quite appropriately, our conception of knowledge goes hand in hand with our conception of ourselves as knowers. To be sure, there is a *superficial* analogy. For, as Gödel points out, we 'verify' axioms by deducing consequences from them concerning areas in which we seem to have more direct 'perception' (clearer intuitions). But we are never told how we know even these, clearer propositions. For example, the 'verifiable' consequences of axioms of higher infinity are (otherwise undecidable) number-theoretic propositions which themselves are 'verifiable' by computation up to any given integer. But the story, to be helpful anywhere, must tell us how we know statements of computational arithmetic—*if they mean what the standard account would have them mean*. And *that* we are not told. So the analogy is at best superficial.

So much for the troubling aspects. More important perhaps, and what I find encouraging, is the evident basic agreement which motivates Gödel's attempt to draw a parallel between mathematics and empirical science. He sees, I think, that something must be said to bridge the chasm, created by his realistic and platonistic interpretation of mathematical propositions, between the entities that form the subject-matter of mathematics and the human knower. Insead of tinkering with the logical form of mathematical propositions or with the nature of the objects known, he postulates a special faculty through which we 'interact' with these objects. We seem to agree on the analysis of the fundamental problem, but clearly disagree about the epistemological issue—about what avenues are open to us through which we may come to know things.

If our account of empirical knowledge is acceptable, it must be in part because it tries to make the connection evident in the case of our theoretical knowledge, where it is not prima facie clear how the causal account is to be filled in. Thus, when we come to mathematics, the absence of a coherent account of how our mathematical intuition is connected with the truth of mathematical propositions renders the over-all account unsatisfactory.

To introduce a speculative historical note, with some foundation in the texts, it might not be unreasonable to suppose that Plato had recourse to the concept of *anamnesis* at least in part to explain how, given the nature of the forms as he depicted them, one could ever have knowledge of them.[11]

[11] 'The soul, then, as being immortal, and having been born again many times, and having seen

The 'combinatorial' view of mathematical truth has epistemological roots. It starts from the proposition that, whatever may be the 'objects' of mathematics, our knowledge is obtained from proofs. Proofs are, or can be (for some, must be), written down or spoken; mathematicians can survey them and come to agree that they *are* proofs. It is largely through these proofs that mathematical knowledge is obtained and transmitted. In short, this aspect of mathematical knowledge—its (essentially linguistic) means of production and transmission—gives their impetus to the class of views that I call 'combinatorial'.

Noticing the role of proofs in the production of knowledge, it seeks the grounds of truth in the proofs themselves. Combinatorial views receive additional impetus from the realization that the platonist casts a shroud of mystery over how knowledge can be obtained at all. Add that realization to the belief that mathematics is a child of our own begetting (mathematical discovery, on these views, is seldom discovery about an independent reality), and it is not surprising that one looks for acts of conception to account for the birth. Many accounts of mathematical truth fall under this rubric. Perhaps almost all. I have mentioned several in passing, and I discussed Hilbert's view in 'On the Infinite' very briefly. The final example I wish to consider is that of conventionalist accounts—the cluster of views that the truths of logic and mathematics are true (or can be made true) in virtue of explicit conventions, where the conventions in question are usually the postulates of the theory. Once more, I will probably do them all an injustice by lumping together a number of views which their proponents would most certainly like to keep apart.

Quine, in his classic paper on this subject,[12] has dealt clearly, convincingly, and decisively with the view that the truths of *logic* are to be accounted for as the products of convention—far better than I could hope to do here. He pointed out that, since we must account for infinitely many truths, the characterization of the eligible sentences as truths must be wholesale rather than retail. But wholesale characterization can proceed only via general principles—and, if we are supposed not to understand any logic at all, we cannot extract the individual instances from the general principles: we would need logic for such a task.

Persuasive as this may be, I wish to add another argument—not because I think this dead horse needs further flogging, but both because Quine's argument is limited to the case of logic and because the principal points I wish to

all things that exist, whether in this world or in the world below, has knowledge of them all' (Plato, *Meno*, 81).

[12] W. V. Quine, 'Truth by Convention', repr. in Benacerraf and Putnam (eds.), *Philosophy of Mathematics*, 2nd edn., 329–54.

bring out do not emerge sufficiently from it. Indeed, Quine grants the conventionalist certain principles I should like to deny him. In resting his case against conventionalism on the need for a wholesale characterization of infinitely many truths, Quine concedes that were there only finitely many truths to be reckoned with, the conventionalist might have a chance to make out his case. He says:

> If truth assignments could be made one by one, rather than an infinite number at a time, the above difficulty would disappear; truths of logic . . . would simply be asserted severally by fiat, and the problem of inferring them from more general conventions would not arise. (p. 344)

Thus, if some way could be found to make sentences of logic wear their truth values upon their sleeves, the objections to the conventionalist account of truth would disappear—for we would have determined truth values for all the sentences, which is all that one could ask.

I wonder, however, what such a sprinkling of the word 'true' would accomplish. Surely it cannot suffice in order to determine a concept of truth to assign values to each and every sentence of the language [suppose now that the language is set theory, in some first-order formalization] (let those with an even number of horseshoes be 'true').

What would make such an assignment of the predicate 'true' the determination of *the concept of truth*? Simply the use of that monosyllable? Tarski has suggested that satisfaction of convention T is a necessary and sufficient condition on a definition of truth for a particular language.[13] A mere (recursive) distribution of truth values can be parlayed into a truth theory that satisfies convention T. We can rest with that provided we are prepared to beg what I think is the main question and ignore the concept of translation that occurs in its (convention T's) formulation. What would be missing, hard as it is to state, is the theoretical apparatus employed by Tarski in providing truth definitions—that is, the analysis of truth in terms of the 'referential' concepts of naming, predication, satisfaction, and quantification. A definition that does not proceed by the customary recursion clauses for the customary grammatical forms may not be adequate, even if it satisfies convention T. The explanation must proceed through reference and satisfaction and, furthermore,

---

[13] Alfred Tarski, 'The Concept of Truth for Formalized Languages', repr. in Tarski, *Logic, Semantics, and Metamathematics* (New York: Oxford University Press, 1956), 152–278. Convention T is stated on pp. 187–8 as follows:

CONVENTION T. A formally correct definition of the symbol 'Tr', formulated in the metalanguage, will be called *an adequate definition of truth* if it has the following consequences:

(α) all sentences which are obtained from the expression '$x \varepsilon$Tr if and only if p' by substituting for the symbol 'x' a structural-descriptive name of any sentence of the language in question and for the symbol 'p' the expression which forms the translation of this sentence into the metalanguage;

(β) the sentence 'for any x, if $x \varepsilon$Tr then $x \varepsilon$S' (in other words 'Tr $\subseteq$ S').

MATHEMATICAL TRUTH

must be supplemented with an account of reference itself. But the defence of this last claim is too involved a matter to take up here.[14]

The Quine of 'Truth by Convention' felt that to determine the truth values of all the contexts that contain a word suffices to determine its reference. That *might* be so, if we already had the concept of truth and chased the reference of the term that interested us down through the truth definition. But there seems to be something patently wrong with trying to fix the concept of truth *itself* in this way. In so doing, we throw away the very crutch which enables that method to work for other concepts. Truth and reference go hand in hand. Our concept of truth, in so far as we have one, proceeds through the mediation of the concepts Tarski has used to define it for the class of languages he has considered—the essence of Tarski's contribution goes much further than convention T, but includes the schemata for the actual definition as well: an analysis of truth for a language that did not proceed through the familiar devices of predication, quantification, etc., should not give us satisfaction.

If this is at all near the mark, then it should be clear why 'combinatorial' views of the nature of mathematical truth fail on my account. They avoid what seems to me to be the necessary route to an account of truth: through the subject-matter of the propositions whose truth is being defined. Motivated by epistemological considerations, they come up with truth conditions whose satisfaction or non-satisfaction mere mortals can ascertain; but the price they pay is their inability to connect these so-called truth conditions with the truth of the propositions for which they are conditions.

Even if it is granted that the truths of first-order logic do not stem from conventions, it might still be claimed that the rest of mathematics (set theory, for logicists; set theory, number theory, and other things for non-logicists) consists of conventions formalized in first-order logic. This view, too, is subject to the objection that such a concept of convention need not bring *truth* along with it.[15] Indeed, it is clear that it does not. For, even ignoring more general objections, once the logic is fixed, it becomes possible that the conventions thus stipulated turn out to be inconsistent. Hence it cannot be maintained that setting down conventions *guarantees* truth. But if it does not *guarantee* truth, what distinguishes those cases in which it provides for it from those in which it does not? Consistency cannot be the answer. To urge it as such is to *mis*construe the significance of the fact that *in*consistency is *proof* that truth has not been attained. The deeper reason once more is that

---

[14] For an excellent presentation of a similar view, see Hartry Field, 'Tarski's Theory of Truth', *Journal of Philosophy*, 69/13 (13 July 1972): 347–75.

[15] Identical arguments will apply to the view, perhaps indistinguishable from this one, that the postulates constitute *implicit definitions* of existing concepts (as opposed to stipulating how new ones are to be understood), if that is advanced to explain how we know the axioms to be true (we learned the language by learning *these* postulates).

postulational stipulation makes no connection between the propositions and their subject-matter—stipulation does not provide for truth. At best, it limits the class of truth definitions (interpretations) consistent with the stipulations. But that is not enough.

To clarify the point, consider Russell's oft-cited dictum: 'The method of "postulating" what we want has many advantages; they are the same as the advantages of theft over honest toil.'[16] On the view I am advancing, that's false. For with theft at least you come away with the loot, whereas implicit definition, conventional postulation, and their cousins are incapable of bringing truth. They are not only morally but practically deficient as well.[17]

---

[16] Bertrand Russell, *Introduction to Mathematical Philosophy* (London: Allen and Unwin, 1919), 71.

[17] This paper was written for presentation at a symposium on mathematical truth, sponsored jointly by the American Philosophical Association, Eastern Division, and the Association for Symbolic Logic, 27 Dec. 1973.

Commentators were Oswaldo Chateaubriand and Saul Kripke, whose comments were not available at the time of publication. Various segments of an early (1967) version of this paper were read at Berkeley, Harvard, Chicago Circle, Johns Hopkins, New York University, Princeton, and Yale. I am grateful for the help I received on these occasions, as well as for many comments from my colleagues at Princeton, both students and faculty. I am particularly indebted to Dick Grandy, Hartry Field, Adam Morton, and Mark Steiner. That these have not resulted in more significant improvements is due entirely to my own stubbornness. The present version is an attempt to summarize the essentials of the longer paper while making minor improvements along the way. The original version was written during 1967–8 with the generous support of the John Simon Guggenheim Foundation and Princeton University. This is gratefully acknowledged.

II

TWO DOGMAS OF EMPIRICISM

W. V. QUINE

Modern empiricism has been conditioned in large part by two dogmas. One is a belief in some fundamental cleavage between truths which are *analytic*, or grounded in meanings independently of matters of fact, and truths which are *synthetic*, or grounded in fact. The other dogma is *reductionism*: the belief that each meaningful statement is equivalent to some logical construct upon terms which refer to immediate experience. Both dogmas, I shall argue, are ill founded.[1] One effect of abandoning them is, as we shall see, a blurring of the supposed boundary between speculative metaphysics and natural science. Another effect is a shift toward pragmatism.

I. BACKGROUND FOR ANALYTICITY

Kant's cleavage between analytic and synthetic truths was foreshadowed in Hume's distinction between relations of ideas and matters of fact and in Leibniz's distinction between truths of reason and truths of fact. Leibniz spoke of the truths of reason as true in all possible worlds. Picturesqueness aside, this is to say that the truths of reason are those which could not possibly be false. In the same vein we hear analytic statements defined as statements whose denials are self-contradictory. But this definition has small explanatory value; for the notion of self-contradictoriness, in the quite broad sense needed for this definition of analyticity, stands in exactly the same need of clarification as does the notion of analyticity itself.[2] The two notions are the two sides of a single dubious coin.

First published in *Philosophical Review*, 60/1 (Jan. 1951): 20–43.

[1] Much of this paper is devoted to a critique of analyticity which I have been urging orally and in correspondence for years past. My debt to the other participants in those discussions, notably Carnap, Church, Goodman, Tarski, and White, is large and indeterminate. White's excellent essay 'The Analytic and the Synthetic: An Untenable Dualism', in his *John Dewey: Philosopher of Science and Freedom* (New York: Dial Press, 1950), says much of what needed to be said on the topic; but in the present paper I touch on some further aspects of the problem. I am grateful to Dr Donald L. Davidson for valuable criticism of the first draft.

[2] See White, 'The Analytic and the Synthetic', 324.

Kant conceived of an analytic statement as one that attributes to its subject no more than is already conceptually contained in the subject. This formulation has two shortcomings: it limits itself to statements of subject–predicate form, and it appeals to a notion of containment which is left at a metaphorical level. But Kant's intent, evident more from the use he makes of the notion of analyticity than from his definition of it, can be restated thus: a statement is analytic when it is true by virtue of meanings and independently of fact. Pursuing this line, let us examine the concept of *meaning* which is presupposed.

We must observe to begin with that meaning is not to be identified with naming, or reference. Consider Frege's example of 'Evening Star' and 'Morning Star'. Understood not merely as a recurrent evening apparition but as a body, the Evening Star is the planet Venus, and the Morning Star is the same. The two singular terms *name* the same thing. But the meanings must be treated as distinct, since the identity 'Evening Star = Morning Star' is a statement of fact established by astronomical observation. If 'Evening Star' and 'Morning Star' were alike in meaning, the identity 'Evening Star = Morning Star' would be analytic.

Again, there is Russell's example of 'Scott' and 'the author of *Waverley*'. Analysis of the meanings of words was by no means sufficient to reveal to George IV that the person named by these two singular terms was one and the same.

The distinction between meaning and naming is no less important at the level of abstract terms. The terms '9' and 'the number of planets' name one and the same abstract entity, but presumably must be regarded as unlike in meaning; for astronomical observation was needed, and not mere reflection on meanings, to determine the sameness of the entity in question.

Thus far we have been considering singular terms. With general terms, or predicates, the situation is somewhat different, but parallel. Whereas a singular term purports to name an entity, abstract or concrete, a general term does not; but a general term is *true of* an entity, or of each of many, or of none. The class of all entities of which a general term is true is called the *extension* of the term. Now, paralleling the contrast between the meaning of a singular term and the entity named, we must distinguish equally between the meaning of a general term and its extension. The general terms 'creature with a heart' and 'creature with a kidney', for example, are perhaps alike in extension but unlike in meaning.

Confusion of meaning with extension in the case of general terms is less common than confusion of meaning with naming in the case of singular terms. It is indeed a commonplace in philosophy to oppose intension (or meaning) to extension, or, in a variant vocabulary, connotation to denotation.

The Aristotelian notion of essence was the forerunner, no doubt, of the modern notion of intension, or meaning. For Aristotle it was essential in men to be rational, accidental to be two-legged. But there is an important difference between this attitude and the doctrine of meaning. From the latter point of view it may indeed be conceded (if only for the sake of argument) that rationality is involved in the meaning of the word 'man' while two-leggedness is not; but two-leggedness may at the same time be viewed as involved in the meaning of 'biped' while rationality is not. Thus, from the point of view of the doctrine of meaning, it makes no sense to say of the actual individual, who is at once a man and a biped, that his rationality is essential and his two-leggedness accidental, or vice versa. Things had essences, for Aristotle, but only linguistic forms have meanings. Meaning is what essence becomes when it is divorced from the object of reference and wedded to the word.

For the theory of meaning the most conspicuous question is as to the nature of its objects: what sort of things are meanings? They are evidently intended to be ideas, somehow—mental ideas for some semanticists, Platonic ideas for others. Objects of either sort are so elusive, not to say debatable, that there seems little hope of erecting a fruitful science about them. It is not even clear, granted meanings, when we have two and when we have one; it is not clear when linguistic forms should be regarded as *synonymous*, or alike in meaning, and when they should not. If a standard of synonymy should be arrived at, we may reasonably expect that the appeal to meanings as entities will not have played a very useful part in the enterprise.

A felt need for meant entities may derive from an earlier failure to appreciate that meaning and reference are distinct. Once the theory of meaning is sharply separated from the theory of reference, it is a short step to recognizing as the business of the theory of meaning simply the synonymy of linguistic forms and the analyticity of statements; meanings themselves, as obscure intermediary entities, may well be abandoned.

The description of analyticity as truth by virtue of meanings started us off in pursuit of a concept of meaning. But now we have abandoned the thought of any special realm of entities called meanings. So the problem of analyticity confronts us anew.

Statements which are analytic by general philosophical acclaim are not, indeed, far to seek. They fall into two classes. Those of the first class, which may be called *logically true*, are typified by

(1) No unmarried man is married.

The relevant feature of this example is that it is not merely true as it stands, but remains true under any and all reinterpretations of 'man' and 'married'. If we suppose a prior inventory of *logical* particles, comprising 'no', 'un-', 'not', 'if', 'then', 'and', etc., then in general a logical truth is a statement which is true and remains true under all reinterpretations of its components other than the logical particles.

But there is also a second class of analytic statements, typified by

(2) No bachelor is married.

The characteristic of such a statement is that it can be turned into a logical truth by putting synonyms for synonyms; thus (2) can be turned into (1) by putting 'unmarried man' for its synonym 'bachelor'. We still lack a proper characterization of this second class of analytic statements, and therewith of analyticity generally, inasmuch as we have had in the above description to lean on a notion of 'synonymy' which is no less in need of clarification than analyticity itself.

In recent years Carnap has tended to explain analyticity by appeal to what he calls state-descriptions.[3] A state-description is any exhaustive assignment of truth values to the atomic, or non-compound, statements of the language. All other statements of the language are, Carnap assumes, built up of their component clauses by means of the familiar logical devices, in such a way that the truth value of any complex statement is fixed for each state-description by specifiable logical laws. A statement is then explained as analytic when it comes out true under every state-description. This account is an adaptation of Leibniz's 'true in all possible worlds'. But note that this version of analyticity serves its purpose only if the atomic statements of the language are, unlike 'John is a bachelor' and 'John is married', mutually independent. Otherwise there would be a state-description which assigned truth to 'John is a bachelor' and falsity to 'John is married', and consequently 'All bachelors are married' would turn out synthetic rather than analytic under the proposed criterion. Thus the criterion of analyticity in terms of state-descriptions serves only for languages devoid of extra-logical synonym-pairs, such as 'bachelor' and 'unmarried man': synonym-pairs of the type which give rise to the 'second class' of analytic statements. The criterion in terms of state-descriptions is a reconstruction, at best, of logical truth.

I do not mean to suggest that Carnap is under any illusions on this point. His simplified model language with its state-descriptions is aimed primarily not at the general problem of analyticity but at another purpose, the clarification of probability and induction. Our problem, however, is analyticity; and here the major difficulty lies not in the first class of analytic statements, the logical truths, but rather in the second class, which depends on the notion of synonymy.

---

[3] R. Carnap, *Meaning and Necessity* (Chicago: University of Chicago Press, 1947), 9 ff.; id., *Logical Foundations of Probability* (Chicago: University of Chicago Press, 1950), 70 ff.

II. DEFINITION

There are those who find it soothing to say that the analytic statements of the second class reduce to those of the first class, the logical truths, by *definition*; 'bachelor', for example, is *defined* as 'unmarried man'. But how do we find that 'bachelor' is defined as 'unmarried man'? Who defined it thus, and when? Are we to appeal to the nearest dictionary, and accept the lexicographer's formulation as law? Clearly this would be to put the cart before the horse. The lexicographer is an empirical scientist, whose business is the recording of antecedent facts; and if he glosses 'bachelor' as 'unmarried man', it is because of his belief that there is a relation of synonymy between these forms, implicit in general or preferred usage prior to his own work. The notion of synonymy presupposed here has still to be clarified, presumably in terms relating to linguistic behaviour. Certainly the 'definition' which is the lexicographer's report of an observed synonymy cannot be taken as the ground of the synonymy.

Definition is not, indeed, an activity exclusively of philologists. Philosophers and scientists frequently have occasion to 'define' a recondite term by paraphrasing it in terms of a more familiar vocabulary. But ordinarily such a definition, like the philologist's, is pure lexicography, affirming a relationship of synonymy antecedent to the exposition in hand.

Just what it means to affirm synonymy, just what the interconnections may be which are necessary and sufficient in order that two linguistic forms be properly describable as synonymous, is far from clear; but, whatever these interconnections may be, ordinarily they are grounded in usage. Definitions reporting selected instances of synonymy come, then, as reports upon usage.

There is also, however, a variant type of definitional activity which does not limit itself to the reporting of pre-existing synonymies. I have in mind what Carnap calls *explication*—an activity to which philosophers are given, and scientists also in their more philosophical moments. In explication the purpose is not merely to paraphrase the *definiendum* into an outright synonym, but actually to improve upon the *definiendum* by refining or supplementing its meaning. But even explication, though not merely reporting a pre-existing synonymy between *definiendum* and *definiens*, does rest, nevertheless, on *other* pre-existing synonymies. The matter may be viewed as follows. Any word worth explicating has some contexts which, as wholes, are clear and precise enough to be useful; and the purpose of explication is to preserve the usage of these favoured contexts while sharpening the usage of other contexts. In order that a given definition be suitable for purposes of explication, therefore, what is required is not that the *definiendum* in its antecedent usage be synonymous with the *definiens*, but just that each of these favoured contexts of the *definiendum*, taken as a whole in its antecedent usage, be synonymous with the corresponding context of the *definiens*.

Two alternative *definientia* may be equally appropriate for the purposes of a given task of explication and yet not be synonymous with each other; for they may serve interchangeably within the favoured contexts, but diverge elsewhere. By cleaving to one of these *definientia* rather than the other, a definition of explicative kind generates, by fiat, a relationship of synonymy between *definiendum* and *definiens* which did not hold before. But such a definition still owes its explicative function, as seen, to pre-existing synonymies.

There does, however, remain still an extreme sort of definition which does not hark back to prior synonymies at all: namely, the explicitly conventional introduction of novel notations for purposes of sheer abbreviation. Here the *definiendum* becomes synonymous with the *definiens* simply because it has been created expressly for the purpose of being synonymous with the *definiens*. Here we have a really transparent case of synonymy created by definition; would that all species of synonymy were as intelligible. For the rest, definition rests on synonymy, rather than explaining it.

The word 'definition' has come to have a dangerously reassuring sound, due no doubt to its frequent occurrence in logical and mathematical writings. We shall do well to digress now into a brief appraisal of the role of definition in formal work.

In logical and mathematical systems either of two mutually antagonistic types of economy may be striven for, and each has its peculiar practical utility. On the one hand we may seek economy of practical expression: ease and brevity in the statement of multifarious relationships. This sort of economy calls usually for distinctive concise notations for a wealth of concepts. Second, however, and oppositely, we may seek economy in grammar and vocabulary; we may try to find a minimum of basic concepts such that, once a distinctive notation has been appropriated to each of them, it becomes possible to express any desired further concept by mere combination and iteration of our basic notations. This second sort of economy is impractical in one way, since a poverty in basic idioms tends to a necessary lengthening of discourse. But it is practical in another way: it greatly simplifies theoretical discourse *about* the language, through minimizing the terms and the forms of construction wherein the language consists.

Both sorts of economy, though prima facie incompatible, are valuable in their separate ways. The custom has consequently arisen of combining both sorts of economy by forging in effect two languages, the one a part of the other. The inclusive language, though redundant in grammar and vocabulary, is economical in message lengths, while the part, called *primitive notation*, is economical in grammar and vocabulary. Whole and part are correlated by rules of translation whereby each idiom not in primitive notation is equated to some complex built up of primitive notation. These rules of translation are the so-called *definitions* which appear in formalized systems. They are best viewed not as adjuncts to one language but as correlations between two languages, the one a part of the other.

But these correlations are not arbitrary. They are supposed to show how the primitive notations can accomplish all purposes, save brevity and convenience, of the redundant language. Hence the *definiendum* and its *definiens* may be expected, in each case, to be related in one or another of the three ways lately noted. The *definiens* may be a faithful paraphrase of the *definiendum* into the narrower notation, preserving a direct synonymy as of antecedent usage; or the *definiens* may, in the spirit of explication, improve upon the antecedent usage of the *definiendum*; or, finally, the *definiendum* may be a newly created notation, newly endowed with meaning here and now.

In formal and informal work alike, thus, we find that definition—except in the extreme case of the explicitly conventional introduction of new notations—hinges on prior relationships of synonymy. Recognizing, then, that the notion of definition does not hold the key to synonymy and analyticity, let us look further into synonymy, and say no more of definition.

III. INTERCHANGEABILITY

A natural suggestion, deserving close examination, is that the synonymy of two linguistic forms consists simply in their interchangeability in all contexts without change of truth value; interchangeability, in Leibniz's phrase, *salva veritate*. Note that synonyms so conceived need not even be free from vagueness, as long as the vaguenesses match.

But it is not quite true that the synonyms 'bachelor' and 'unmarried man' are everywhere interchangeable *salva veritate*. Truths which become false under substitution of 'unmarried man' for 'bachelor' are easily constructed with help of 'bachelor of arts' or 'bachelor's buttons'. Also with help of quotation, thus:

'Bachelor' has less than ten letters.

Such counter-instances can, however, perhaps be set aside by treating the phrases 'bachelor of arts' and 'bachelor's buttons' and the quotation 'Bachelor' each as a single indivisible word and then stipulating that the interchangeability *salva veritate* which is to be the touchstone of synonymy is not supposed to apply to fragmentary occurrences inside of a word. This account of synonymy, supposing it acceptable on other counts, has indeed the drawback of appealing to a prior conception of 'word' which can be counted on to present difficulties of formulation in its turn. Nevertheless, some progress might be claimed in having reduced the problem of synonymy to a problem of wordhood. Let us pursue this line a bit, taking 'word' for granted.

The question remains whether interchangeability *salva veritate* (apart from occurrences within words) is a strong enough condition for synonymy, or whether, on the contrary, some non-synonymous expressions might be thus interchangeable. Now let us be clear that we are not concerned here with synonymy in the sense of complete identity in psychological associations or poetic quality; indeed, no two expressions are synonymous in such a sense. We are concerned only with what may be called *cognitive synonymy*. Just what this is cannot be said without successfully finishing the present study; but we know something about it from the need which arose for it in connection with analyticity in Section I. The sort of synonymy needed there was merely such that any analytic statement could be turned into a logical truth by putting synonyms for synonyms. Turning the tables and assuming analyticity, indeed, we could explain cognitive synonymy of terms as follows (keeping to the familiar example): to say that 'bachelor' and 'unmarried man' are cognitively synonymous is to say no more nor less than that the statement:

(3) All and only bachelors are unmarried men

is analytic.[4]

What we need is an account of cognitive synonymy not presupposing analyticity—if we are to explain analyticity conversely with help of cognitive synonymy, as undertaken in Section I. And indeed, such an independent account of cognitive synonymy is at present up for consideration: namely, interchangeability *salva veritate* everywhere except within words. The question before us, to resume the thread at last, is whether such interchangeability is a sufficient condition for cognitive synonymy. We can quickly assure ourselves that it is, by examples of the following sort. The statement

(4) Necessarily all and only bachelors are bachelors

is evidently true, even supposing 'necessarily' so narrowly construed as to be truly applicable only to analytic statements. Then, *if* 'bachelor' and 'unmarried man' are interchangeable *salva veritate*, the result

(5) Necessarily, all and only bachelors are unmarried men

---

[4] This is cognitive synonymy in a primary, broad sense. Carnap (*Meaning and Necessity*, 56 ff.) and C. I. Lewis (*Analysis of Knowledge and Valuation* (La Salle, Ill.: Open Court, 1946), 83 ff. have suggested how, once this notion is at hand, a narrower sense of cognitive synonymy which is preferable for some purposes can in turn be derived. But this special ramification of concept-building lies aside from the present purposes, and must not be confused with the broad sort of cognitive synonymy here concerned.

TWO DOGMAS OF EMPIRICISM

of putting 'unmarried man' for an occurrence of 'bachelor' in (4) must, like (4), be true. But to say that (5) is true is to say that (3) is analytic, and hence that 'bachelor' and 'unmarried men' are cognitively synonymous.

Let us see what there is about the above argument that gives it its air of hocus-pocus. The condition of interchangeability *salva veritate* varies in its force with variations in the richness of the language at hand. The above argument supposes we are working with a language rich enough to contain the adverb 'necessarily', this adverb being so construed as to yield truth when and only when applied to an analytic statement. But can we condone a language which contains such an adverb? Does the adverb really make sense? To suppose that it does is to suppose that we have already made satisfactory sense of 'analytic'. Then what are we so hard at work on right now?

Our argument is not flatly circular, but something like it. It has the form, figuratively speaking, of a closed curve in space.

Interchangeability *salva veritate* is meaningless until relativized to a language whose extent is specified in relevant respects. Suppose now we consider a language containing just the following materials. There is an indefinitely large stock of one- and many-place predicates, mostly having to do with extra-logical subject-matter. The rest of the language is logical. The atomic sentences consist each of a predicate followed by one or more variables; and the complex sentences are built up of atomic ones by truth functions and quantification. In effect, such a language enjoys the benefits also of descriptions and class names, and indeed singular terms generally, these being contextually definable in known ways.[5] Such a language can be adequate to classical mathematics, and indeed to scientific discourse generally, except in so far as the latter involves debatable devices such as modal adverbs and contrary-to-fact conditionals. Now a language of this type is *extensional*, in this sense: any two predicates which *agree extensionally* (i.e., are true of the same objects) are interchangeable *salva veritate*.

In an extensional language, therefore, interchangeability *salva veritate* is no assurance of cognitive synonymy of the desired type. That 'bachelor' and 'unmarried man' are interchangeable *salva veritate* in an extensional language assures us of no more than that (3) is true. There is no assurance here that the extensional agreement of 'bachelor' and 'unmarried man' rests on meaning rather than merely on accidental matters of fact, as does extensional agreement of 'creature with a heart' and 'creature with a kidney'.

For most purposes extensional agreement is the nearest approximation to synonymy we need care about. But the fact remains that extensional agreement falls far short of cognitive synonymy of the type required for explaining

---

[5] See, e.g., my *Mathematical Logic* (New York: Norton, 1940; Cambridge, Mass: Harvard University Press, 1947), sects. 24, 26, 27; or *Methods of Logic* (New York: Holt, 1950), sect. 37 ff.

analyticity in the manner of Section I. The type of cognitive synonymy required there is such as to equate the synonymy of 'bachelor' and 'unmarried man' with the analyticity of (3), not merely with the truth of (3).

So we must recognize that interchangeability *salva veritate*, if construed in relation to an extensional language, is not a sufficient condition of cognitive synonymy in the sense needed for deriving analyticity in the manner of Section I. If a language contains an intensional adverb 'necessarily' in the sense lately noted, or other particles to the same effect, then interchangeability *salva veritate* in such a language does afford a sufficient condition of cognitive synonymy; but such a language is intelligible only if the notion of analyticity is already clearly understood in advance.

The effort to explain cognitive synonymy first, for the sake of deriving analyticity from it afterward as in Section I, is perhaps the wrong approach. Instead, we might try explaining analyticity somehow without appeal to cognitive synonymy. Afterward we could doubtless derive cognitive synonymy from analyticity satisfactorily enough if desired. We have seen that cognitive synonymy of 'bachelor' and 'unmarried man' can be explained as analyticity of (3). The same explanation works for any pair of one-place predicates, of course, and it can be extended in obvious fashion to many-place predicates. Other syntactical categories can also be accommodated in fairly parallel fashion. Singular terms may be said to be cognitively synonymous when the statement of identity formed by putting ' = ' between them is analytic. Statements may be said simply to be cognitively synonymous when their biconditional (the result of joining them by 'if and only if') is analytic.[6] If we care to lump all categories into a single formulation, at the expense of assuming again the notion of 'word' which was appealed to early in this section, we can describe any two linguistic forms as cognitively synonymous when the two forms are interchangeable (apart from occurrences within 'words') *salva* (no longer *veritate* but) *analyticitate*. Certain technical questions arise, indeed, over cases of ambiguity or homonymy; let us not pause for them, however, for we are already digressing. Let us rather turn our backs on the problem of synonymy and address ourselves anew to that of analyticity.

IV. SEMANTICAL RULES

Analyticity at first seemed most naturally definable by appeal to a realm of meanings. On refinement, the appeal to meanings gave way to an appeal to synonymy or definition. But definition turned out to be a will-o'-the-wisp,

---

[6] The 'if and only if' itself is intended in the truth-functional sense. See Carnap, *Meaning and Necessity*, 14.

and synonymy turned out to be best understood only by dint of a prior appeal to analyticity itself. So we are back at the problem of analyticity.

I do not know whether the statement 'Everything green is extended' is analytic. Now does my indecision over this example really betray an incomplete understanding, an incomplete grasp of the 'meanings', of 'green' and 'extended'? I think not. The trouble is not with 'green' or 'extended', but with 'analytic'.

It is often hinted that the difficulty in separating analytic statements from synthetic ones in ordinary language is due to the vagueness of ordinary language, and that the distinction is clear when we have a precise artificial language with explicit 'semantical rules'. This, however, as I shall now attempt to show, is a confusion.

The notion of analyticity about which we are worrying is a purported relation between statements and languages: a statement S is said to be *analytic for* a language L, and the problem is to make sense of this relation generally—that is, for variables 'S' and 'L'. The point that I want to make is that the gravity of this problem is not perceptibly less for artificial languages than for natural ones. The problem of making sense of the idiom 'S is analytic for L', with variable 'S' and 'L', retains its stubbornness even if we limit the range of the variable 'L' to artificial languages. Let me now try to make this point evident.

For artificial languages and semantical rules we look naturally to the writings of Carnap. His semantical rules take various forms, and to make my point, I shall have to distinguish certain of the forms. Let us suppose, to begin with, an artificial language L_0 whose semantical rules have the form explicitly of a specification, by recursion or otherwise, of all the analytic statements of L_0. The rules tell us that such and such statements, and only those, are the analytic statements of L_0. Now here the difficulty is simply that the rules contain the word 'analytic', which we do not understand! We understand what expressions the rules attribute analyticity to, but we do not understand what the rules attribute analyticity to those expressions. In short, before we can understand a rule which begins 'A statement S is analytic for language L_0 if and only if . . .', we must understand the general relative term 'analytic for'; we must understand 'S is analytic for L' where 'S' and 'L' are variables.

Alternatively, we may indeed view the so-called rule as a conventional definition of a new simple symbol 'analytic-for-L_0', which might better be written untendentiously as 'K' so as not to seem to throw light on the interesting word 'analytic'. Obviously, any number of classes K, M, N, etc. of statements of L_0 can be specified for various purposes or for no purpose; what does it mean to say that K, as against M, N, etc., is the class of the 'analytic' statements of L_0?

By saying what statements are analytic for L_0, we explain 'analytic for-L_0' but not 'analytic', not 'analytic for'. We do not begin to explain the idiom 'S is analytic for L' with variable 'S' and 'L', even though we be content to limit the range of 'L' to the realm of artificial languages.

Actually we do know enough about the intended significance of 'analytic' to know that analytic statements are supposed to be true. Let us then turn to a second form of semantical rule, which says not that such and such statements are analytic, but simply that such and such statements are included among the truths. Such a rule is not subject to the criticism of containing the un-understood word 'analytic'; and we may grant for the sake of argument that there is no difficulty over the broader term 'true'. A semantical rule of this second type, a rule of truth, is not supposed to specify all the truths of the language; it merely stipulates, recursively or otherwise, a certain multitude of statements which, along with others unspecified, are to count as true. Such a rule may be conceded to be quite clear. Derivatively, afterward, analyticity can be demarcated thus: a statement is analytic if it is (not merely true but) true according to the semantical rule.

Still there is really no progress. Instead of appealing to an unexplained word 'analytic', we are now appealing to an unexplained phrase 'semantical rule'. Not every true statement which says that the statements of some class are true can count as a semantical rule—otherwise *all* truths would be 'analytic' in the sense of being true according to semantical rules. Semantical rules are distinguishable, apparently, only by the fact of appearing on a page under the heading 'Semantical Rules'; and this heading is itself then meaningless.

We can say indeed that a statement is *analytic-for-L_0* if and only if it is true according to such and such specifically appended 'semantical rules', but then we find ourselves back at essentially the same case which was originally discussed: 'S is analytic-for-L_0 if and only if . . .'. Once we seek to explain 'S is analytic for L' generally for variable 'L' (even allowing limitation of 'L' to artificial languages), the explanation 'true according to the semantical rules of L' is unavailing; for the relative term 'semantical rule of' is as much in need of clarification, at least, as 'analytic for'.

It might conceivably be protested that an artificial language L (unlike a natural one) is a language in the ordinary sense *plus* a set of explicit semantical rules—the whole constituting, let us say, an ordered pair; and that the semantical rules of L then are specifiable simply as the second component of the pair L. But, by the same token and more simply, we might construe an artificial language L outright as an ordered pair whose second component is the class of its analytic statements; and then the analytic statements of L become specifiable simply as the statements in the second component of L. Or better still, we might just stop tugging at our bootstraps altogether.

Not all the explanations of analyticity known to Carnap and his readers have been covered explicitly in the above considerations, but the extension to other forms is not hard to see. Just one additional factor should be mentioned which sometimes enters: sometimes the semantical rules are in effect rules of translation into ordinary language, in which case the analytic statements of the artificial language are in effect recognized as such from the analyticity of their specified translations in ordinary language. Here, certainly, there can be no thought of an illumination of the problem of analyticity from the side of the artificial language.

From the point of view of the problem of analyticity the notion of an artificial language with semantical rules is a *feu follet par excellence*. Semantical rules determining the analytic statements of an artificial language are of interest only in so far as we already understand the notion of analyticity; they are of no help in gaining this understanding.

Appeal to hypothetical languages of an artificially simple kind could conceivably be useful in clarifying analyticity, if the mental or behavioural or cultural factors relevant to analyticity—whatever they may be—were somehow sketched into the simplified model. But a model which takes analyticity merely as in irreducible character is unlikely to throw light on the problem of explicating analyticity.

It is obvious that truth in general depends on both language and extra-linguistic fact. The statement 'Brutus killed Caesar' would be false if the world had been different in certain ways, but it would also be false if the word 'killed' happened rather to have the sense of 'begat'. Hence the temptation to suppose in general that the truth of a statement is somehow analysable into a linguistic component and a factual component. Given this supposition, it next seems reasonable that in some statements the factual component should be null; and these are the analytic statements. But, for all its a priori reasonableness, a boundary between analytic and synthetic statements simply has not been drawn. That there is such a distinction to be drawn at all is an unempirical dogma of empiricists, a metaphysical article of faith.

V. THE VERIFICATION THEORY AND REDUCTIONISM

In the course of these sombre reflections we have taken a dim view first of the notion of meaning, then of the notion of cognitive synonymy, and finally of the notion of analyticity. But what, it may be asked, of the verification theory of meaning? This phrase has established itself so firmly as a catchword of empiricism that we should be very unscientific indeed not to look beneath it for a possible key to the problem of meaning and the associated problems.

The verification theory of meaning, which has been conspicuous in the literature from Peirce onward, is that the meaning of a statement is the method of empirically confirming or infirming it. An analytic statement is that limiting case which is confirmed no matter what.

As urged in Section I, we can as well pass over the question of meanings as entities, and move straight to sameness of meaning, or synonymy. Then what the verification theory says is that statements are synonymous if and only if they are alike in point of method of empirical confirmation or infirmation.

This is an account of cognitive synonymy not of linguistic forms generally, but of statements.[7] However, from the concept of synonymy of statements we could derive the concept of synonymy for other linguistic forms, by considerations somewhat similar to those at the end of Section III. Assuming the notion of 'word', indeed, we could explain any two forms as synonymous when the putting of the one form for an occurrence of the other in any statement (apart from occurrences within 'words') yields a synonymous statement. Finally, given the concept of synonymy thus for linguistic forms generally, we could define analyticity in terms of synonymy and logical truth, as in Section I. For that matter, we could define analyticity more simply in terms of just synonymy of statements together with logical truth; it is not necessary to appeal to synonymy of linguistic forms other than statements. For a statement may be described as analytic simply when it is synonymous with a logically true statement.

So, if the verification theory can be accepted as an adequate account of statement synonymy, the notion of analyticity is saved after all. However, let us reflect. Statement synonymy is said to be likeness of method of empirical confirmation or infirmation. Just what are these methods which are to be compared for likeness? What, in other words, is the nature of the relationship between a statement and the experiences which contribute to or detract from its confirmation?

The most naïve view of the relationship is that it is one of direct report. This is *radical reductionism*. Every meaningful statement is held to be translatable into a statement (true or false) about immediate experience. Radical reductionism, in one form or another, well antedates the verification theory of meaning explicitly so-called. Thus Locke and Hume held that every idea must either originate directly in sense experience or else be compounded of ideas thus originating; and taking a hint from Tooke,[8] we might rephrase this doctrine in semantical jargon by saying that a term, to be significant at all,

---

[7] The doctrine can indeed be formulated with terms rather than statements as the units. Thus C. I. Lewis describes the meaning of a term as '*a criterion in mind*, by reference to which one is able to apply or refuse to apply the expression in question in the case of presented, or imagined, things or situations' (*Analysis of Knowledge*, 133).

[8] John Horne Tooke, *The Diversions of Purley* (London, 1776; Boston, 1806), vol. 1, ch. 2.

must be either a name of a sense-datum or a compound of such names or an abbreviation of such a compound. So stated, the doctrine remains ambiguous as between sense-data as sensory events and sense-data as sensory qualities; and it remains vague as to the admissible ways of compounding. Moreover, the doctrine is unnecessarily and intolerably restrictive in the term-by-term critique which it imposes. More reasonably, and without yet exceeding the limits of what I have called radical reductionism, we may take full statements as our significant units—thus demanding that our statements as wholes be translatable into sense-datum language, but not that they be translatable term by term.

This emendation would unquestionably have been welcome to Locke and Hume and Tooke, but historically it had to await two intermediate developments. One of these developments was the increasing emphasis on verification or confirmation, which came with the explicitly so called verification theory of meaning. The objects of verification of confirmation being statements, this emphasis gave the statement an ascendancy over the word or term as unit of significant discourse. The other development, consequent upon the first, was Russell's discovery of the concept of incomplete symbols defined in use.

Radical reductionism, conceived now with statements as units, set itself the task of specifying a sense-datum language and showing how to translate the rest of significant discourse, statement by statement, into it. Carnap embarked on this project in the *Aufbau*.[9]

The language which Carnap adopted as his starting-point was not a sense-datum language in the narrowest conceivable sense, for it included also the notations of logic, up through higher set theory. In effect it included the whole language of pure mathematics. The ontology implicit in it (i.e., the range of values of its variables) embraced not only sensory events but classes, classes of classes, and so on. Empiricists there are who would boggle at such prodigality. Carnap's starting-point is very parsimonious, however, in its extra-logical or sensory part. In a series of constructions in which he exploits the resources of modern logic with much ingenuity, he succeeds in defining a wide array of important additional sensory concepts which, but for his constructions, one would not have dreamed were definable on so slender a basis. Carnap was the first empiricist who, not content with asserting the reducibility of science to terms of immediate experience, took serious steps toward carrying out the reduction.

Even supposing Carnap's starting-point satisfactory, his constructions were, as he himself stressed, only a fragment of the full programme. The

[9] R. Carnap, *Der logische Aufbau der Welt* (Berlin: Weltkreis-Verlag, 1928); trans. R. A. George as *The Logical Structure of the World* (Berkeley: University of California Press, 1967).

construction of even the simplest statements about the physical world was left in a sketchy state. Carnap's suggestions on this subject were, despite their sketchiness, very suggestive. He explained spatio-temporal point-instants as quadruples of real numbers, and envisaged assignment of sense qualities to point-instants according to certain canons. Roughly summarized, the plan was that qualities should be assigned to point-instants in such a way as to achieve the laziest world compatible with our experience. The principle of least action was to be our guide in constructing a world from experience.

Carnap did not seem to recognize, however, that his treatment of physical objects fell short of reduction not merely through sketchiness, but in principle. Statements of the form 'Quality q is at point-instant $x; y; z; t$' were, according to his canons, to be apportioned truth values in such a way as to maximize and minimize certain over-all features, and with growth of experience the truth values were to be progressively revised in the same spirit. I think this is a good schematization (deliberately oversimplified, to be sure) of what science really does; but it provides no indication, not even the sketchiest, of how a statement of the form 'Quality q is at $x; y; z; t$' could ever be translated into Carnap's initial language of sense-data and logic. The connective 'is at' remains an added, undefined connective; the canons counsel us in its use but not in its elimination.

Carnap seems to have appreciated this point afterwards; for in his later writings he abandoned all notion of the translatability of statements about the physical world into statements about immediate experience. Reductionism in its radical form has long since ceased to figure in Carnap's philosophy.

But the dogma of reductionism has, in a subtler and more tenuous form, continued to influence the thought of empiricists. The notion lingers that to each statement, or each synthetic statement, there is associated a unique range of possible sensory events such that the occurrence of any of them would add to the likelihood of truth of the statement, and that there is associated also another unique range of possible sensory events whose occurrence would detract from that likelihood. This notion is of course implicit in the verification theory of meaning.

The dogma of reductionism survives in the supposition that each statement, taken in isolation from its fellows, can admit of confirmation or infirmation at all. My counter-suggestion, issuing essentially from Carnap's doctrine of the physical world in the *Aufbau*, is that our statements about the external world face the tribunal of sense experience not individually but only as a corporate body.

The dogma of reductionism, even in its attenuated form, is intimately connected with the other dogma: that there is a cleavage between the analytic and the synthetic. We have found ourselves led, indeed from the latter problem to the former through the verification theory of meaning. More directly, the one dogma clearly supports the other in this way: as long as it is taken to be significant in general to speak of the confirmation and infirmation of a statement, it seems significant to speak also of a limiting kind of statement which is vacuously confirmed, *ipso facto*, come what may; and such a statement is analytic.

The two dogmas are, indeed, at root identical. We lately reflected that in general the truth of statements does obviously depend both upon language and upon extra-linguistic fact; and we noted that this obvious circumstance carries in its train, not logically but all too naturally, a feeling that the truth of a statement is somehow analysable into a linguistic component and a factual component. The factual component must, if we are empiricists, boil down to a range of confirmatory experiences. In the extreme case where the linguistic component is all that matters, a true statement is analytic. But I hope we are now impressed with how stubbornly the distinction between analytic and synthetic has resisted any straightforward drawing. I am impressed also, apart from prefabricated examples of black and white balls in an urn, with how baffling the problem has always been of arriving at any explicit theory of the empirical confirmation of a synthetic statement. My present suggestion is that it is nonsense, and the root of much nonsense, to speak of a linguistic component and a factual component in the truth of any individual statement. Taken collectively, science has its double dependence upon language and experience; but this duality is not significantly traceable to the statements of science taken one by one.

Russell's concept of definition in use was, as remarked, an advance over the impossible term-by-term empiricism of Locke and Hume. The statement, rather than the term, came with Russell to be recognized as the unit accountable to an empiricist critique. But what I am now urging is that even in taking the statement as unit we have drawn our grid too finely. The unit of empirical significance is the whole of science.

VI. EMPIRICISM WITHOUT THE DOGMAS

The totality of our so-called knowledge or beliefs, from the most casual matters of geography and history to the profoundest laws of atomic physics or even of pure mathematics and logic, is a man-made fabric which impinges on experience only along the edges. Or, to change the figure, total science is like a field of force whose boundary conditions are experience. A conflict with experience at the periphery occasions readjustments in the interior of the field. Truth values have to be redistributed over some of our statements.

Re-evaluation of some statements entails re-evaluation of others, because of their logical interconnections—the logical laws being in turn simply certain further statements of the system, certain further elements of the field. Having re-evaluated one statement, we must re-evaluate some others, whether they be statements logically connected with the first or whether they be the statements of logical connections themselves. But the total field is so undetermined by its boundary conditions, experience, that there is much latitude of choice as to what statements to re-evaluate in the light of any single contrary experience. No particular experiences are linked with any particular statements in the interior of the field, except indirectly through considerations of equilibrium affecting the field as a whole.

If this view is right, it is misleading to speak of the empirical content of an individual statement—especially if it be a statement at all remote from the experiential periphery of the field. Furthermore, it becomes folly to seek a boundary between synthetic statements, which hold contingently on experience, and analytic statements, which hold come what may. Any statement can be held true come what may, if we make drastic enough adjustments elsewhere in the system. Even a statement very close to the periphery can be held true in the face of recalcitrant experience by pleading hallucination or by amending certain statements of the kind called logical laws. Conversely, by the same token, no statement is immune to revision. Revision even of the logical law of the excluded middle has been proposed as a means of simplifying quantum mechanics; and what difference is there in principle between such a shift and the shift whereby Kepler superseded Ptolemy or Einstein Newton, or Darwin Aristotle?

For vividness I have been speaking in terms of varying distances from a sensory periphery. Let me try now to clarify this notion without metaphor. Certain statements, though *about* physical objects and not sense experience, seem peculiarly germane to sense experience—and in a selective way: some statements to some experiences, others to others. Such statements, especially germane to particular experiences, I picture as near the periphery. But in this relation of 'germaneness' I envisage nothing more than a loose association reflecting the relative likelihood, in practice, of our choosing one statement rather than another for revision in the event of recalcitrant experience. For example, we can imagine recalcitrant experiences to which we would surely be inclined to accommodate our system by re-evaluating just the statement that there are brick houses on Elm Street, together with related statements on the same topic. We can imagine other recalcitrant experiences to which we would be inclined to accommodate our system by re-evaluating just the statement that there are no centaurs, along with kindred statements. A recalcitrant experience can, I have already urged, be accommodated by any of various

TWO DOGMAS OF EMPIRICISM

alternative re-evaluations in various alternative quarters of the total system; but, in the cases which we are now imagining, our natural tendency to disturb the total system as little as possible would lead us to focus our revisions upon these specific statements concerning brick houses or centaurs. These statements are felt, therefore, to have a sharper empirical reference than highly theoretical statements of physics or logic or ontology. The latter statements may be thought of as relatively centrally located within the total network, meaning merely that little preferential connection with any particular sense-data obtrudes itself.

As an empiricist, I continue to think of the conceptual scheme of science as a tool, ultimately, for predicting future experience in the light of past experience. Physical objects are conceptually imported into the situation as convenient intermediaries—not by definition in terms of experience, but simply as irreducible posits comparable, epistemologically, to the gods of Homer. Let me interject that for my part I do, *qua* lay physicist, believe in physical objects and not in Homer's gods; and I consider it a scientific error to believe otherwise. But in point of epistemological footing the physical objects and the gods differ only in degree and not in kind. Both sorts of entities enter our conception only as cultural posits. The myth of physical objects is epistemologically superior to most in that it has proved more efficacious than other myths as a device for working a manageable structure into the flux of experience.

Imagine, for the sake of analogy, that we are given the rational numbers. We develop an algebraic theory for reasoning about them, but we find it inconveniently complex, because certain functions such as square root lack values for some arguments. Then it is discovered that the rules of our algebra can be much simplified by conceptually augmenting our ontology with some mythical entities, to be called irrational numbers. All we continue to be really interested in, first and last, are rational numbers; but we find that we can commonly get from one law about rational numbers to another much more quickly and simply by pretending that the irrational numbers are there too.

I think this a fair account of the introduction of irrational numbers and other extensions of the number system. The fact that the mythical status of irrational numbers eventually gave way to the Dedekind–Russell version of them as certain infinite classes of ratios is irrelevant to my analogy. That version is impossible anyway as long as reality is limited to the rational numbers and not extended to classes of them.

Now I suggest that experience is analogous to the rational numbers and that the physical objects, in analogy to the irrational numbers are posits which serve merely to simplify our treatment of experience. The physical objects are no more reducible to experience than the irrational numbers to rational numbers, but their incorporation into the theory enables us to get more easily from one statement about experience to another.

The salient differences between the positing of physical objects and the positing of irrational numbers are, I think, just two. First, the factor of simplification is more overwhelming in the case of physical objects than in the numerical case. Second, the positing of physical objects is far more archaic, being indeed coeval, I expect, with language itself. For language is social, and so depends for its development upon intersubjective reference.

Positing does not stop with macroscopic physical objects. Objects at the atomic level and beyond are posited to make the laws of macroscopic objects, and ultimately the laws of experience, simpler and more manageable; and we need not expect or demand full definition of atomic and subatomic entities in terms of macroscopic ones, any more than definition of macroscopic things in terms of sense-data. Science is a continuation of common sense, and it continues the common-sense expedient of swelling ontology to simplify theory.

Physical objects, small and large, are not the only posits. Forces are another example; and indeed we are told nowadays that the boundary between energy and matter is obsolete. Moreover, the abstract entities which are the substance of mathematics—ultimately classes and classes of classes and so on up—are another posit in the same spirit. Epistemologically, these are myths on the same footing with physical objects and gods, neither better nor worse except for differences in the degree to which they expedite our dealings with sense experiences.

The over-all algebra of rational and irrational numbers is under determined by the algebra of rational numbers, but is smoother and more convenient; and it includes the algebra of rational numbers as a jagged or gerrymandered part. Total science, mathematical and natural and human, is similarly, but more extremely, underdetermined by experience. The edge of the system must be kept squared with experience; the rest, with all its elaborate myths or fictions, has as its objective the simplicity of laws.

Ontological questions, under this view, are on a par with questions of natural science. Consider the question whether to countenance classes as entities. This, as I have argued elsewhere,[10] is the question whether to quantify with respect to variables which take classes as values. Now Carnap has maintained[11] that this is a question not of matters of fact but of choosing a convenient language form, a convenient conceptual scheme or framework for science. With this I agree, but only on the proviso that the same be conceded regarding scientific hypotheses generally. Carnap has recognized[12] that he is

---

[10] e.g., in 'Notes on Existence and Necessity', *Journal of Philosophy*, 40 (1943): 113–27.

[11] R. Carnap, 'Empiricism, Semantics, and Ontology', *Revue internationale de philosophie*, 4 (1950): 20–40.   [12] Ibid. 32 n.

TWO DOGMAS OF EMPIRICISM 51

able to preserve a double standard for ontological questions and scientific hypotheses only by assuming an absolute distinction between the analytic and the synthetic; and I need not say again that this is a distinction which I reject.

Some issues do, I grant, seem more a question of convenient conceptual scheme and others more a question of brute fact. The issue over there being classes seems more a question of convenient conceptual scheme; the issue over there being centaurs, or brick houses on Elm Street, seems more a question of fact. But I have been urging that this difference is only one of degree, and that it turns upon our vaguely pragmatic inclination to adjust one strand of the fabric of science rather than another in accommodating some particular recalcitrant experience. Conservatism figures in such choices, and so does the quest for simplicity.

Carnap, Lewis, and others take a pragmatic stand on the question of choosing between language forms, scientific frameworks; but their pragmatism leaves off at the imagined boundary between the analytic and the synthetic. In repudiating such a boundary, I espouse a more thorough pragmatism. Each man is given a scientific heritage plus a continuing barrage of sensory stimulation; and the considerations which guide him in warping his scientific heritage to fit his continuing sensory promptings are, where rational, pragmatic.

III

ACCESS AND INFERENCE

W. D. HART

One might characterize two styles of epistemology by distinguishing what is known from the means by which knowledge is acquired. One sort of epistemology begins with a catalogue of means of learning, and asks what knowledge those means can provide. Such an epistemology runs a sceptical risk of discovering that we know a good deal less than we thought we did. Another sort of epistemology begins by assuming that by and large we know what we think we know, and asks what means are required to provide such knowledge. This second sort runs a risk of discovering our psychologies to be rather richer than we thought they were. The former puts epistemology before metaphysics, while the second founds epistemology on metaphysics. But there is a third strategy: perhaps we can so juggle means and ends simultaneously as to preserve most of what we know and a plausible psychology while also attaining harmony between the two. Such is my hope.

It is striking that despite his talent for exposition, Quine is so little understood. I shall begin by outlining some of Quine's characteristic moves. Consider 'On What There Is' and the sixth section of 'Two Dogmas of Empiricism'.[1] The first thing to keep firmly in mind is that these passages contain an *argument*; they are more than just the assertion of a view. This argument proceeds through two stages, and the first stage is epistemological, not at all semantical. The first stage of Quine's argument has two premisses. I call the first premiss Duhem's thesis: beyond a minimal level, no scientific hypothesis is tested individually; only relatively large and heterogeneous bodies of hypotheses are tested against experiment and observation.

Duhem's thesis is supported by examining the history of natural science. Consider, for example, testing Kepler's hypothesis that the planets travel in elliptical orbits about the sun against Tycho Brahe's meticulous observations.

First published in *Proceedings of the Aristotelian Society*, suppl. vol. 53 (1979): 153–65. I have somewhat revised the original paper. Reproduced by courtesy of the Editor of the Aristotelian Society: © 1979.

[1] W. V. O. Quine, 'On What There Is' and 'Two Dogmas of Empiricism', in *From a Logical Point of View*, 2nd edn. (Cambridge, Mass.: Harvard University Press, 1961), 1–19, 20–46. The latter is reproduced here as Chapter II.

Here we need to know that a planet (once) was at some point on the straight prolongation of the line of sight from Tycho to the point in, say, the sky at which the planet appeared to Tycho. In other words, Kepler's hypothesis alone does not suffice to yield Tycho's observations; it requires as well at least the very different hypothesis that light rays *in vacuo* travel in straight lines. Suppose we could survey the large and heterogeneous body of scientific hypotheses (and initial conditions) sufficient to yield an interesting scientific prediction, but that this prediction is observed to be false. It follows that the conjunction of hypotheses (and initial conditions) is false. Nothing in this result establishes *which* hypothesis (or initial condition) fails; refutation so far attaches only to the whole conjunction. But if only large and heterogeneous bodies of hypotheses are ever tested against experience, then there will never be grounds forcing refutation on one rather than another of the hypotheses. In symmetrical justice, should observation bear out the prediction, then there are no grounds for assigning the resultant confirmation to some at the exclusion of others of the hypotheses required to account for that happy prediction.

I call the second premiss of Quine's argument his 'indispensability thesis': beyond a minimal level, we do not know how to do natural science without mathematics. In logicians' jargon: any reasonably sophisticated natural scientific theory could be formalized only as an extension of some part of mathematics. That Quine's indispensability thesis is true should be evident to anyone with even the most elementary scientific education. On the other hand, *why* it is true may be a deeper matter. Conservation principles appear to play a crucial role in scientific explanation. We ask, 'Where did the drive, force, energy of this event come from, how did it evolve, and where did it go?' Without conservation of something, that question makes no sense; indeed, perhaps causation requires conservation. At any rate, conservation is always of amounts of quantities, and once upon a time, mathematics was defined as the science of quantity. So perhaps it is inevitable that if natural science is to provide sophisticated explanations, then it will incorporate some mathematics. (More on these themes elsewhere.)

It follows from Duhem's thesis and Quine's indispensability thesis that such mathematics as is required by natural science for making true predictions (and by Quine's thesis, such mathematics there is) is confirmed thereby; there are no grounds for confining such confirmation to the natural science and denying it to the mathematical science. I call this conclusion of the first stage of Quine's argument his epistemology for mathematics'. There are a few points about it worth noting.

Quine's epistemology for mathematics says nothing whatsoever about meaning. He draws a semantical conclusion only by adding a further premiss:

verificationism. The old slogan had it that meaning is method of verification; if so, then only what is confirmed has meaning. So if only large and heterogeneous bodies of hypotheses can be confirmed, there can be no grounds for assigning distinct meanings to distinct scientific hypotheses; there are no propositions. The verificationist crux here is the view that the unit of meaning is the smallest unit of confirmation. Thus, for example, Kepler's hypotheses of elliptical orbits cannot be assigned a distinct meaning from that of the hypothesis that light travels in straight lines. Paradoxically enough, our 'heterogeneous' hypotheses merge in meaning.

This semantic holism completes the second stage of Quine's argument: except for those special stimulus meanings described later in *Word and Object*, there are no propositions. Understood properly, this is a remarkable thesis: the unit of meaning is the smallest unit capable of confirmation (verificationism); these are large (but not, by compactness, infinite) and heterogeneous in the case of scientific hypotheses (Duhem's thesis); so distinct meanings for such hypotheses cannot be distinguished (semantic holism). And since science requires mathematics (indispensability thesis), meanings for the mathematical results required for confirmation of scientific hypotheses likewise cannot be distinguished from meanings for those hypotheses (there is no analytic/synthetic distinction). (Note how verificationism is here extended to mathematics, a move Quine shares with the later Wittgenstein and latterly Dummett.)

It is clear that this second and semantic stage of Quine's argument depends crucially on verificationism; without it, we get only epistemology for mathematics. Dummett's objections to Quine are primarily, if not exclusively, directed at his semantic holism. But Quine and Dummett share an impulse toward verificationism. I find this perplexing. Much of the first half of this philosophical century was given over to logical positivism, now remembered in no small part for its verifiability theories of meaning. I should have hoped that if there were any settled philosophical opinions, one of them might be that verification is hopeless. (Perhaps that is because I was brought up on Carl Hempel's excellent essay, 'Problems and Changes in the Empiricist Criterion of Meaning.'[2]) But even in the last quarter of this century, it would seem that verificationism will not lie down.

Here is an argument which may be novel to some. The verificationist axe was the principle that what is not verifiable is not meaningful; that is, what is meaningful is also verifiable. Since what is true is meaningful and what is verifiable can be known, it follows that what is true can be known. This is a

---

[2] Carl G. Hempel, 'Problems and Changes in the Empiricist Criterion of Meaning', in *Semantics and the Philosophy of Language*, ed. Leonard Linsky (Urbana: University of Illinois Press, 1952), 163–85.

weak consequence of verificationism and a weak thesis of idealism (since its denial, some truths cannot be known, asserts a strong independence of truth from knowability). Add to propositional calculus a weak modal logic: if a conditional and its antecedent are both necessarily true, then so is its consequent; and permit inference of the necessitation of anything proved in this system. As for knowledge, suppose first that anything known is true. Assume second a weak form of the view that logical consequences of what is known are known: if a conjunction is known, so are its conjuncts. Our last assumption is the weak consequence of verificationism (which mixes modal and epistemic operators): what is true can be known. From these assumptions alone, it follows that everything true is known.[3] This result being preposterous, it follows that one of our assumptions is false. Since I believe the rest (and in particular that if a conjunction is known, so are its conjuncts), I conclude that the weak consequence of verificationism, and with it verificationism, is false. (I am thereby committed to the view that there are truths which

---

[3] In logical crochets, the argument goes thus: Suppose for *reductio* that p & ¬Kp. Then ◊ K (p & ¬Kp) by the weak consequence of verificationism. Our second knowledge assumption allows us to distribute the 'K' across the ampersand, and our weak modal logic is strong enough to allow us to do this inside the diamond; thus, ◊(Kp & K ¬Kp). Our first knowledge assumption allows us to drop the outer 'K' from the second conjunct, and our weak modal logic is strong enough to allow us to do so inside the diamond; thus, ◊ (Kp & ¬Kp). This says that a contradiction might be true, an *absurdum* in any modal logic, even ours.

This unjustly neglected logical gem has a curious history. It was first published by F. B. Fitch in 'A Logical Analysis of some Value Concepts', *Journal of Symbolic Logic*, 28 (1963): 135–42. Fitch credits the argument to an anonymous referee of a paper he sent to the *Journal of Symbolic Logic* in 1945 but never published. I would like to know the identity of that referee.

Incidentally, on an intuitionist reading, it just might be that every truth is known. For being in an intuitionist position to assert that $(\forall x)(Fx \rightarrow Gx)$ requires a method which, given an object and a proof that it is F, yields a proof that it is G. In the present instance this means: suppose we are given a sentence (statement, proposition, thought, what have you) and a proof that it is true. Read the proof; thereby you come to know that the sentence is true. Reflecting on your recent learning, you recognize that the sentence is now known by you; this shows that the truth is known. If this argument is intuitionistically acceptable (it is reminiscent of intuitionistic defences of a version of the axiom of choice), then I think that fact reflects poorly on intuitionism; surely we have good inductive grounds for believing that there are truths as yet unknown.

In the years since this paper was first published in 1979, a literature on this argument has grown up. See Dorothy Edgington, 'The Paradox of Knowability', *Mind*, 94 (1985): 557–68, and Wlodzimierz Rabinowicz, 'Intuitionistic Truth', *Journal of Philosophical Logic*, 14 (1985): 191–228. Timothy Williamson in particular has made a hobby of the argument. Here is just a sample: 'Intuitionism Disproved?', *Analysis*, 42 (1982): 203–8; 'On Knowledge and the Unknowable', *Analysis*, 47 (1987): 154–8; 'Anthropocentrism and Truth', *Philosophia*, 17 (1987): 33–53; 'On the Paradox of Knowability', *Mind*, 96 (1987), 256–61; 'Assertion, Denial and Some Cancellation Rules in Modal Logic', *Journal of Philosophical Logic*, 17 (1988): 299–318; 'Knowability and Constructivism', *Philosophical Quarterly*, 38 (1988): 422–32; 'Verification, Falsification, and Cancellation in KT', *Notre Dame Journal of Formal Logic*, 31 (1990): 286–90; 'Two Incomplete Anti-Realist Modal Epistemic Logics', *Journal of Symbolic Logic*, 55 (1990): 297–314; 'Fitch and Intuitionistic Knowability', *Analysis*, 50 (1990): 182–7; 'On Intuitionistic Modal Epistemic Logic', *Journal of Philosophical Logic*, 21 (1992): 63–89.

cannot possibly be known, a result which on pain of paradox could not be proved constructively; I am an unreconstructed realist.)

Let us forget meaning for the moment and return to knowledge. The first stage of Quine's argument does *not* conclude that certain mathematics is confirmed *only* by being required by natural science for true prediction and correct explanation; it is compatible with there being other routes to confirmation. I think it is hopeless to wish for a *proof* that there are no other routes; the matter seems instead an experimental question. That is, one tries to devise other ways mathematics might be confirmed; one surveys such proposals in the literature; and one thinks through whether they stand up to criticism; if they all fail persistently and for good reason, then eventually one has amassed solid inductive support for thinking that only Quine's epistemology for mathematics will do. Such now seems to me to be the case. (Note that, unlike those who discover in us something like a capacity to perceive single abstracta common to many concreta, Quine invokes here no mysterious new faculties for justifying mathematical belief.)

There are two common objections to Quine's epistemology which deserve mention. First, Quine's account deals only with such mathematics as is applied in natural science. What about the rest? Here we might lift an idea from Gödel.[4] It is clear that hypotheses in natural science are not confirmed only by experiment and observation; a lower-level hypothesis so confirmed can in turn transmit its confirmation to higher-level hypotheses which explain it. For example, Tycho's observations confirm Kepler's (rather odd) laws of planetary motion; those laws are in turn in part explained by Newton's laws of motion and gravitation; and in the standard inductive volte-face, Kepler's laws thus transmit confirmation to Newton's. Analogously, such arithmetic and analysis (and so on) as is required by natural science for true prediction and correct explanation is confirmed thereby; such arithmetic and analysis are organized, systematized, and sometimes even *explained*[5] by more

[4] Kurt Gödel, 'What is Cantor's Continuum Problem?', in P. Benacerraf and H. Putnam (eds.), *Philosophy of Mathematics: Selected Readings*, 2nd edn. (Cambridge: Cambridge University Press, 1983), 470–85, esp. 476–7.

[5] Not all proofs of a theorem explain why it is true, although good ones should. Gauss, e.g., proved the law of quadratic reciprocity in many different ways, because although each proof showed that the theorem was true, none explained why; I am told that it required the quite abstract class field theory of this century to explain the law. (If the view here suggested is true, then Hempel's deductive-nomological model of explanation is not sufficient to characterize explanation in mathematics; for all correct proofs conform to a deductive-nomological model, but some are not explanations.)

Turning from mathematics to science in general, I doubt that the notions of confirmation and explanation are independent. It is well known that triviality ensues from the thesis that what entails an observation is confirmed by it. See Nelson Goodman, *Fact, Fiction, and Forecast* (London: Athlone Press, 1954), 70; entailment is not sufficient for confirmation. More likely, the best explanation of an observation is confirmed by it. See Gilbert H. Harman, 'The Inference to the Best Explanation', *Philosophical Review*, 74 (1965): 88–95. While Hempel's deductive-nomological

abstract mathematical theories (like number theory, topology, algebra, measure theory, and even (some) set theory); and in that way more concrete mathematics transmits confirmation to less. For this analogy to yield confirmation to *all* of received mathematics, each of its bits would have to fit somewhere into a tier of illuminating generalizations ultimately grounded somewhere in mathematics applied in natural science. While that is by no means obviously true, it nevertheless coheres with what seems to be a commonly held conception of the history of mathematics. (On the other hand, perhaps we do not have reason to believe wild-eyed claims about, say, large cardinals which cannot be fitted into such a tier.)

Second, there is the matter of proof. Quine's account makes no mention of it, but surely the phenomenology of mathematics is overwhelmingly one of acquiring knowledge by proving theorems. But this objection is superficial; proofs do not spring from nothing. While we still do not command a clear view of what proofs *in medias res* really are, it is no accident that the paradigm of a mathematical theory is axiomatic. Because you cannot get something for nothing, proofs of theorems in axiomatic systems always trace their lineage back to unproved axioms which are precisely that: unproved. (This point stands even if a potentially infinite regress of proof is possible.) Deduction of theorems from axioms issues in knowledge of theorems only if the axioms are known; so, since the axioms are not proved, proof cannot be the only, or even the basic, source of mathematical knowledge. When, as for us it eventually must, the regress of proof gives out, it must in good conscience be asked how the axioms are known. It is here that Quine has an answer: the axioms are confirmed by best accounting for those of their consequences required by natural science for true prediction and correct explanation. (This view clearly owes much to the inductive justification for the principles of mathematics sketched by Russell on the first page of the preface to volume 1 of *Principia Mathematica*.)

Since Dummett argues against Quine, and since Dummett is not always easy to follow, it may not be amiss if I try to summarize Dummett's argument and respond to it. In this I shall follow Colin McGinn's illuminating essay 'Truth and Use'.[6] My summary must perforce by very brief; for more, see McGinn's essay.

---

model is probably right in claiming that entailment is necessary for explanation, it is probably not sufficient; as a famous example has it, one may deduce the height of a tree from the length of its shadow, the angle between its shadow and the line of sight from the end of the shadow to the treetop, and some laws of trigonometry and optics, but that does not explain the tree's height. If some mathematics is confirmed by observation, then that is because it is inextricably part of what best explains such observation.

[6] Colin McGinn, 'Truth and Use', in Mark Platts (ed.), *Reference, Truth and Reality: Essays in the Philosophy of Language* (London: Routledge and Kegan Paul, 1980), 19–40.

Dummett argues that general considerations from the theory of meaning suffice to refute realism, that is, the 'view' that there is something in virtue of which some of what we say is true (or false) quite independently of limitations on our capacity to come to know the truth value of what we say; the theory of meaning shows that conditions for truth cannot outstrip our capacity to learn. Suppose we have a materially adequate truth theory for a natural language L, that is, a finitely axiomatized theory *à la* Tarski which yields one of the familiar T-sentences for each sentence of L. If this truth theory is to provide a theory of meaning for L, then granted that speakers of L know what sentences of L mean, the truth theory should in some way express this knowledge. Moreover, this knowledge should in some way mesh smoothly with speakers' capacity to *use* L. For any natural language is essentially an instrument of *communication*. Consequently, as is now widely held, meaning must be publicly accessible—that is, both manifested in and recoverable from observable linguistic use; and what is so manifested and recovered ought to be some sort of knowledge had by speakers of L and encoded in the truth theory for L. Dummett argues that no knowledge-transcendent species of truth can sustain the required link through knowledge to use.

We cannot in general identify knowledge of a sentence's truth conditions with the capacity to state them. For it the speaker uses the sentence to do so, we have a circle; and if he uses another sentence, we are launched on a regress. At this point, Dummett proposes that knowledge of truth conditions be associated with a capacity to learn the sentence's truth value. If only the capacity to recognize a sentence's truth value makes the required connection between a sentence's truth conditions, a speaker's knowledge of these conditions, and his ability to use that sentence in communication, then a realist, knowledge-transcendent species of truth can play no part in an adequate theory of meaning. According to Dummett, it is not enough that a theory of understanding (knowledge of meaning) merely record what a speaker knows, for example, that 'Snow is white' is true if and only if snow is white. It must also supply such an explanation of this knowledge that meaning shall be manifested in and recoverable from observable linguistic use. According to Dummett, there could be no such way to come to understand or to communicate by using a sentence whose truth value we could not learn.

This argument seems objectionably reductionist to me. We perceive some objects and we perform some tests on them in order to learn whether certain predicates are true or false of them. These are the elements from which some empiricists try to construct a capacity to understand certain subject–predicate sentences. In such an empiricist scheme, observation of objects and tests performed on them are our basic, perhaps only, evidence for all that we are justified in believing. What Dummett, and many before him, cannot stomach is that we might think or mean by what we say or understand by what another says something whose content goes beyond the evidence for such an assertion. As it were, there are no real inductions; how could we get the idea of objects we cannot perceive? I do not know; but then neither do I know how we get ideas of objects we do perceive. (Empiricist semantics steals plausibility from the unexpressed notion that it is quite plain how we acquire ideas of objects we perceive. But that notion should be suspect. Crabs see stones, but what ideas of stones do they have? Ideas are *not* faint copies of impressions.) But still, I claim, it is a fact that in explaining what we do perceive, we come up with thoughts and ideas of things we do not, and sometimes cannot, perceive. What we observe to be true is evidence that what would best explain it is also true; but that evidence does not in general constitute the meaning of that for which it is evidence. To insist otherwise is reductionism, verificationism, and flies in the face of subtle and indirect scientific argument. The history of philosophy is replete with reductionist programmes whose proponents insisted stridently that their constructions *must* be possible. In fact, not one was ever executed; how long does it take for the message to get through?

Suppose then that we can come up with evidence-transcendent conceptions. How do they manifest themselves in the use of language? Is there any mode of manifestation alternative to Dummett's proposal, a capacity to recognize as such what would satisfy such a conception? McGinn says yes: the capacity to communicate, to understand, and be understood by others sharing such a conception. Suppose another says to you, 'The continuum has $\aleph_1$ members'. What matters for understanding is not the capacity to find the continuum and count up its members, or even the ability to recognize a proof of the continuum hypothesis. What matters in the first instance is what Dummett says is not enough, that you should know that this sentence is true (in a canonical way) if and only if the continuum has $\aleph_1$ members, and all of the great deal more that goes into being able to interpret another. Certainly it would be interesting to know how you acquired such ideas and how you came to associate them with his words; but ignorance on that score should not make you doubt that you understand him, that he is communicating by exploiting your knowledge of English. But, Dummett may ask, in what does this understanding consist? I suspect 'consist'. If it is colourless, then Butler's answer will do: it is what it is and not another thing. If it is an insistence on a reduction, then I reject it. What someone perceives and does can provide evidence for what he understands; but once again, I distinguish evidence for the truth of sentence from its meaning.

In a sense, then, it is true that epistemological holism fails to give an 'account' of understanding, learning, knowledge, recognition, or communication. It fails to reduce them to other somehow more privileged phenomena, and it is none the worse for that. Of course Quine, set as he is against 'pernicious mentalism', cheerfully expects the eventual passing of such myths. But I am as realist about the mind as I am about mathematics, so I am sanguine about understanding even if there is nothing else in which it consists. Because I distinguish meaning from evidence, I accept epistemological holism and underdetermination of theory by evidence, I reject semantic holism, and I distinguish meanings among heterogeneous hypotheses.

Quine's epistemological holism supplemented by Gödel's ideas gives us reason to believe that (many of) the received theorems of classical mathematics are true. This is not quite yet knowledge of abstract entities, unless some of those theorems say that there are abstract entities. Oddly enough, this seems both obviously true and hotly debated at length. So consider Euclid's famous result, 'There are infinitely many primes'. This, I assume, is a result that Quine's construction gives us reason to believe. Moreover, it says that there are numbers, namely, infinitely many prime ones. This can be dressed up in talk of objectual quantification, ontological commitment, and how truth derives from satisfaction, but the point seems too patent to need belabouring yet again. So what needs showing is that numbers are abstract objects. (What, do you ask, is an abstract object? If it contributes to answering 'What is a physical object?' to say 'Read physics', then it contributes to answering your question to say 'Read mathematics'.) I find it obvious that numbers are abstract, but here is an argument. First, it seems clear to me that everything is mental, physical, or abstract. Hume argues persuasively that no physical object exists necessarily.[7] (That is not to say there might have been nothing physical; rather, each physical thing might not have been.) It seems if anything clearer that each mental entity might not have been; indeed, perhaps once there were no minds at all. But numbers, according to tradition, exist necessarily if they exist at all. Since (by Quine's argument) they do exist, they do so necessarily. Therefore, they are neither mental nor physical, so they are abstract. (Of course Quine, set as he is against modality, will disdain this argument; but I am as realist about modality as I am about minds and mathematics.)

In the first instance, we want numbers to account naturally for mathematical truth. Since 'true' is not equivocal, we want a uniform account of truth, in terms of satisfaction of open and closed sentences by sequences of objects, to go for natural science and mathematics. But we want mathematical truth to account smoothly for observed natural phenomena; that is Quine's construction. So there is an analogy between inferring the existence of

---

[7] David Hume, *Dialogues concerning Natural Religion*, ch. IX.

electromagnetic fields from the fact that the hypothesis of such fields is the best way to account for observed phenomena like light and inferring the existence of numbers from the fact that beyond a minimal level, no scientific explanation of anything is possible without mathematical results stating, on the natural view of them and truth, that numbers exist; in neither case does it much matter whether one has no epistemic access to the naked inferred entity.

We thus have reason to believe, and I think enough to know, that there are abstract objects. But I suspect that this is not what some opponents of platonism were after. It is hard to be sure, but it seems to me that they want something rather more intimate than an inductive inference to the conclusion that there are numbers; they want to grasp them. It seems to me that their thinking is dominated by what one might call a 'direct object grammar of knowledge'.

Ever since Ryle made it, we have become accustomed to a distinction between knowing how and knowing that. But there is also knowing objects, and there may even be an inarticulate tradition that would like to reduce propositional knowledge to knowing objects, or perhaps make the second a necessary condition of the first; can one know that a is F if one does not know a? At any rate, it seems to me that some think knowing that there are numbers is not enough; it looks to me as though they think we ought to know abstract objects.

Quine once wrote, 'numbers are known only by their laws';[8] we do not grasp numbers. That this should be so is, I think, reasonably clear; it is a virtue of Quine's epistemology for mathematics that it invokes no marvellous faculty by which we discern numbers. Perception is the only means by which we become acquainted with objects, and, I think, we know only objects with which we are acquainted. As Grice argues persuasively, perception is essentially causal.[9] Because they are very abstract, numbers are causally inert. (Perhaps we can go a bit deeper. Quine says that causation is energy flow.[10] But whatever acquires energy thereby has a mass-equivalent; that is from special relativity. Only concreta have mass. So, since numbers are abstract, they are causally inert.) Thus, we do not know numbers.

But if we do not know numbers, should we therefore doubt that we know that there are numbers. On the contrary, the case of numbers demonstrates that knowing that certain objects exist, or even individual facts about individual numbers, does not require even the possibility of knowing any of

[8] W. V. O. Quine, 'Ontological Relativity', in *Ontological Relativity and Other Essays* (New York: Columbia University Press, 1969), 44.

[9] H. P. Grice, 'The Causal Theory of Perception', in Robert J. Swartz (ed.), *Perceiving, Sensing and Knowing* (Garden City, NY: Doubleday, 1965), 438–72.

[10] W. V. O. Quine, *The Roots of Reference* (La Salle, Ill.: Open Court, 1973), 4–8.

those objects. One may perfectly well come by inference to know that an object exists, and if so, whether one can be acquainted with it as well is quite beside the point. But there are a couple of residual points of interest here.

First, in another context Quine and others have revived an old distinction between what are now often called knowledge *de re* and knowledge *de dicto*. There are at present two principal treatments of this distinction in the literature. One, due to David Kaplan, sought to account for *de re* knowledge (in part) by requiring the knower to interact causally with the *res* of which he has *de re* knowledge.[11] The other, due to Brian Loar, sought the account along more description-theoretic lines by supposing certain singular terms to specify both the *res* known and a part of the content of what the *de re* knower is said to know.[12] The subject is complex, subtle, and difficult, but I suspect that Kaplan's view handles his fascinating problem of the shortest spy better than Loar's. If, then, *de re* knowledge does require causal commerce with the *res*, then since numbers are causally inert, it would follow that there is no *de re* knowledge of numbers. Mildly interesting perhaps, but hardly a problem: mathematical knowledge, even of singular propositions, is never *de re*.

Second, a few causal conceptions of reference have recently been bruited about the literature. If such views are intended to require that a person's utterance of a singular term on an occasion must trace back through a causal chain to an object (or, in Quine's terms, some of the energy of that act of utterance flows ultimately from that object) in order either that the person or his singular term refer on that occasion to that object, then since I believe that some people and some singular terms refer to causally inert numbers, I am therefore forced to conclude that such causal theories of reference are false of numbers. How reference to numbers is possible I do not know; but none of these remarks apply to historical chain theories in which initial dubbings need not involve causal commerce with the referent.

It is the tendency (but only that) of these two points that mathematical knowledge is what Russell might have called 'knowledge by description', not knowledge by acquaintance; numbers, as Quine says, are known only by their laws. If we are to preserve knowledge of the received mathematical truths, understood as saying the Platonic things they seem to say (and such should always be the presumption of any inquiry into meaning), while also preserving credibility for the theory of our learning faculties, then some such result seems all but inevitable.

[11] David Kaplan, 'Quantifying In', *Synthese*, 19 (Dec. 1968): 178–214.
[12] Brian Loar, 'Reference and Propositional Attitudes', *Philosophical Review*, 81 (1972): 43–62.

IV

THE PHILOSOPHICAL BASIS OF INTUITIONISTIC LOGIC

MICHAEL DUMMETT

The question with which I am here concerned is: What plausible rationale can there be for repudiating, within mathematical reasoning, the canons of classical logic in favour of those of intuitionistic logic? I am, thus, not concerned with justifications of intuitionistic mathematics from an eclectic point of view, that is, from one which would admit intuitionistic mathematics as a legitimate and interesting form of mathematics alongside classical mathematics: I am concerned only with the standpoint of the intuitionists themselves, namely, that classical mathematics employs forms of reasoning which are not valid on any legitimate construal of mathematical statements (save, occasionally, by accident, as it were, under a quite unintended reinterpretation). Nor am I concerned with exegesis of the writings of Brouwer or of Heyting: the question is what forms of justification of intuitionistic mathematics will stand up, not what particular writers, however eminent, had in mind. And, finally, I am concerned only with the most fundamental feature of intuitionistic mathematics, its underlying logic, and not with the other respects (such as the theory of free choice sequences) in which it differs from classical mathematics. It will therefore be possible to conduct the discussion wholly at the level of elementary number theory. Since we are, in effect, solely concerned with the logical constants—with the sentential operators and the first-order quantifiers—our interest lies only with the most general features of the notion of a mathematical construction, although it will be seen that we need to consider these in a somewhat delicate way.

Any justification for adopting one logic rather than another as the logic for mathematics must turn on questions of *meaning*. It would be impossible to contrive such a justification which took meaning for granted, and represented the question as turning on knowledge or certainty. We are certain of the truth of a statement when we have conclusive grounds for it and are certain that the grounds which we have *are* valid grounds for it and *are* conclusive. If classical arguments for mathematical statements are called in question, this cannot possibly be because it is thought that we are, in general, unable to tell with certainty whether an argument is classically valid, unless it is also intuitionistically valid: rather, it must be that what is being put in doubt is whether arguments which are valid by classical but not by intuitionistic criteria are absolutely valid, that is, whether they really conclusively establish their conclusions as true. Even if it were held that classical arguments, while not in general absolutely valid, nevertheless always conferred a high probability on their conclusions, it would be wrong to characterize the motive for employing only intuitionistic arguments as lying in a desire to attain knowledge in place of mere probable opinion in mathematics, since the very thesis that the use of classical arguments did not lead to knowledge would represent the crucial departure from the classical conception, beside which the question of whether or not one continued to make use of classical arguments as mere probabilistic reasoning is comparatively insignificant. (In any case, within standard intuitionistic mathematics, there is no reason whatever why the existence of a classical proof of it should render a statement probable, since if, for example, it is a statement of analysis, its being a classical theorem does not prevent it from being intuitionistically disprovable.)

So far as I am able to see, there are just two lines of argument for repudiating classical reasoning in mathematics in favour of intuitionistic reasoning. The first runs along the following lines. The meaning of a mathematical statement determines and is exhaustively determined by its *use*. The meaning of such a statement cannot be, or contain as an ingredient, anything which is not manifest in the use made of it, lying solely in the mind of the individual who apprehends that meaning: if two individuals agree completely about the use to be made of the statement, then they agree about its meaning. The reason is that the meaning of a statement consists solely in its role as an instrument of communication between individuals, just as the powers of a chess piece consist solely in its role in the game according to the rules. An individual cannot communicate what he cannot be observed to communicate: if one individual associated with a mathematical symbol or formula some mental content, where the association did not lie in the use he made of the symbol or formula, then he could not convey that content by means of the symbol or formula, for his audience would be unaware of the association and would have no means of becoming aware of it.

The argument may be expressed in terms of the *knowledge* of meaning, that is of understanding. A model of meaning is a model of understanding, that is, a representation of what it is that is known when an individual knows the meaning. Now knowledge of the meaning of a particular symbol or expression is frequently verbalizable knowledge, that is, knowledge which consists in the ability to state the rules in accordance with which the expression or symbol is used or the way in which it may be replaced by an equivalent expression or sequence of symbols. But to suppose that, in general, a knowledge of meaning consisted in verbalizable knowledge would involve an infinite regress: if a grasp of the meaning of an expression consisted, in general, in the ability to *state* its meaning, then it would be impossible for anyone to learn a language who was not already equipped with a fairly extensive language. Hence that knowledge which, in general, constitutes the understanding of the language of mathematics must be implicit knowledge. Implicit knowledge cannot, however, meaningfully be ascribed to someone unless it is possible to say in what the manifestation of that knowledge consists: there must be an observable difference between the behaviour or capacities of someone who is said to have that knowledge and someone who is said to lack it. Hence it follows, once more, that a grasp of the meaning of a mathematical statement must, in general, consist of a capacity to use that statement in a certain way, or to respond in a certain way to its use by others.

Another approach is via the idea of learning mathematics. When we learn a mathematical notation, or mathematical expressions, or, more generally, the language of a mathematical theory, what we learn to do is to make use of the statements of that language: we learn when they may be established by computation and how to carry out the relevant computations; we learn from what they may be inferred and what may be inferred from them, that is, what role they play in mathematical proofs and how they can be applied in extra-mathematical contexts; and perhaps we learn also what plausible arguments can render them probable. These things are all that we are shown when we are learning the meanings of the expressions of the language of the mathematical theory in question, because they are all that we can be shown; and, likewise, our proficiency in making the correct use of the statements and expressions of the language is all that others have from which to judge whether or not we have acquired a grasp of their meanings. Hence it can only be in the capacity to make a correct use of the statements of the language that a grasp of their meanings, and those of the symbols and expressions which they contain, can consist. To suppose that there is an ingredient of meaning which transcends the use that is made of that which carries the meaning is to suppose that someone might have learned all that is directly taught when the language of a mathematical theory is taught to him, and might then behave in every way like someone who understood that language, and yet not actually understand it, or understand it only incorrectly. But to suppose this is to make meaning ineffable, that is, in principle incommunicable. If this is possible, then no one individual ever has a guarantee that he is understood by any other individual; for all he knows, or can ever know, everyone else may attach to his words or to the symbols which he employs a meaning quite different from that which he attaches to them. A notion of meaning so private to the individual is one that has become completely irrelevant to mathematics as it is actually practised, namely, as a body of theory on which many individuals are corporately engaged, an inquiry within which each can communicate his results to others.

It might seem that an approach to meaning which regarded it as exhaustively determined by use would rule out any form of revisionism. If *use* constitutes meaning, then, it might seem, *use* is beyond criticism: there can be no place for rejecting any established mathematical practice, such as the use of certain forms of argument or modes of proof, since that practice, together with all others which are generally accepted, is simply constitutive of the meanings of our mathematical statements, and we surely have the right to make our statements mean whatever we choose that they shall mean. Such an attitude is one possible development of the thesis that *use* exhaustively determines meaning: it is, however, one which can, ultimately, be supported only by the adoption of a holistic view of language. On such a view, it is illegitimate to ask after the content of any single statement, or even after that of any one theory, say a mathematical or a physical theory; the significance of each statement or of each deductively systematized body of statements is modified by the multiple connections which it has, direct and remote, with other statements in other areas of our language taken as a whole, and so there is no adequate way of understanding the statement short of knowing the entire language. Or, rather, even this image is false to the facts: it is not that a statement or even a theory has, as it were, a primal meaning which then gets modified by the interconnections that are established with other statements and other theories; rather, its meaning simply consists in the place which it occupies in the complicated network which constitutes the totality of our linguistic practices. The only thing to which a definite content may be attributed is the totality of all that we are, at a given time, prepared to assert; and there can be no simple model of the content which that totality of assertions embodies; nothing short of a complete knowledge of the language can reveal it.

Frequently such a holistic view is modified to the extent of admitting a class of observation statements which can be regarded as more or less directly registering our immediate experience, and hence as each carrying a determinate individual content. These observation statements lie, in Quine's famous image of language, at the periphery of the articulated structure formed by all the sentences of our language, where alone experience

INTUITIONISTIC LOGIC 67

impinges. To these peripheral sentences, meanings may be ascribed in a more or less straightforward manner, in terms of the observational stimuli which prompt assent to and dissent from them. No comparable model of meaning is available for the sentences which lie further towards the interior of the structure: an understanding of them consists solely in a grasp of their place in the structure as a whole and their interaction with its other constituent sentences. Thus, on such a view, we may accept a mathematical theory, and admit its theorems as true, only because we find in practice that it serves as a convenient substructure deep in the interior of the complex structure which forms the total theory: there can be no question of giving a representation of the truth conditions of the statements of the mathematical theory under which they may be judged individually as acceptable, or otherwise, in isolation from the rest of language.

Such a conception bears an evident analogy with Hilbert's view of classical mathematics; or, more accurately, with Boole's view of his logical calculus. For Hilbert, a definite individual content, according to which they may be individually judged as correct or incorrect, may legitimately be ascribed only to a very narrow range of statements of elementary number theory: these correspond to the observation statements of the holistic conception of language. All other statements of mathematics are devoid of such a content, and serve only as auxiliaries, though psychologically indispensable auxiliaries, to the recognition as correct of the finitistic statements which alone are individually meaningful. The other mathematical statements are not, on such a view, devoid of significance: but their significance lies wholly in the role which they play within the mathematical theories to which they belong, and which are themselves significant precisely because they enable us to establish the correctness of finitistic statements. Boole likewise distinguished, amongst the formulas of his logical calculus, those which were interpretable from those which were uninterpretable: a deduction might lead from some interpretable formulae as premisses, via uninterpretable formulae as intermediate steps, to a conclusion which was once more interpretable.

The immediately obvious difficulty about such a manner of construing a mathematical, or any other, theory is to know how it can be justified. How can we be sure that the statements or formulae to which we ascribe a content, and which are derived by such a means, are true? The difference between Hilbert and Boole, in this respect, was that Hilbert took the demand for justification seriously, and saw the business of answering it as the prime task for his philosophy of mathematics, while Boole simply ignored the question. Of course, the most obvious way to find a justification is to extend the interpretation to all the statements or formulae with which we are concerned, and, in the case of Boole's calculus, this is very readily done, and indeed yields a great simplification of the calculus. Even in Hilbert's case, the consistency proof, once found, does yield an interpretation of the infinitistic statements, though one which is relative to the particular proof in which they occur, not one uniform for all contexts. Without such a justification, the operation of the mechanism of the theory or the language remains quite opaque to us; and it is because the holist is oblivious of the demand for justification, or of the unease which the lack of one causes us, that I said that he is to be compared to Boole rather than to Hilbert. In his case, the question would become: With what right do we feel an assurance that the observation statements deduced with the help of the complex theories, mathematical, scientific and otherwise, embedded in the interior of the total linguistic structure, are true, when these observation statements are interpreted in terms of their stimulus meanings? To this the holist attempts no answer, save a generalized appeal to induction: these theories have 'worked' in the past, in the sense of having for the most part yielded true observation statements, and so we have confidence that they will continue to work in the future.

The path of thought which leads from the thesis that use exhaustively determines meaning to an acceptance of intuitionistic logic as the correct logic for mathematics is one which rejects a holistic view of mathematics, and insists that each statement of any mathematical theory must have a determinate individual content. A grasp of this content cannot, in general, consist of a piece of verbalizable knowledge, but must be capable of being fully manifested by the use of the statement: but that does not imply that every aspect of its existing use is sacrosanct. An existing practice in the use of a certain fragment of language is capable of being subjected to criticism if it is impossible to systematize it, that is, to frame a model whereby each sentence carries a determinate content which can, in turn, be explained in terms of the use of that sentence. What makes it possible that such a practice may prove to be incoherent and therefore in need of revision is that there are different aspects to the use of a sentence; if the whole practice is to be capable of systematization in the present sense, there must be a certain harmony between these different aspects. This is already apparent from the holistic examples cited. One aspect of the use of observation statements lies in the propensities we have acquired to assent to and dissent from them under certain types of stimuli; another lies in the possibility of deducing them by means of non-observational statements, including highly theoretical ones. If the linguistic system as a whole is to be coherent, there must be harmony between these two aspects: it must not be possible to deduce observation statements from which the perceptual stimuli require dissent. Indeed, if the observation statements are to retain their status as observation statements, a stronger demand must be made: of an observation statement deduced by means of theory, it must hold that we can place ourselves in a situation in which stimuli occur which require assent to it. This condition is thus a demand that, in a certain sense, the language as a whole be a conservative extension of that fragment of the language containing only observation statements. In just the same way, Hilbert's philosophy of mathematics requires that classical number theory, or even classical analysis, be a conservative extension of finitistic number theory.

For utterances considered quite generally, the bifurcation between the two aspects of their use lies in the distinction between the conventions governing the occasions on which the utterance is appropriately made and those governing both the responses of the hearer and what the speaker commits himself to by making the utterance: schematically, between the *conditions for* the utterance and the *consequences of* it. Where, as in mathematics, the utterances with which we are concerned are *statements*, that is, utterances by means of which assertions can be effected, this becomes the distinction between the grounds on which the statement can be asserted and its inferential consequences, the conclusions that can be inferred from it. Plainly, the requirement of harmony between these in respect of some type of statement is the requirement that the addition of statements of that type to the language produces a conservative extension of the language; that is, that it is not possible, by going via statements of this type as intermediaries, to deduce from premisses not of that type conclusions, also not of that type, which could not have been deduced before. In the case of the logical constants, a loose way of putting the requirement is to say that there must be a harmony between the introduction and elimination rules; but, of course, this is not accurate, since the whole system has to be considered (in classical logic, for example, it is possible to infer a disjunctive statement, say by double-negation elimination, without appeal to the rule of disjunction introduction). An alternative way of viewing the dichotomy between the two principal aspects of the use of statements is as a contrast between *direct* and *indirect* means of establishing them. So far as a logically complex statement is concerned, the introduction rules governing the logical constants occurring in the statement display the most direct means of establishing the statement, step by step in accordance with its logical structure; but the statement may be accepted on the basis of a complicated deduction which relies also on elimination rules, and we require a harmony which obtains only if a statement that has been indirectly established always could (in some sense of 'could') have been established directly. Here again the demand is that the admission of the more complex inferences yield a conservative extension of the language. When only introduction rules are used, the inference involves only statements of logical complexity no greater than that of the conclusion: we require that the derivation of a statement by inferences involving statements of greater logical complexity shall be possible only when its derivation by the more direct means is in some sense already possible.

On any molecular view of language—any view on which individual sentences carry a content which belongs to them in accordance with the way they are compounded out of their own constituents, independently of other sentences of the language not involving those constituents—there must be some demand for harmony between the various aspects of the use of sentences, and hence some possibility of criticizing or rejecting existing practice when it does not display the required harmony. Exactly what the harmony is which is demanded depends upon the theory of meaning accepted for the language, that is, the general model of that in which the content of an individual sentence consists; that is why I rendered the above remarks vague by the insertion of phrases like 'in some sense'. It will always be legitimate to demand, of any expression or form of sentence belonging to the language, that its addition to the language should yield a conservative extension; but, in order to make the notion of a conservative extension precise, we need to appeal to some concept such as that of truth or that of being assertible or capable in principle of being established, or the like; and just which concept is to be selected, and how it is to be explained, will depend upon the theory of meaning that is adopted.

A theory of meaning, at least of the kind with which we are mostly familiar, seizes upon some one general feature of sentences (at least of assertoric sentences, which is all we need be concerned with when considering the language of mathematics) as central: the notion of the content of an individual sentence is then to be explained in terms of this central feature. The selection of some one such feature of sentences as central to the theory of meaning is what is registered by philosophical dicta of the form 'Meaning is . . .'—for example, 'The meaning of a sentence is the method of its verification', 'The meaning of a sentence is determined by its truth conditions', etc. (The slogan 'Meaning is use' is, however, of a different character: the 'use' of a sentence is not, in this sense, a *single* feature; the slogan simply restricts the *kind* of feature that may legitimately be appealed to as constituting or determining meaning.) The justification for thus selecting some one single feature of sentences as central—as being that in which their individual meanings consist—is that it is hoped that every other feature of the use of sentences can be derived, in a uniform manner, from this central one. If, for example, the notion of truth is taken as central to the theory of meaning, then the meanings of individual expressions will consist in the manner in which they contribute to determining the truth conditions of sentences in which they occur; but this conception of meaning will be justified only if it is possible, for an arbitrary assertoric sentence whose truth conditions are taken as known, to describe, in terms of the notion of truth, our actual practice in the use of such a sentence; that is, to give a general characterization of the linguistic practice of making assertions, of the conditions under which they are made, and the responses which they elicit. Obviously, we are very far from being able to construct such a general theory of the use of sentences, of the practice of speaking a language; equally obviously, it is likely that, if we ever do attain such an account, it will involve a considerable modification of the ideal pattern under which the account will take a quite general form, irrespective of the individual content of the sentence as given in terms of whatever is taken as the central notion of the theory of meaning. But it is only to the extent that we shall eventually be able to approximate to such a pattern that it is possible to give substance to the claim that it is in terms of some *one* feature, such as truth or verification, that the individual meanings of sentences and of their component expressions are to be given.

It is the multiplicity of the different features of the use of sentences, and the consequent legitimacy of the demand, given a molecular view of language, for harmony between them, that makes it possible to criticize existing practice, to call in question uses that are actually made of sentences of the language. The thesis with which we started, that use exhaustively determines meaning, does not, therefore, conflict with a revisionary attitude to some aspect of language: what it does do is to restrict the selection of the feature of sentences which is to be treated as central to the theory of meaning. On a platonistic interpretation of a mathematical theory, the central notion is that of truth: a grasp of the meaning of a sentence of the language of the theory consists in a knowledge of what it is for that sentence to be true. Since, in general, the sentences of the language will not be ones whose truth value we are capable of effectively deciding, the condition for the truth of such a sentence will be one which we are not, in general, capable of recognizing as obtaining whenever it obtains, or of getting ourselves into a position in which we can so recognize it. Nevertheless, on the theory of meaning which underlies platonism, an individual's grasp of the meaning of such a sentence consists in his knowledge of what the condition is which has to obtain for the sentence to be true, even though the condition is one which he cannot, in general, recognize as obtaining when it does obtain.

Such a conception violates the principle that use exhaustively determines meaning; or, at least, if it does not, a strong case can be put up that it does, and it is this case which constitutes the first type of ground which appears to exist for repudiating classical in favour of intuitionistic logic for mathematics. For, if the knowledge that constitutes a grasp of the meaning of a sentence has to be capable of being manifested in actual linguistic practice, it is quite obscure in what the knowledge of the condition under which a sentence is true can consist, when that condition is not one which is always capable of being recognized as obtaining. In particular cases, of course, there may be no problem, namely when the knowledge in question may be taken as verbalizable knowledge; that is when the speaker is able to *state*, in other words, what the condition is for the truth of the sentence; but, as we have already noted, this cannot be the general case. An ability to state to express the content of the sentence in other words. We accept such a capacity as evidence of a grasp of the meaning of the original sentence on the presumption that the speaker understands the words in which he is stating its truth condition; but at some point it must be possible to break out of the circle: even if it were always possible to find an equivalent, understanding plainly cannot in general consist in the ability to find a synonymous expression. Thus the knowledge in which, on the platonistic view, a grasp of the meaning of a mathematical statement consists must, in general, be implicit knowledge, knowledge which does not reside in the capacity to state that which is known. But, at least on the thesis that use exhaustively determines meaning, and perhaps on any view whatever, the ascription of implicit knowledge to someone is meaningful only if he is capable, in suitable circumstances, of fully manifesting that knowledge. (Compare Wittgenstein's question why a dog cannot be said to expect that his master will come home next week.) When the sentence is one which we have a method for effectively deciding, there is again no problem: a grasp of the condition under which the sentence is true may be said to be manifested by a mastery of the decision procedure, for the individual may, by that means, get himself into a position in which he can recognize that the condition for the truth of the sentence obtains or does not obtain, and we may reasonably suppose that, in this position, he displays by his linguistic behaviour his recognition that the sentence is, respectively, true or false. But, when the sentence is one which is not in this way effectively decidable, as is the case with the vast majority of sentences of any interesting mathematical theory, the situation is different. Since the sentence is, by hypothesis, effectively undecidable, the condition which must, in general, obtain for it to be true is not one which we are capable of recognizing whenever it obtains, or of getting ourselves in a position to do so. Hence any behaviour which displays a capacity for acknowledging the sentence as being true in all cases in which the condition for its truth can be recognized as obtaining will fall short of being a full manifestation of the knowledge of the condition for its truth: it shows only that the condition can be recognized in certain cases, not that we have a grasp of what, in general, it is for that condition to obtain even in those cases when we are incapable of recognizing that it does. It is, in fact, plain that the knowledge which is being ascribed to one who is said to understand the sentence is knowledge which transcends the capacity to manifest that knowledge by the way in which the sentence is used. The platonistic theory of meaning cannot be a theory in which meaning is fully determined by use.

If to know the meaning of a mathematical statement is to grasp its use; if we learn the meaning by learning the use, and our knowledge of its meaning is a knowledge which we must be capable of manifesting by the use we make of it: then the notion of *truth*, considered as a feature which each mathematical statement either determinately possesses or determinately lacks, independently of our means of recognizing its truth value, cannot be the central notion for a theory of the meanings of mathematical statements. Rather, we have to look at those things which are actually features of the use which we learn to make of mathematical statements. What we actually learn to do, when we learn some part of the language of mathematics, is to recognize, for each statement, what counts as establishing that statement as true or as false. In the case of very simple statements, we learn some computation procedure which decides their truth or falsity: for more complex statements, we learn to recognize what is to be counted as a proof or a disproof of them. That is the practice of which we acquire a mastery: and it is in the mastery of that practice that our grasp of the meanings of the statements must consist. We must, therefore, replace the notion of truth, as the central notion of the theory of meaning for mathematical statements, by the notion of *proof*: a grasp of the meaning of a statement consists in a capacity to recognize a proof of it when one is presented to us, and a grasp of the meaning of any expression smaller than a sentence must consist in a knowledge of the way in which its presence in a sentence contributes to determining what is to count as a proof of that sentence. This does not mean that we are obliged uncritically to accept the canons of proof as conventionally acknowledged. On the contrary, as soon as we construe the logical constants in terms of this conception of meaning, we become aware that certain forms of reasoning which are conventionally accepted are devoid of justification. Just because the conception of meaning in terms of proof is as much a molecular, as opposed to holistic, theory of meaning as that of meaning in terms of truth conditions, forms of inference stand in need of justification, and are open to being rejected as unjustified. Our mathematical practice has been disfigured by a false conception of what our understanding of mathematical theories consisted in.

This sketch of one possible route to an account of why, within mathematics, classical logic must be abandoned in favour of intuitionistic logic obviously leans heavily upon Wittgensteinian ideas about language. Precisely because it rests upon taking with full seriousness the view of language as an instrument of social communication, it looks very unlike traditional intuitionist accounts, which, notoriously, accord a minimum of importance to language or to symbolism as a means of transmitting thought, and are constantly disposed to slide in the direction of solipsism. However, I said at the outset that my concern in this paper was not in the least with the exegesis of actual intuitionist writings: however little it may gibe with the view of the intuitionists themselves, the considerations that I have sketched appear to me to form one possible type of argument in favour of adopting an intuitionistic version of mathematics in place of a classical one (at least as far as the logic employed is concerned), and, moreover, an argument of considerable power. I shall not take the time here to attempt an evaluation of the argument, which would necessitate inquiring how the platonist might reply to it, and how the debate between them would then proceed: my interest lies, rather, in asking whether this is the only legitimate route to the adoption of an intuitionistic logic for mathematics.

Now the first thing that ought to strike us about the form of argument which I have sketched is that it is virtually independent of any considerations relating specifically to the *mathematical* character of the statements under discussion. The argument involved only certain considerations within the theory of meaning of a high degree of generality, and could, therefore, just as well have been applied to any statements whatever, in whatever area of language. The argument told in favour of replacing, as the central notion for the theory of meaning, the condition under which a statement is true, whether we know or can know when that condition obtains, by the condition under which we acknowledge the statement as conclusively established, a condition which we must, by the nature of the case, be capable of effectively recognizing whenever it obtains. Since we were concerned with mathematical statements, which we recognize as true by means of a proof (or, in simple cases, a computation), this meant replacing the notion of truth by that of proof: evidently, the appropriate generalization of this, for statements of an arbitrary kind, would be the replacement of the notion of truth, as the central notion of the theory of meaning, by that of verification; to know the meaning of a statement is, on such a view, to be capable of recognizing whatever counts as verifying the statement, that is, as conclusively establishing it as true. Here, of course, the verification would not ordinarily consist in the bare occurrence of some sequence of sense experiences, as on the positivist conception of the verification of a statement. In the mathematical case, that which establishes a statement as true is the production of a deductive argument terminating in that statement as conclusion; in the general case, a statement will, in general, also be established as true by a process of reasoning, though here the reasoning will not usually be purely deductive in character, and the premisses of the argument will be based on observation; only for a restricted class of statements—the observation statements—will their verification be of a purely observational kind, without the mediation of any chain of reasoning or any other mental, linguistic, or symbolic process.

If follows that, in so far as an intuitionist position in the philosophy of mathematics (or, at least, the acceptance of an intuitionistic logic for mathematics) is supported by an argument of this first type, similar, though not necessarily identical, revisions must be made in the logic accepted for statements of other kinds. What is involved is a thesis in the theory of meaning of the highest possible level of generality. Such a thesis is vulnerable in many places: if it should prove that it cannot be coherently applied to any one region of discourse, to any one class of statements, then the thesis cannot be generally true, and the general argument in favour of it must be fallacious. Construed in this way, therefore, a position in the philosophy of mathematics will be capable of being undermined by considerations which have nothing directly to do with mathematics at all.

Is there, then, any alternative defence of the rejection, for mathematics, of classical in favour of intuitionistic logic? Is there any such defence which turns on the fact that we are dealing with *mathematical* statements in particular, and leaves it entirely open whether or not we wish to extend the argument to statements of any other general class?

Such a defence must start from some thesis about mathematical statements the analogue of which we are free to reject for statements of other kinds. It is plain what this thesis must be: namely, the celebrated thesis that mathematical statements do not relate to an objective mathematical reality existing independently of us. The adoption of such a view apparently leaves us free either to reject or to adopt an analogous view for statements of any other kind. For instance, if we are realists about the physical universe, then we may contrast mathematical statements with statements ascribing physical properties to material objects: on this combination of views, material-object statements do relate to an objective reality existing independently of ourselves, and are rendered true or false, independently of our knowledge of their truth values or of our ability to attain such knowledge or the particular means, if any, by which we do so, by that independently existing reality; the assertion that mathematical statements relate to no such external reality gains its substance by contrast with the physical case. Unlike material objects, mathematical objects are, on this thesis, creations of the human mind: they are objects of thought, not merely in the sense that they can be thought about, but in the sense that their being is to be thought of; for them, *esse est concipi*.

On such a view, a conception of meaning as determined by truth conditions is available for any statements which do relate to an independently existing reality, for then we may legitimately assume, of each such statement, that it possesses a determinate truth value, true or false, independently of our knowledge, according as it does or does not agree with the constitution of that external reality which it is about. But, when the statements of some class do not relate to such an external reality, the supposition that each of them possesses such a determinate truth value is empty, and we therefore cannot regard them as being given meanings by associating truth conditions with them; we have, in such a case, *faute de mieux*, to take them as having been given meaning in a different way, namely, by associating with them conditions of a different kind—conditions that we are capable of recognizing when they obtain—namely, those conditions under which we take their assertion or their denial as being conclusively justified.

The first type of justification of intuitionistic logic which we considered conformed to Kreisel's dictum, 'The point is not the existence of mathematical objects, but the objectivity of mathematical truth': it bore directly upon the claim that mathematical statements possess objective truth values, without raising the question of the ontological status of mathematical objects or the metaphysical character of mathematical reality. But a justification of the second type violates the dictum: it makes the question whether mathematical statements possess objective truth values depend upon a prior decision as to the being of mathematical objects. And the difficulty about it lies in knowing on what we are to base the premiss that mathematical objects are the creations of human thought in advance of deciding what is the correct model for the meanings of mathematical statements or what is the correct conception of truth as relating to them. It appears that, on this view, before deciding whether a grasp of the meaning of a mathematical statement is to be considered as consisting in a knowledge of what has to be the case for it to be true or in a capacity to recognize a proof of it when one is presented, we have first to resolve the metaphysical question whether mathematical objects—natural numbers, for example—are, as on the constructivist view, creations of the human mind or, as on the platonist view, independently existing abstract objects. And the puzzle is to know on what basis we could possibly resolve this metaphysical question, at a stage at which we do not even know what model to use for our understanding of mathematical statements. We are, after all, being asked to choose between two metaphors, two pictures. The platonist metaphor assimilates mathematical inquiry to the investigations of the astronomer: mathematical structures, like galaxies, exist, independently of us, in a realm of reality which we do not inhabit but which those of us who have the skill are capable of observing and reporting on. The constructivist metaphor assimilates mathematical activity to that of the artificer fashioning objects in accordance with the creative power of his imagination. Neither metaphor seems, at first sight, especially apt, nor one more apt than the other: the activities of the mathematician seem strikingly unlike those either of the astronomer or of the artist. What basis can exist for deciding which metaphor is to be preferred? How are we to know in which respects the metaphors are to be taken seriously, how the pictures are to be used?

Preliminary reflection suggests that the metaphysical question ought not to be answered first: we cannot, as the second type of approach would have us do, *first* decide the ontological status of mathematical objects, and then, with that as premiss, deduce the character of mathematical truth or the correct model of meaning for mathematical statements. Rather, we have first to decide on the correct model of meaning—either an intuitionistic one, on the basis of an argument of the first type, or a platonistic one, on the basis of some rebuttal of it; and then one or other picture of the metaphysical character of mathematical reality will force itself on us. If we have decided upon a model of the meanings of mathematical statements according to which we have to repudiate a notion of truth considered as determinately attaching, or failing to attach, to such statements independently of whether we can now, or ever will be able to, prove or disprove them, then we shall be unable to use the picture of mathematical reality as external to us and waiting to be discovered. Instead, we shall inevitably adopt the picture of that reality as being the product of our thought, or, at least, as coming into existence only as it is thought. Conversely, if we admit a notion of truth as attaching objectively to our mathematical statements independently of our knowledge, then, likewise, the picture of mathematical reality as existing, like the galaxies, independently of our observation of it will force itself on us in an equally irresistible manner. But, when we approach the matter in this way, there is no puzzle over the interpretation of these metaphors: psychologically inescapable as they may be, their non-metaphorical content will consist entirely in the two contrasting models of the meanings of mathematical statements, and the issue between them will become simply the issue as to which of these two models is correct. If, however, a view as to the ontological status of mathematical objects is to be treated as a *premiss* for deciding between the two models of meaning, then the metaphors cannot without circularity be explained solely by reference to those models; and it is obscure how else they are to be explained.

These considerations appear, at first sight, to be reinforced by reflection upon Frege's dictum, 'Only in the context of a sentence does a name stand for anything'. We cannot refer to an object save in the course of saying something about it. Hence, any thesis concerning the ontological status of objects of a given kind must be, at the same time, a thesis about what makes a statement involving reference to such objects true—in other words, a thesis about what properties an object of that kind can have. Thus, to say that fictional characters are the creations of the imagination is to say that a statement about a fictional character can be true only if it is imagined as being true, that a fictional character can have only those properties which it is part of the story that he has; to say that something is an object of sense—that for it *esse est percipi*—is to say that it has only those properties it is perceived as having: in both cases, the ontological thesis is a ground for rejecting the law of excluding middle as applied to statements about those objects. Thus we cannot separate the question of the ontological status of a class of objects from the question of the correct notion of truth for statements about those objects, that is, of the kind of thing in virtue of which such statements are true, when they are true. This conclusion corroborates the idea that an answer to the former question cannot serve as a premiss for an answer to the latter one.

Nevertheless, the position is not so straightforward as all this would make it appear. From the possibility of an argument of the first type for the use of intuitionistic logic in mathematics, it is evident that a model of the meanings of mathematical statements in terms of proof rather than of truth need not rest upon any particular view about the ontological character of mathematical objects. There is no substantial disagreement between the two models of meaning so long as we are dealing only with decidable statements: the crucial divergence occurs when we consider ones which are not effectively decidable, and the linguistic operation which first enables us to frame effectively undecidable mathematical statements is that of quantification over infinite totalities, in the first place over the totality of natural numbers. Now suppose someone who has, on whatever grounds, been convinced by the platonist claim that we do not create the natural numbers, and yet that reference to natural numbers is not a mere *façon de parler*, but is a genuine instance of reference to objects: he believes, with the platonist, that natural numbers are abstract objects, existing timelessly and independently of our knowledge of them. Such a person may, nevertheless, when he comes to consider the meaning of existential and universal quantification over the natural numbers, be convinced by a line of reasoning such as that which I sketched as constituting the first type of justification for replacing classical by intuitionistic logic. He may come to the conclusion that quantification over a denumerable totality cannot be construed in terms of our grasp of the conditions under which a quantified statement is true, but must, rather, be understood in terms of our ability to recognize a proof or disproof of such a statement. He will therefore reject a classical logic for number-theoretic statements in general, admitting only intuitionistically valid arguments involving them. Such a person would be accepting a platonistic view of the existence of mathematical objects (at least the objects of number theory), but rejecting a platonistic view of the objectivity of mathematical statements.

Our question is, rather, whether the opposite combination of views is possible: whether one may consistently hold that natural numbers are the creations of human thought, yet believe that there is a notion of truth under which each number-theoretic statement is determinately either true or false, and that it is in terms of our grasp of their truth conditions that our understanding of number-theoretic statements is to be explained. If such a combination is possible, then, it appears, there can be no route from the ontological thesis that mathematical objects are the creations of our thought to the model of the meanings of mathematical statements which underlies the adoption of an intuitionistic logic.

This is not the only question before us: for, even if these two views cannot be consistently combined, it would not follow that the ontological thesis could serve as a premiss for the constructivist view of the meanings of mathematical statements; our difficulty was to understand how the ontological thesis could have any substance if it were not merely a picture encapsulating that conception of meaning. The answer is surely this: that, while it is surely correct that a thesis about the ontological status of objects of a given kind, for example, natural numbers, must be understood as a thesis about that in which the truth of certain statements about those objects consists, it need not be taken as, in the first place, a thesis about the entire class of such statements; it may, instead, be understood as a thesis only about some restricted subclass of such statements, those which are basic to the very possibility of making reference to those objects. Thus, for example, the thesis that natural numbers are creations of human thought may be taken as a thesis about the sort of thing which makes a numerical equation or inequality true, or, more generally, a statement formed from such equations by the sentential operators and bounded quantification. To say that the only notion of truth we can have for number-theoretic statements generally is that which equates truth with our capacity to prove a statement is to prejudge the issue about the correct model of meaning for such statements, and therefore cannot serve as a premiss for the constructivist view of meaning. But to say that, for decidable number-theoretic statements, truth consists in provability, is not in itself to prejudge the question in what the truth of undecidable statements, involving unbounded quantification, consists: and hence the possibility is open that a view about the one might serve as a premiss for a view about the other. Our problem is to discover whether it can do so in fact: whether there is any legitimate route from the thesis that natural numbers are creations of human thought, construed as a thesis about the sort of thing which makes decidable number-theoretic statements true, to a view of the meanings of number-theoretic statements generally which would require the adoption for them of an intuitionistic rather than a classical logic.

In order to resolve this question, it is necessary for us to take a rather closer look at the notion of truth for mathematical statements, as understood intuitionistically. The most obvious suggestion that comes to mind in this connection is that the intuitionistic notion of truth conforms, just as does the classical notion, to Tarski's schema:

(T) $\qquad\qquad\qquad S$ is true iff A,

where an instance of the schema is to be formed by replacing 'A' by some number-theoretic statement and 'S' by a canonical name of that sentence, as, for example in,

'There are infinitely many twin primes' is true iff there are infinitely many twin primes.

It is necessary to admit counter-examples to the schema (T) in any case in which we wish to hold that there exist sentences which are neither true nor false: for if we replace 'A' by such a sentence, the left-hand side of the biconditional becomes false (on the assumption that, if the negation of a sentence is true, that sentence is false), although, by hypothesis, the right-hand side is not false. But, in intuitionistic logic, that semantic principle holds good which stands to the double negation of the law of excluded middle as the law of bivalence stands to the law of excluded middle itself: it is inconsistent to assert of any statement that it is neither true nor false; and hence there seems no obstacle to admitting the correctness of the schema (T). Of course, in doing so, we must construe the statement which appears on the right-hand side of any instance of the schema in an intuitionistic manner. Provided we do this, a truth definition for the sentences of an intuitionistic language, say that of Heyting arithmetic, may be constructed precisely on Tarski's lines, and will yield, as a consequence, each instance of the schema (T).

However, notoriously, such an approach leaves many philosophical problems unresolved. The truth definition tells us, for example, that

'598017 + 246532 = 844549' is true

just in the case in which 598017 + 246532 = 844549. We may perform the computation, and discover that 598017 + 246532 does indeed equal 844549; but does that mean that the equation was already true before the computation was performed, or that it would have been true even if the computation had never been performed? The truth definition leaves such questions quite unanswered, because it does not provide for inflections of tense or mood of the predicate 'is true': it has been introduced only as a predicate as devoid of tense as are all ordinary mathematical predicates; but its role in our language does not reveal why such inflections of tense or even of mood should be forbidden.

INTUITIONISTIC LOGIC

These difficulties raise their heads as soon as we make the attempt to introduce tense into mathematics, as intuitionism provides us with some inclination to do; this can be seen from the problems surrounding the theory of the creative subject. These problems are well brought out in Troelstra's discussion of the topic. It is evident that we ought to admit as an axiom

(α) $\qquad (\vdash_n A) \to A;$

if we know that, at any stage, A has been (or will be) proved, then we are certainly entitled to assert A. But ought we to admit the converse in the form

(β) $\qquad A \to \exists n\, (\vdash_n A)?$

Its double negation

(γ) $\qquad A \to \neg\neg \exists n\, (\vdash_n A)$

is certainly acceptable: if we know that A is true, then we shall certainly never be able to assert, at least on purely mathematical grounds, that it will never be proved. But can we equate truth with the obtaining of a proof at some stage, in the past or in the future, as the equivalence

(δ) $\qquad A \leftrightarrow \exists n\, (\vdash_n A)$

requires us to do? (To speak of 'truth' here seems legitimate, since, while Tarski's truth predicate is a predicate of sentences, the sentential operator to which it corresponds is a redundant one, which can be inserted before or deleted from in front of any clause without change of truth value.)

If we accept the axiom (β), and hence the equivalence (δ), we run into certain difficulties, on which Troelstra comments. The operator '$\exists n\, (\vdash_n \ldots)$' becomes a redundant truth operator, and hence may be distributed across any logical constant, as in

(ε) $\qquad (\vdash_k \forall m\, A(m)) \to \forall m\, \exists n\, (\vdash_n A(m)).$

As Troelstra observes, this appears to have the consequence that, if we have once proved a universally quantified statement, we are in some way committed to producing, at some time in the future, individual proofs of all its instances, whereas, palpably, we are under no such constraint. The solution to which he inclines is that proposed by Kreisel, namely, that the operator '$\vdash_n$' must be so construed that a proof, at stage n, of a universally quantified statement counts as being, at the same time, a proof of each instance, so that we could assert the stronger thesis

(ζ) $\qquad (\vdash_k \forall m\, A(m)) \to \forall m\, (\vdash_k A(m)).$

(Troelstra in fact recommends this interpretation on separate grounds, as enabling us to escape a paradox about constructive functions: he himself points out, however, that this paradox can alternatively be avoided by introducing distinctions of level which seem intrinsically plausible.) The difficulty about this solution is that it must be extended to every recognized logical consequence. From

(η) $$(m \leq n \:\&\: (\vdash_m A)) \to (\vdash_n A),$$

we have

(θ) $$(n = \max(m, k) \:\&\: (\vdash_m A) \:\&\: (\vdash_k C)) \to ((\vdash_n A) \:\&\: (\vdash_n C)),$$

while from (δ) we obtain

(ι) $$(\vdash_m A) \:\&\: (\vdash_k (A \to B)) \to \exists n (\vdash_n B).$$

We could in the same way complain that this committed us, whenever we had proved a statement A and had recognized some other statement B as being a consequence of A, to actually drawing that consequence some time in the future; and, if our interpretation of the operator '$\vdash_n$' is to be capable of dealing with this difficulty in the same way as with the special case of instances of a universally quantified statement, we should have to allow that a proof that a theorem had a certain consequence was, at the same time, a proof of that consequence, and, likewise, that a proof of a statement already known to have a certain consequence was, at the same time, a proof of that consequence; we should, that is, have to accept the law

(κ) $$(n = \max(m, k) \:\&\: (\vdash_m A) \:\&\: (\vdash_k (A \to B))) \to (\vdash_n B).$$

We should thus so have to construe the notion of proof that a proof of a statement is taken as simultaneously constituting a proof of anything that has already been recognized as a consequence of that statement. We can, no doubt, escape having to say that it is simultaneously a proof of whatever, in a platonistic sense, as a matter of fact is an intuitionistic consequence of the statement: but when are we to be said to have recognized that one statement is a consequence of another? If a proof of a universally quantified statement is simultaneously a proof of all its instances, it is difficult to see how we can avoid conceding that a demonstration of the validity of a schema of first-order predicate logic is simultaneously a demonstration of the truth of all its instances, or an acceptance of the induction schema simultaneously an acceptance of all cases of induction. The resulting notion of proof would be far removed indeed from actual mathematical experience, and could not be explained as no more than an idealization of it.

The trouble with all this is that, as a representation of actual mathematical experience, we are operating with too simplified a notion of proof. The axiom (η) is acceptable in the sense that, prescinding from the occasional accident, once a theorem has been proved, it always remains *available* to be subsequently appealed to: but the idea that, having acknowledged the two premisses of a *modus ponens*, we have *thereby* recognized the truth of the conclusion, is plausible only in a case in which we are simultaneously bearing in mind the truth of the two premisses. To have once proved a statement is not thereafter to be continuously aware of its truth: if it were, then we should indeed always know the logical consequences of everything which we know, and should have no need of proof.

Acceptance of axiom (β) leads to the conclusion that we shall eventually prove every logical consequence of everything we prove. This, as a representation of the intuitionist notion of proof, is an improvement upon Beth trees, as normally presented: for these are set up in such a way that, at any stage (node), every logical consequence of statements true at that stage is already true; the Beth trees are adapted only to situations, such as those involving free choice sequences, where new information is coming in that is not derived from the information we have at earlier stages. But the idea that we shall eventually establish every logical consequence of everything we know is implausible and arbitrary: and it cannot be rescued by construing each proof as, implicitly, a proof also of the consequences of the statement proved, save at the cost of perverting the whole conception. If we wish to do so, there seems no reason why we should not take the stages represented by the numerical subscripts as punctuated by proofs, however short the stages thereby become, and the notion of proof as relating only to what is quite explicitly proved, so that, at each stage, one and only one new statement is proved, and consider what axioms hold under the resulting interpretation of the symbol '$\vdash_n$'. It thus appears that, under this interpretation, the axiom (β) must be rejected in favour of the weaker axiom (γ).

Looked at another way, however, the stronger axiom (β) seems entirely acceptable. If, that is, we interpret the implication sign in its intuitionistic sense, the axiom merely says that, given a proof of A, we can effectively find a proof that A was proved at some stage; and this seems totally innocuous and banal. But, if axiom (β) is innocuous, how did we arrive at our earlier difficulties? The only possibility seems to be that our logical laws are themselves at fault. For instance, the law

(λ) $$\forall x A(x) \to A(m)$$

leads, via axiom (β), to the conclusion

(μ) $$\forall x A(x) \to \exists n \, (\vdash_n A(m)),$$

which appears, on the present interpretation of '$\vdash_n$', to say that we shall explicitly prove every instance of every universally quantified statement which we prove; so perhaps the error lies in the law (λ) itself. A law such as (λ) is ordinarily justified by saying that, given a proof of $\forall xA(x)$, we can, for each m, effectively find a proof of $A(m)$. If this is to remain a sufficient justification of (μ), then (μ) must be construed as saying that, given a proof of $\forall xA(x)$, we can effectively find a proof that $A(m)$ will be proved at some stage. How can we do this, for given m? Obviously, by proving $A(m)$ and noting the stage at which we do so. This means, then, that the existentially quantified statement

(ν) $\quad\quad\quad\quad\quad\quad\quad\quad \exists n\, (\vdash_n A(m))$

is to be so understood that its assertion does not amount to a claim that we shall, as a matter of fact, prove $A(m)$ at some stage n, but only that we are capable of bringing it about that $A(m)$ is proved at some stage. Our difficulties thus appear to have arisen from understanding the existential quantifier in (β) in an excessively classical or realistic manner, namely, as meaning that there will in fact be a stage n at which the statement is proved rather than as meaning that we have an effective means, if we choose to apply it, of making it the case that there is such a stage. The point here is that it is not merely a question of interpreting the existential quantifier intuitionistically rather than classically in the sense that we can assert that there is a stage n at which a statement will be proved only if we have an effective means for identifying a particular such stage. Rather, if quantification over temporal stages is to be introduced into mathematical statements, then it must be treated like quantification over mathematical objects and mathematical constructions: the assertion that there is a stage n at which such-and-such will hold is justified provided that we possess an enduring capability of bringing about such a stage, regardless of whether we ever exercise this capability or not.

The confusions concerning the theory of the creative subject which we have been engaged in disentangling arose in part from a perfectly legitimate desire, to relate the intuitionistic truth of a mathematical statement with a use of the logical constants which is alien to intuitionistic mathematics. Troelstra's difficulties sprang from his desire to construe the expression '$\exists n\, (\vdash_n A)$' as meaning that A would in fact be proved at some stage: but, whether we interpret the existential quantifier classically or constructively, such a construal of it fails to gibe with the way it and the other logical constants are construed within ordinary mathematical statements, and hence, however we try to modify our notion of a statement's being proved, we shall not obtain anything equivalent to the mathematical statement A itself. Nevertheless, the desire to express the condition for the intuitionistic truth of a mathematical statement in terms which do not presuppose an understanding of the intuitionistic logical constants as used within mathematical statements is entirely licit. Indeed, if it were impossible to do so, intuitionists would have no way of conveying to platonist mathematicians what it was that they were about: we should have a situation quite different from that which in fact obtains, namely, one in which some people found it natural to extend basic computational mathematics in a classical direction, and others found it natural to extend it in an intuitionistic direction, and neither could gain a glimmering of what the other was at. That we are not in this situation is because intuitionists and platonists can find a common ground, namely, statements, both mathematical and non-mathematical, which are, in the view of both, decidable, and about whose meaning there is therefore no serious dispute and which both sides agree obey a classical logic. Each party can, accordingly, by use of and reference to these unproblematic statements, explain to the other what his conception of meaning is for those mathematical statements which are in dispute. Such an explanation may not be accepted as legitimate by the other side (the whole point of the intuitionist position is that undecidable mathematical statements cannot legitimately be given a meaning by laying down truth conditions for them in the platonistic manner): but at least the conception of meaning held by each party is not wholly opaque to the other.

This dispute between platonists and intuitionists is a dispute over whether or not a realist interpretation is legitimate for mathematical statements: and the situation I have just indicated is quite characteristic for disputes concerning the legitimacy of a realist interpretation of some class of statements, and is what allows a *dispute* to take place at all. Typically, in such a dispute there is some auxiliary class of statements about which both sides agree that a realist interpretation is possible (depending upon the grounds offered by the anti-realists for rejecting a realist interpretation for statements of the disputed class, this auxiliary class may or may not consist of statements agreed to be effectively decidable); and typically, it is in terms of the truth conditions of statements of this auxiliary class that the anti-realist frames his conception of meaning, his non-classical notion of truth, for statements of the disputed class, while the realist very often appeals to statements of the auxiliary class as providing an analogy for his conception of meaning for statements of the disputed class. Thus, when the dispute concerns statements about the future, statements about the present will form the auxiliary class; when it concerns statements about material objects, the auxiliary class will consist of sense-data statements; when the dispute concerns statements about character traits, the auxiliary class will consist of statements about actual or hypothetical behaviour; and so on.

If the intuitionistic notion of truth for mathematical statements can be explained only by a Tarski-type truth definition which takes for granted the meanings of the intuitionistic logical constants, then the intuitionist notion of truth, and hence of meaning, cannot be so much as conveyed to anyone who does not accept it already, and no debate between intuitionists and platonists is possible, because they cannot communicate with one another. It is therefore wholly legitimate, and, indeed, essential, to frame the condition for the intuitionistic truth of a mathematical statement in terms which are intelligible to a platonist and do not beg any questions, because they employ only notions which are not in dispute.

The obvious way to do this is to say that a mathematical statement is intuitionistically true if there exists an (intuitionistic) proof of it, where the existence of a proof does not consist in its platonic existence in a realm outside space and time, but in our actual possession of it. Such a notion of truth, obvious as it is, already departs at once from that supplied by the analogue of the Tarski-type truth definition, since the predicate 'is true', thus explained, is significantly tensed: a statement not now true may later become true. For this reason, when 'true' is so construed, the schema (T) is incorrect: for the negation of the right-hand side of any instance will be a mathematical statement, while the negation of the left-hand side will be a non-mathematical statement, to the effect that we do not as yet possess a proof of a certain mathematical statement, and hence the two sides cannot be equivalent. We might, indeed, seek to restore the equivalence by replacing 'is true' on the left-hand side by 'is or will be true': but this would lead us back into the difficulties we encountered with the theory of the creative subject, and I shall not further explore it.

What does require exploration is the notion of proof being appealed to, and that also of the existence of a proof. It has often and, I think, correctly been held that the notion of proof needs to be specialized if it is to supply a non-circular account of the meanings of the intuitionistic logical constants. It is possible to see this by considering disjunction and existential quantification. The standard explanation of disjunction is that a construction is a proof of $A \vee B$ just in case it is a proof either of A or of B. Despite this, it is not normally considered legitimate to assert a disjunction, say, in the course of a proof, only when we actually have a proof of one or other disjunct. For instance, it would be quite in order to assert that

$10^{10^{10}} + 1$ is either prime or composite

without being able to say which alternative held good, and to derive some theorem by means of an argument by cases. What makes this legitimate, on the standard intuitionist view, is that we have a method which is in principle effective for deciding which of the two alternatives is correct: if we were to take the trouble to apply this method, the appeal to an argument by cases could be dispensed with. Generally speaking, therefore, if we take a statement as being true only when we actually possess a proof of it, an assertion of a disjunctive statement will not amount to a claim that it is true, but only to a claim that we have a means, effective in principle, for obtaining a proof of it. This means, however, that we have to distinguish between a proof proper, a proof in the sense of 'proof' used in the explanations of the logical constants, and a cogent argument. In the course of a cogent argument for the assertibility of a mathematical statement, a disjunction of which we do not possess an actual proof may be asserted, and an argument by cases based upon this disjunction. This argument will not itself be a proof, since any initial segment of a proof must again be a proof: it merely indicates an effective method by which we might obtain a proof of the theorem if we cared to apply it. We thus appear to require a distinction between a proof proper—a canonical proof— and the sort of argument which will normally appear in a mathematical article or textbook, an argument which we may call a 'demonstration'. A demonstration is just as cogent a ground for the assertion of its conclusion as is a canonical proof, and is related to it in this way: that a demonstration of a proposition provides an effective means for finding a canonical proof. But it is in terms of the notion of a canonical proof that the meanings of the logical constants are given. Exactly similar remarks apply to the existential quantifier.

There is some awkwardness about this way of looking at disjunction and existential quantification, namely, in the divorce between the notions of truth and of assertibility. It might be replied that the significance of the act of assertion is not, in general, uniquely determined by the notion of truth: for instance, even when we take the notion of truth for mathematical statements as given, it still needs to be stipulated whether the assertion of a mathematical statement amounts to a claim to have a proof of it, or whether it may legitimately be based on what Polya calls a 'plausible argument' of a non-apodeictic kind. (We can imagine people whose mathematics wholly resembles ours, save that they do not construe an assertion as embodying a claim to have more than a plausible argument.) It nevertheless remains that, if the truth of a mathematical statement consists in our possession of a canonical proof of it, while its assertion need be based on possession of no more than a demonstration, we are forced to embrace the awkward conclusion that it may be legitimate to assert a statement even though it is *known* not to be true. However, if the sign of disjunction and the existential quantifier were the only logical constants whose explanation appeared to call for a distinction between canonical proofs and demonstrations, the distinction might be avoided altogether by modifying their explanations, to allow that a proof of a disjunction consisted in any construction of which we could recognize that it would effectively yield a proof of one or other disjunct, and similarly for existential quantification: we should then be able to say that a statement could be asserted only when it was (known to be) true.

However, the distinction is unavoidable if the explanations of universal quantification, implication, and negation are to escape circularity. The standard explanation of implication is that a proof of $A \to B$ is a construction of which we can recognize that, applied to any proof of A, it would yield a proof of B. It is plain that the notion of proof being used here cannot be one which admits unrestricted use of *modus ponens*: for, if it did, the explanation would be quite empty. We could admit anything we liked as constituting a proof of $A \to B$, and it would remain the case that, given such a proof, we had an effective method of converting any proof of A into a proof of B, namely, by adding the proof of $A \to B$ and performing a single inference by *modus ponens*. Obviously, this is not what is intended: what is intended is that the proof of $A \to B$ should supply a means of converting a proof of A into a proof of B without appeal to *modus ponens*—at least, without appeal to any *modus ponens* containing $A \to B$ as a premiss. The kind of proof in terms of which the explanation of implication is being given is, therefore, one of a restricted kind. On the assumption that we have, or can effectively obtain, a proof of $A \to B$ of this restricted kind, an inference from $A \to B$ by *modus ponens* is justified, because it is in principle unnecessary. The same must, by parity of reasoning, hold good for any other application of *modus ponens* in the main (though not in any subordinate) deduction of any proof. Thus, if the intuitionistic explanation of implication is to escape, not merely circularity, but total vacuousness, there must be a restricted type of proof—canonical proof—in terms of which the explanation is given, and which does not admit *modus ponens* save in subordinate deductions. Arguments employing *modus ponens* will be perfectly valid and compelling, but they will, again, not be proofs in this restricted sense: they will be demonstrations, related to canonical proofs as supplying a means effective in principle for finding canonical proofs. Exactly similar remarks apply to universal quantification *vis-à-vis* universal instantiation and to negation *vis-à-vis* the rule *ex falso quodlibet*: the explanations of these operators presuppose a restricted type of proof in which the corresponding elimination rules do not occur within the main deduction.

What exactly the notion of a canonical proof amounts to is obscure. The deletion of elimination rules from a canonical proof suggests a comparison with the notion of a normalized deduction. On the other hand, Brouwer's celebrated remarks about fully analysed proofs in connection with the bar theorem do not suggest that such a proof is one from which unnecessary detours have been cut out—the proof of the bar theorem consists in great part in cutting out such detours from a proof taken already to be in 'fully analysed' form. Rather, Brouwer's idea appears to be that, in a fully analysed proof, all operations on which the proof depends will actually have been carried out. That is why such a proof may be an infinite structure: a proof of a universally quantified statement will be an operation which, applied to each natural number, will yield a proof of the corresponding instance; and, if this operation is carried out for each natural number, we shall have proofs of denumerably many statements. The conception of the mental construction which is the fully analysed proof as being an infinite structure must, of course, be interpreted in the light of the intuitionist view that all infinity is potential infinity: the mental construction consists of a grasp of general principles according to which any finite segment of the proof could be explicitly constructed. The direction of analysis runs counter to the direction of deduction; while one could not be convinced by an actually infinite proof-structure (because one would never reach the conclusion), one may be convinced by a potentially infinite one, because its infinity consists in our grasp of the principles governing its analysis. Indeed, it might reasonably be said that the standard intuitionistic meanings of the universal and conditional quantifiers involve that a proof is such a potentially infinite structure. Nevertheless, the notion of a fully analysed proof, that is, of the result of applying every operation involved in the proof, is far from clear, because it is obscure what the effect of the analysis would be on conditionals and negative statements. We can systematically display the results of applying the operation which constitutes a proof of a statement involving universal quantification over the natural numbers, because we can generate each natural number in sequence. But the corresponding application of the operation which constitutes the proof of a statement of the form $A \to B$ would consist in running through all putative canonical proofs of A and either showing, in each case, that it was not a proof of A or transforming it into a proof of B: and, at least without a firm grasp upon the notion of a canonical proof, we have no idea how to generate all the possible candidates for being a proof of A.

The notion of canonical proof thus lies in some obscurity; and this state of affairs is not indefinitely tolerable, because, unless it is possible to find a coherent and relatively sharp explanation of the notion, the viability of the intuitionist explanations of the logical constants must remain in doubt. But, for present purposes, it does not matter just how the notion of canonical proof is to be explained; all that matters is that we require some distinction between canonical proofs and demonstrations, related to one another in the way that has been stated. Granted that such a distinction is necessary, there is no motivation for refusing to apply it to the case of disjunctions and existential statements.

Let us now ask whether we want the intuitionistic truth of a mathematical statement to consist in the existence of a canonical proof or of a demonstration. If by the 'existence' of a proof or demonstration we mean that we have actually explicitly carried one out, then either choice leaves us with certain counter-intuitive consequences. On either view, naturally, a valid rule of inference will not always lead from true premisses to a true conclusion, namely, if we have not explicitly drawn the inference: this will always be so on any view which equates truth with our actual possession of some kind of proof. If we take the stricter line, and hold a statement to be true only when we possess a canonical proof of it, then, as we have seen, we shall have to allow that a statement may be asserted even though it is known not to be true. If, on the other hand, we allow that a statement is true when we possess merely a demonstration of it, then truth will not distribute over disjunction: we may possess a demonstration of $A \vee B$ without having a demonstration either of A or of B. Now, granted, once we have admitted a significant tense for the predicate 'is true', then, as we have noted, the schema (T) cannot be maintained as in all cases correct: but our instinct is to permit as little divergence from it as possible, and it is for this reason that we are uneasy about a notion of truth which is not distributive over disjunction or existential quantification.

A natural emendation is to relax slightly the requirement that a proof or demonstration should have been explicitly given. The question is how far we may consistently go along this path. If we say merely that a mathematical statement is true just in case we are aware that we have an effective means of obtaining a canonical proof of it, this will not be significantly different from equating truth with our actual possession of a demonstration. It might be allowed that there would be some cases when we had demonstrated the premisses of, say, an inference by *modus ponens* in which we were aware that we could draw the conclusion, though we had not quite explicitly done so; but there will naturally be others in which we were not aware of this, that is, had not noticed it; if it were not so, we could never discover new demonstrations. It is therefore tempting to go one step further, and say that a statement is true provided that we are in fact in possession of a means of obtaining a canonical proof of it, whether or not we are aware of the fact. Would such a step be a betrayal of intuitionist principles?

In which cases would it be correct to say that we possess an effective means of finding a canonical proof of a statement, although we do not know that we have such a means? Unless we are to suppose that we can attain so sharp a notion of a canonical proof that it would be possible to enumerate effectively all putative such proofs of a given statement (the supposition whose implausibility causes our difficulty over the notion of a fully analysed proof), there is only one such case: that in which we possess a demonstration of a disjunctive or existential statement. Such a demonstration provides us with what we recognize as an effective means (in principle) for finding a canonical proof of the disjunctive or existential statement demonstrated. Such a canonical proof, when found, will be a proof of one or other disjunct, or of one instance of the existentially quantified statement: but we cannot, in general, tell which. For example, when $A(x)$ is a decidable predicate, the decision procedure constitutes a demonstration of the disjunction '$A(\bar{n}) \vee \neg A(\bar{n})$', for specific n; but, until we apply the procedure, we do not know which of the two disjuncts we can prove. It is very difficult for us to resist the temptation to suppose that there is already, unknown to us, a determinate answer to the question which of the two disjuncts we should obtain a proof of, were we to apply the decision procedure; that, for example, that it is already the case either that, if we were to test it out, we should find that $10^{10^{10}} + 1$ is prime, or that, if we were to test it out, we should find that it was composite. What is involved here is the passage from a subjunctive conditional of the form:

$A \to (B \vee C)$

to a disjunction of subjunctive conditionals of the form

$(A \to B) \vee (A \to C)$.

Where the conditional is interpreted intuitionistically, this transition is, of course, invalid: but the subjunctive conditional of natural language does not coincide with the conditional of intuitionistic mathematics. It is, indeed, the case that the transition is not in general valid for the subjunctive conditional of natural language either: but, when we reflect on the cases in which the inference fails, it is difficult to avoid thinking that the present case is not one of them.

There are two obvious kinds of counter-example to this form of inference for ordinary subjunctive conditionals: perhaps they are really two subvarieties of a single type. One is the case in which the antecedent A requires supplementation before it will yield a determinate one of the disjuncts B and C. For instance, we may safely agree that if Fidel Castro were to meet President Nixon, he would either insult him or speak politely to him; but it might not be determinately true, of either of those things, that he would do it, since it might depend upon some so far unspecified further condition, such as whether the meeting took place in Cuba or outside. Schematically, this kind of case is one in which we can assert:

$A \to (B \vee C)$,
$(A \mathbin{\&} Q) \to B$,
$(A \mathbin{\&} \neg Q) \to C$,

but in which the subjunctive antecedent A neither implies nor presupposes either Q or its negation; in such a case, we cannot assert either $A \to B$ or $A \to C$. The other kind of counter-example is that in which we do not consider the disjuncts to be determined by anything at all: no supplementation of the antecedent would be sufficient to decide between them in advance. If an electron were to pass through that screen, it would have passed through one aperture or the other: but there is nothing at all which will determine in advance through which it would pass. Similar cases will arise, for those who believe in free will in the traditional sense, in respect of human actions.

If we were to carry out the decision procedure for determining the primality or otherwise of some specific large number N, we should either obtain the result that N is prime or obtain the result that N is composite. Is this, or is it not, a case in which we may conclude that it either holds good that, if we were to carry out the procedure, we should find that N is prime, or that, if we were to carry out the procedure, we should find that N is composite? The difficulty of resisting the conclusion that it is such a case stems from the fact that it does not display either of the characteristics found in the two readily admitted types of counter-example to the form of inference we are considering. No further circumstance could be relevant to the result of the procedure— this is part of what is meant by calling it a computation; and, since at each step the outcome of the procedure is determined, how can we deny that the overall outcome is determinate also?

If we yield to this line of thought, then we must hold that every statement formed by applying a decidable predicate to a specific natural number already has a definite truth value, true or false, although we may not know it. And, if we hold this, it makes no difference whether we chose at the outset to say that natural numbers are creations of the human mind or that they are eternally existing abstract objects. Whichever we say, our decision how to interpret undecidable statements of number theory, and, in the first place, statements of the forms $\forall x\, A(x)$ and $\exists x\, A(x)$, where $A(x)$ is decidable, will be independent of our view about the ontological status of natural numbers. For on this view of the truth of mathematical statements, each decidable number-theoretic statement will already be determinately true or false, independently of our knowledge, just as it is on a platonistic view: any thesis about the ontological character of natural numbers will then be quite irrelevant to the interpretation of the quantifiers. As we noted, it would be possible for someone to be prepared to regard natural numbers as timeless abstract objects, and to regard decidable predicates as being determinately true or false of them, and yet to be convinced by an argument of the first type, based on quite general considerations concerning meaning, that unbounded quantification over natural numbers was not an operation which in all cases preserved the property of possessing a determinate truth value, and therefore to fall back upon a constructivist interpretation of it. Conversely, if someone who thought of the natural numbers as creations of human thought also believed, for the reasons just indicated, that each decidable predicate was determinately true or false of each of them, he might accept a classical interpretation of the quantifiers. He would do so if he was unconvinced by the general considerations about meaning which we reviewed, that is, by the first type of argument for the adoption of an intuitionistic logic for mathematics: the fact that he was prepared to concede that the natural numbers come into existence only in virtue of our thinking about them would play no part in his reflections on the meanings of the quantifiers. Dedekind, who declared that mathematical structures are free creations of the human mind, but nevertheless appears to have construed statements about them in a wholly platonistic manner, may perhaps be an instance of just such a combination of ideas.

One who rejects the idea that there is already a determinate outcome for the application, to any specific case, of an effective procedure is, however, in a completely different position. If someone holds that the only acceptable sense in which a mathematical statement, even one that is effectively decidable, can be said to be true is that in which this means that we presently possess an actual proof or demonstration of it, then a classical interpretation of unbounded quantification over the natural numbers is simply unavailable to him. As is frequently remarked, the classical or platonistic conception is that such quantification represents an infinite conjunction or disjunction: the truth value of the quantified statement is determined as the infinite sum or product of the truth values of the denumerably many instances. Whether or not this be regarded as an acceptable means of determining the meaning of these operators, the explanation presupposes that all the instances of the quantified statement themselves already possess determinate truth values: if they do not, it is impossible to take the infinite sum or product of these. But if, for example, we do not hold that such a predicate as 'x is odd $\to x$ is not perfect' already has a determinate application to each natural number, though we do not know it, then it is just not open to him to think that, by attaching a quantifier to this predicate, we obtain a statement that is determinately true or false.

One question which we asked earlier was this: Can the thesis that natural numbers are creations of human thought be taken as a premiss for the adoption of an intuitionistic logic for number-theoretic statements? And another question was: What content can be given to the thesis that natural numbers are creations of human thought that does not prejudge the question what is the correct notion of truth for number-theoretic statements in general? The tentative answer which we gave to this latter question was that the thesis might be taken as relating to the appropriate notion of truth for a restricted class of number-theoretic statements, say, numerical equations or, more generally, decidable statements. From what we have said about the intuitionistic notion of truth for mathematical statements, it has now become apparent that there is one way in which the thesis that natural numbers are creations of the human mind might be taken, namely, as relating precisely to the appropriate notion of truth for decidable statements of arithmetic, which would provide a ground for rejecting a platonistic interpretation of number-theoretic statements generally, without appeal to any general thesis concerning the notion of meaning. This way of taking the thesis would amount to holding that there is no notion of truth applicable even to numerical equations save that in which a statement is true when we have actually performed a computation (or effected a proof) which justifies that statement. Such a claim must rest, as we have seen, on the most resolute scepticism concerning subjunctive conditionals: it must deny that there exists any proposition which is now true about what the result of a computation which has not yet been performed would be if it were to be performed. Anyone who can hang on to a view as hard-headed as this has no temptation at all to accept a platonistic view of number-theoretic statements involving unbounded quantification: he has a rationale for an intuitionistic interpretation of them which rests upon considerations relating solely to mathematics and demanding no extension to other realms of discourse (save in so far as the subjunctive conditional is involved in explanations of the meanings of statements in these other realms). But, for anyone who is not prepared to be quite as hard-headed as that, the route to a defence of an intuitionistic interpretation of mathematical statements which begins from the ontological status of mathematical objects is closed: the only path that he can take to this goal is that which I sketched at the outset: one turning on the answers given to general questions in the theory of meaning.

V

MATHEMATICAL INTUITION

CHARLES PARSONS

In a much quoted passage, Gödel writes:

> But, despite their remoteness from sense-experience, we do have something like a perception of the objects of set theory, as is seen from the fact that the axioms force themselves upon us as being true. I don't see any reason why we should have less confidence in this kind of perception, i.e. in mathematical intuition, than in sense-perception.[1]

If we leave aside its specific reference to set theory, the passage is a classic expression of what might be called the philosophical conception of mathematical intuition. As I see it, the principal mark of this conception is an analogy between sense perception as a cognitive relation to the physical world and 'something like a perception' giving a similar relation to mathematical objects, and perhaps other abstract entities. If it is to be central to the philosophy of mathematics, it should play a role like that of sense perception in our knowledge of the everyday world and of physics.

My aim in this paper is to begin a reasoned explication of this conception. I shall argue that something answering to it does in fact exist. However, this positive result is very limited in scope, and we shall already see some limitations of the conception. Unlike Gödel, I shall not focus on set theory, where the conception of intuition has special difficulties, which I have discussed elsewhere.[2] One is more likely to make progress by concentrating on the simplest case, such as elementary geometry or arithmetic. I shall concentrate on the latter, but look at it from a somewhat geometric point of view.

First published in the *Proceedings of the Aristotelian Society*, 80 (1979–80): 145–68. Reproduced by courtesy of the Editor of the Aristotelian Society: © 1979–80.

[1] Kurt Gödel, 'What is Cantor's Continuum Problem?', in Paul Benacerraf and Hilary Putnam (eds.), *Philosophy of Mathematics* 2nd edn. (Cambridge: Cambridge University Press, 1983), 483–4. This passage and others cited below are from a supplement added to this edition of the paper, which first appeared in 1947.

[2] Charles Parsons, 'What is the Iterative Conception of Set?', in R. E. Butts and Jaakko Hintikka (eds.), *Logic, Foundations of Mathematics, and Computability Theory* (Dordrecht: Reidel, 1977), 339–45.

I

When Gödel speaks of something like a perception *of* the objects of set theory, he expresses something central to the conception I am examining: mathematical intuition has a certain *de re* character; it involves a relation of a person to (presumably mathematical) *objects*. The vocabulary of sense perception contains locutions expressing relations to physical objects or events: *a* sees *x*, *a* hears *x*, *a* smells *x*, *a* perceives *x*, etc. Just how literally this is to be carried over into the concept of mathematical intuition is one of the trickiest questions about it.

For some perceptual verbs, notably 'see' and 'perceive', we can contrast such object-relational uses with uses with sentence complements, which we can call propositional attitude uses. Which type of use is more fundamental has been controversial, but the *existence* of the object-relational uses is obvious. The matter is otherwise with mathematical intuition, and philosophers have not expressed themselves very clearly on the point. However, we can find both kinds of use in the philosophical literature.[3] To abbreviate reference to the object-relational or propositional attitude use of 'intuit', I shall talk of intuition *of* and intuition *that*.

We find some unclarity already in the above-cited passage of Gödel: that there is 'something like a perception of the objects of set theory' is, he says, 'seen from the fact that the axioms force themselves on us as being true'. Here he seems to conclude from the evident character of certain *statements*, which we might express as intuitions *that*, to the existence of intuitions *of*. The premiss may be disputed, but even if it is granted, the *inference* seems to be a *non sequitur*. What Gödel says in the next paragraph by way of explanation (and probably qualification) is quite obscure.

Intuition *that* is of course a very traditional rationalistic theme. It might be taken to subsume almost any conception of the evidence or self-evidence of truths of reason, where this is taken not to be derived from habit, practice, or convention. Just for this reason, the analogy with perception does not enter the picture until it is used for an *account* of such rational evidence, or perhaps to mark clearly the distinction between a proposition's being genuinely evident and its merely seeming obviously true. At this point the analogy is likely to be developed in the direction of intuition *of*, simply because the presence of an object is so central to perception.

I suggest that we can find such a picture in Descartes, for whom *clear and distinct perception* is certainly mainly a propositional attitude.[4] Two

---

[3] The relevance of the distinction to mathematical intuition is pointed out by Mark Steiner in *Mathematical Knowledge* (Ithaca, NY: Cornell University Press, 1975), 131. Steiner maintains that no one would defend object-relational mathematical intuition. That seems to me clearly false.

[4] Descartes seems to rely on perception *of* in his explanations, e.g., the explanation of clarity in

important philosophers of the past who seem more directly committed to intuition *of* where the objects involved may be mathematical are Kant and Husserl. In Kant, intuition as a propositional attitude plays no explicit role. By definition, an intuition is a singular representation, that is a representation of a single object.[5] When Kant in the *Critique of Pure Reason* says that it is through intuition that knowledge has 'immediate relation' to objects (A19 = B33), this immediacy seems to be a direct presence of the object to the mind, as in perception. At all events, intuition gives 'immediate evidence' to propositions of, for example, geometry.[6] Thus intuition *that* seems to be present in Kant, although his official use of 'intuition' is only for intuition *of*.

Husserl's discussions of 'categorial intuition' in the *Logische Untersuchungen* and of 'intuition of essences' in the *Ideen* represent a sustained and interesting attempt to develop a theory of rational evidence based on an analogy with perception, in which the feature of perception as being of an object is central. Husserl understands rational evidence in general as intuition, and undertakes to give a unified account of intuition *of* and intuition *that*.

Both Kant's and Husserl's conceptions have had some influence on discussions of the foundations of mathematics in this century. Kant's influence is more visible and pervasive. Hilbert's conception of the intuitive character of finitary mathematics is explicitly based on a Kantian conception of pure intuition, though perhaps more on Kant's theory of geometry than on his theory of arithmetic. Intuitionism also owes much to Kant, particularly to the notion of time as the form of inner sense. Husserl's ideas have not had nearly so much influence, but he did have an impact on Weyl and Gödel.[7]

Principles, I, 40. Cf. Descartes, *Philosophical Writings*, trans. E. Anscombe and P. Geach (Indianapolis: Bobbs–Merrill, 1971), 190.

[5] Immanuel Kant, *Logic*, §1 (Academy edn., IX, 91).

[6] My interpretation of the immediacy of intuition in relation to *objects* is controversial. See my 'Kant's Philosophy of Arithmetic', in S. Morgenbesser, P. Suppes, and M. White (eds.), *Philosophy, Science, and Method: Essays in Honor of Ernest Nagel* (New York: St Martin's Press, 1969), esp. 569–71, and Jaakko Hintikka, 'Kantian Intuitions', *Inquiry*, 15 (1972): 341–5. However, I do not see how there could be controversy about the fact that according to Kant intuition (in particular a priori intuition) confers *evidence* that is immediate. *This* immediacy can surely not be reduced to singularity, as Hintikka proposes for the other dimension of immediacy.

[7] Concerning Hilbert, Gödel writes: 'What Hilbert means by "Anschauung" is substantially Kant's space-time intuition confined, however, to configurations of a finite number of discrete objects' (from note h added to 'On an Extension of Finitary Mathematics which has not yet been Used', unpublished English translation, emended by the author with some additional notes, of 'Über eine noch nicht benützte Erweiterung des finiten Standpunktes', *Dialectica*, 12 (1958): 280–7. As if to stress the difference with the notion of intuition in 'What is Cantor's Continuum Problem?', the term 'Anschauung' is translated as 'concrete intuition'.

The aspect of intuitionism which was most original and may have proved most fruitful, seeing the meaning of a mathematical statement as constituted by what would be a proof of it, does not seem to owe its inspiration to Kant. In recent discussions of the foundations of intuitionism, the very concept of intuition seems to drop out.

II

The idea of 'something like a perception' of mathematical objects seems at first sight outrageous. If mathematical objects are given to us in a way similar to that in which physical objects are given to our senses, should it not be *obvious* that this is so? But the history of philosophical discussion about mathematics shows that it is not. Whatever mysteries and philosophical puzzles there may be about perception, it works to a large extent as a straightforward empirical concept. We can make a lot of assured judgements about when we perceive something, and confidence about the description of our experience can often survive doubt about what it is an experience of. Thus the proposition that I now *see* before me a typewriter with paper in it is one that I expect that no other philosopher, were he in this room now, would dispute except on the basis of sceptical arguments, and many of these would not touch weaker statements such as that it *looks* to me as if I see these things. There is a phenomenological datum here that is as close to being undisputed as anything is in philosophy.

It is hard to maintain that the case is the same for mathematical objects. Is it *obvious* that there is an experience of intuiting the number 7 or a triangle, or at least of its 'looking' as if I were intuiting 7 or a triangle? Are there any experiences we can appeal to here that are anywhere near as undisputed as my present experience of seeing my typewriter? If we don't know what to point to, isn't that already a serious disanalogy between sense perception and whatever consciousness we have of mathematical objects?

This embarrassment is connected with an obvious disanalogy. In normal cases of perception, there is a physical action of the object perceived on our sense-organs. Our perception is, as it were, founded on this action, and there are serious philosophical reasons for holding that such a causal relation is a necessary condition for perceiving an object.[8] It would be implausible to suppose that in *mathematical* intuition there is a causal action of a mathematical object on us (presumably on the mind). Moreover, this is no part of the view of the upholders of mathematical intuition that I have mentioned, though it is sometimes included in popular conceptions of 'platonism'.

At this point we find qualifications in accounts of mathematical intuition, which raise the question just how close an analogy with perception is intended. Gödel says that 'mathematical intuition need not be conceived of as a

---

[8] In attempting to develop an analogy between perception and knowledge of abstract objects, Husserl is helped by his phenomenological perspective. The causal foundation of perception is not part of the subject-matter of phenomenology, even in the form it takes in the *Logische Untersuchungen* (hereafter LU). Husserl does undertake to show that in categorial intuition there is something analogous to sensations in sense perception. In my view, he lapses into obscurity in explaining this (LU VI, §56). I am not sure to what extent this can be cleared up.

faculty giving an *immediate* knowledge of the objects concerned'.[9] Husserl is even prepared to call categorial intuition 'perception' (*Wahrnehmung*),[10] but he contrasts sense perception as *schlicht*, in which the object is 'immediately given', with categorial intuition which is *founded* in other 'acts', such as ordinary perceptions and imaginings.[11]

Kant expresses puzzlement about how intuition can be a priori. In the *Prolegomena*, after introducing the notion of pure intuition, he writes (§8):

> An intuition is such a representation as would immediately depend on the presence (*Gegenwart*) of the object. Hence it seems impossible to intuit spontaneously (*ursprünglich*) *a priori* because intuition would in that event have to take place without either a former or a present object to refer to, and in consequence could not be an intuition... But how can the intuition of an object precede the object itself?[12]

Here (and elsewhere) Kant does not explicitly express a view about intuition of *mathematical* objects. It is clear from the context that by 'object' he means *real* object, in practice physical object. So the question is how it is possible for a priori intuition to be 'of' physical objects that are not given a priori.

In §9 Kant claims that the puzzle is resolved by the fact that a priori intuition contains only the form of our sensibility. It is a nice question just what this does to the characterization of intuition that gives rise to the puzzle. Clearly, in the a priori case, the causal dependence of the intuition on the object has to go. Whether and how the *phenomenological* presence of an object is preserved is a further question, as is the question whether the object thus present is a physical or a purely mathematical object. The former is not ruled out by the a priori character of pure intuition, since the 'presence' might be that characteristic of *imagination* rather than sense. In fact, a number of passages in Kant indicate that just that is his position.

We might find a difficulty for the idea of intuition of mathematical objects in what, following Leibniz, might be called their incompleteness. I do not

---

[9] Gödel, 'What is Cantor's Continuum Problem?', 271; in Benacerraf and Putnam (eds.), *Philosophy of Mathematics*, 2nd edn. (Cambridge: Cambridge University Press, 1983), 483–4.
[10] LU VI, §45. [11] Ibid., §46.
[12] Kant's puzzle is related to the dilemma about mathematical truth posed by Paul Benacerraf in 'Mathematical Truth', reprinted here as Chapter I. According to Benacerraf, our best theory of mathematical *truth* (Tarski's) involves postulating mathematical objects, while our best account of *knowledge* requires causal relations of the objects of knowledge to us; but mathematical objects are acausal.

One can present Kant's problem as a similar dilemma: mathematical truth requires applicability to the physical world. But our best account of mathematical knowledge makes it rest on intuition, which requires the prior presence of the object. But this contradicts the a priori character of mathematics.

This is of interest because it is a form of the dilemma that does not require that the semantics of mathematics involve mathematical objects (which it seems one might avoid by a modal interpretation of quantifiers). But of course it depends on other assumptions, in particular that mathematics is a priori.

need to go into this much, because it has been much discussed, not least by me.[13] The properties and relations of mathematical objects that play a role in mathematical reasoning are those determined by the basic relations of some system or structure to which all the objects involved belong, such as the natural numbers, Euclidean or some other space, a given group, field, or other such structure, or the universe of sets or some model thereof. It seems that the properties and relations of mathematical objects about which there is a 'fact of the matter' are either in some way expressible in terms of the basic relations of this structure or else are 'external relations' which are independent of the choice of a system of objects to realize the structure.

Consider, for example, the natural numbers, with 0 and the successor function S as giving the relevant structure (perhaps with other functions such as addition, if we give ourselves no second-order apparatus). Examples of the former type are number-theoretic properties such as being prime or being the sum of four squares. External relations include those arising in counting other objects, and such properties as being believed by me to be prime. Such relations will not in general be definable in the language of number theory, even higher-order, but they are in general definable in terms of the basic relations and others that do not depend on the choice of a system of objects and relations to realize the structure.

Now the question is, how can mathematical intuition place objects 'before our minds' when these objects are not identifiable individually at all? For example, unless one is presupposing a structure including numbers and sets, it seems indeterminate whether the number 2 is identical to the one-element set $\{\{\Lambda\}\}$, the two-element set $\{\Lambda,\{\Lambda\}\}$, or neither.[14] How can this be if numbers and sets are objects of mathematical intuition? Can such intuition be a significant source of mathematical knowledge if it does not determine the answers to such simple questions?

One could press the matter further and urge the possibility of an interpretation of mathematics which dispenses with distinctively mathematical objects. One such possibility is a nominalistic reconstrual of such objects. Another, more promising as an approach to the whole of mathematics, is a modal interpretation of quantifiers in which, roughly, statements of the existence of

---

[13] Charles Parsons, 'Frege's Theory of Number', in Max Black (ed.), *Philosophy in America* (London: Allen and Unwin, 1965); id., 'Ontology and Mathematics', *Philosophical Review*, 80 (1971): 151–76, at 154–7; id., 'Quine on the Philosophy of Mathematics', in L. Hahn and P. A. Schilpp (eds.), *The Philosophy of W. V. Quine* (La Salle, Ill.: Open Court, 1986), 369–95. Among others see esp. Paul Bernays, 'Mathematische Existenz und Widerspruchsfreiheit' (1950), in *Abhandlungen zur Philosophie der Mathematik* (Darmstadt: Wissenschaftliche Buchgesellschaft, 1976), 92–106; and Paul Benacerraf, 'What Numbers Could Not Be', *Philosophical Review*, 74 (1965): 47–73.

[14] The first follows from Zermelo's proposal for a set-theoretic construal of numbers, the second from von Neumann's, the third if set theory takes the natural numbers as individuals.

a mathematical object satisfying some condition are rendered as statements of the *possible* existence of an object satisfying purely structural conditions.[15]

These difficulties are at bottom one. What is really essential to mathematical objects is the relations constituting the structure to which they belong. Accordingly, in the end there is no objective ground for preferring one realization over another as 'the' intended domain of objects, in particular for rejecting concrete (nominalistic) realizations if they are available. Moreover, actual, as opposed to merely possible, realization of a structure adds nothing mathematically relevant. Both these points need *some* qualification, first because often actually given realizations of a structure presuppose some more comprehensive structure, such as the natural numbers of sets, and because in discussions of potential totalities, something like a distinction between actual and potential existence can be made. However, the exactly right way to put these points need not concern us here.

III

I propose to show that there is at least a limited application of the notion of mathematical intuition *of* which is able to meet these objections. First, let us review briefly the reasons why one might introduce the concept. Intuition *that* becomes a persuasive idea when one reflects on the obviousness of elementary truths of mathematics. Two alternative views have had influential advocates in this century: conventionalism, the view that at least some mathematical propositions are true by convention, and a form of empiricism according to which mathematics is continuous with science and the axioms of mathematics have a status similar to that of high-level theoretical hypotheses. Both these views have unattractive features. Conventionalism has been much criticized, and I need not repeat the criticisms here.

The empiricist view, even in the subtle and complex form it takes in the work of Professor Quine, seems subject to the objection that it leaves unaccounted for precisely the *obviousness* of elementary mathematics (and perhaps also of logic). It seeks to meet the difficulties of early empiricist views of mathematics by assimilating mathematics to the theoretical part of science. But there are great differences: first, the 'topic-neutrality' of logic, which receives considerable recognition in Quine's writings, although he insists that it

---

[15] Hilary Putnam, 'Mathematics without Foundations', *Journal of Philosophy*, 84 (1967): 5–22, reprinted here as Chapter VIII; my 'Ontology and Mathematics', 158–64; Charles S. Chihara, *Ontology and the Vicious Circle Principle* (Ithaca, NY: Cornell University Press, 1973), 191.

Applied to arithmetic and other more elementary parts of mathematics, the modal interpretation of quantifiers may serve to defuse scruples about abstract objects. In 'Quine on the Philosophy of Mathematics', I argue that this is not the case for higher set theory.

depends on a specification of the logical constants that is at bottom arbitrary; second, the very close connection of mathematics and logic, where the potential field of application of mathematics is as wide as that of logic, in spite of the fact that the existence of mathematical objects makes mathematics not strictly topic-neutral; third, the existence of very general principles that are universally regarded as obvious, where on an empiricist view one would expect them to be bold hypotheses, about which a prudent scientist would maintain reserve, keeping in mind that experience might not bear them out; fourth, the fact that differences about logic and elementary mathematics, such as the issues raised by intuitionism, are naturally explained as differences about *meaning*. Quine recognizes this by the role that logic plays in his theory of translation, but the obviousness of logic is an unexamined premiss of that theory.

Some version of the pre-Quinean view of logic as true by virtue of meaning may be the most promising way of addressing the difficulties of the Quinean view of *logic*. There is no a priori reason why the conception of intuition we are examining should play a role in working out such a view. In the case of arithmetic, the situation is different because, unlike logic, it has ontological commitments. That a structure such as the natural numbers should exist, or at least should be *possible* in some mathematically relevant way, is hard to make out as true by virtue of the meanings of arithmetical or other expressions.

Just at this point, the idea of intuition *of* suggests itself. We are taking as a gross fact about arithmetic, that a considerable body of arithmetical truths is known to us in some more direct way than is the case for the knowledge we acquire by empirical reasoning. And this knowledge takes the form of truths about certain objects—the natural numbers. What is more natural than the hypothesis that we have direct knowledge of these truths because the objects they are about are given to us in some direct way? The model we offer of this givenness is the manner in which a physical body is given to us in perception.

IV

As applied to the natural numbers, this picture is oversimplified. However, I propose to meet the difficulties by a strategy suggested by Kant's conception of pure intuition as giving the *form* of empirical intuition and by Husserl's thesis that categorial intuition is *founded* on sensible intuition. The quasi-perceptual manner in which mathematical objects can be given to us is in a certain way exemplified by situations of *ordinary* perception *or imagination* of realizations (sometimes partial) of the structures involved.

MATHEMATICAL INTUITION

Elsewhere I have presented an account of arithmetical intuition.[16] However, the presentation was tied to a modal interpretation of quantifiers, and the idea was intended to have greater generality. The following exposition is intended to make some other aspects of the earlier account more explicit.

It is well to follow Hilbert and to begin by considering the 'syntax' of a 'language' with a single basic symbol '|' (stroke), whose well-formed expressions are just arbitrary strings containing just this symbol, that is, |, ||, |||, ... This sequence of strings is isomorphic to the natural numbers, if one takes '|' as 0 and the operation of adding one more '|' on the right as the successor operation. This yields an interpretation of arithmetic as a kind of geometry of strings of strokes. At first sight the interpretation leaves out the concept of *number*, that is the role of natural numbers as cardinals and ordinals.

Ordinary perception of a string of strokes would have to be perception of a *token*, but we naturally think of such symbols as types. Beginning with the notion of a token being *a* stroke, we can recast the explanation of the stroke-language in such a way that types are not presupposed as objects. Two strings are 'of the same type' if they *can* be placed side by side so that strokes correspond one to one.[17] The use of 'can' in the criterion for sameness of type may be non-essential; someone, such as an actualist nominalist, could argue that some other type of empirical test is sufficient, or he might appeal to an inductive definition: two strings are of the same type if they are both single strokes, or if the strings consisting of all but the right-most stroke in each are of the same type. However, shortly we shall face a much stronger temptation to use modality.

That one can go this far (and indeed much farther) in doing syntax nominalistically is not news. What is less widely appreciated is that we have here the basis of an explanation of types, which first of all makes them no more mysterious than other objects, in spite of their 'abstractness', and secondly makes it quite reasonable to say that they are *given* in a way analogous to that in which middle-sized physical objects are given. Indeed, ordinary language recognizes this, in that we speak of hearing or seeing *words* and *sentences*, where what is clearly meant are types.

Of course a perception of a string of stroke-tokens is not by itself an 'intuition' of a stroke-string type. One has to approach it with the *concept* of a type, first of all to have the capacity to recognize other tokens as of the same type or not. Something more than the mere capacity is involved, which might be described as seeing something *as* the type. But this much is present in ordinary perception as well. One can of course see an object without recognizing it as this or that, but when it does occur, such recognition is part of normal

---

[16] Parsons, 'Ontology and Mathematics', section III.
[17] Ibid. 159–60, to which I refer the reader for details.

perception, and when one sees an object, one at least recognizes it under *some* description that permits re-identification.

One might object that in the case of ordinary perception, the *Auffassung* as an object, even of a particular kind, is entirely spontaneous and natural, whereas what I want to call 'intuiting' a symbol-type is a conscious exercise of a conceptual apparatus which may be quite artificial. I agree that this may be true in this case and is certainly true in some. However, in some cases, taking what is given as a type is quite spontaneous and natural. The most obvious is the understanding of natural language: the hearer is without reflection ready to re-identify the type (in the linguistic, not the acoustic, sense). Typically, the hearer of an utterance has a more explicit conception of *what was uttered* (e.g., what words) than he has of an objective identification of the *event* of the utterance.[18] I believe that the same is true of some other kinds of universals, such as sense qualities and shapes.[19] Indeed, in all these cases it seems not to violate ordinary language to talk of perception of the universal *as an object*, where an instance of it is present. This is not just an overblown way of talking of perceiving an instance *as* an instance (e.g., seeing something red *as* being red), because the identification of the universal can be firmer and more explicit than the identification of the object that is an instance of it.

These observations should begin to dispel the widespread impression that mathematical intuition is a 'special' faculty, which perhaps comes into play only in doing pure mathematics. At least one type of essentially mathematical intuition, of symbol-and expression-types, is perfectly ordinary and recognized as such by ordinary language. If a positive account of mathematical intuition is to get anywhere, it has to make clear, as its advocates intended, that mathematical intuition is not an isolated epistemological concept, to be applied only to pure mathematics, but must be so closely related to the concepts by which we describe perception and our knowledge of the physical world that the 'faculty' involved will be seen to be at work when one is not consciously doing mathematics.

---

[18] Of course when we talk of what was uttered, and even more of what was *said*, this is often best understood in a way that invites regimentation in terms of *propositions*. One might then offer a similar argument for the claim that propositions are objects of intuition. However, such considerations cannot get us past the well-known doubts about the objectivity of propositions. A response to an auditory stimulus can count as intuition of a *sentence* because we can attribute to the hearer a reasonably sharp concept of *same sentence*.

[19] Such a view is not necessarily incompatible with all versions of nominalism. The British empiricists sometimes understood sense qualities as universals, but admitted them as 'simple ideas' rather than 'abstract ideas'. Similarly, the *qualia* admitted by Nelson Goodman in *The Structure of Appearance* (Cambridge, Mass.: Harvard University Press, 1951) are universals. Neither on the empiricist's view, nor Goodman's, nor on the view I suggest, should sense qualities be understood as a kind of *attribute* in the sense of something denoted by a nominalized predicate.

V

The preceding discussion indicates that we should be careful in talking, with Husserl, of 'intuition' of a type as founded on perception of a token. In ordinary cases there will be perception in the full sense, which requires physical presence and action on the senses. Ordinary talk of hearing *words* normally carries this implication. In many cases the token will be pushed into the background by the type, but that does not make the former not an object of perception. However, even in normal cases the background and further experience that are necessary to the perception's being of something physically *real* are irrelevant to its being of the *form* given by the type. In most cases, physical reality is important not for taking in the type, but for further considerations: what is likely to be of interest about the words is that they were spoken by a speaker at a certain time, or stand written in a certain book.

Perceptions and imaginings, as founding such intuitions, play a paradigmatic role. It is through this that intuition of a type can give rise to propositional knowledge about the type, that is, intuition *that*. A simple case is singular propositions about types, such as that ||| is the successor of ||. We see this to be true on the basis of a single intuition, but of course in its implications for tokens it is a general proposition. Let a be the token of ||| above; let b be the token of || above; the statement implies that if c and d are respectively of the same type as a and b, then c consists of a part of the same type as b, and one additional stroke on the right. We can of course buttress the statement that ||| is the successor of || by considering arbitrary tokens of the relevant types and verifying the above consequence. But we have to verify it in the same way, by instances that we take as paradigmatic. This situation is not peculiar to our artificial framework. The same is true of calculations done on paper and of formal proofs, such as the deductions done in elementary logic courses.

A more problematic situation arises when we consider general propositions about *types*, which have in their scope indefinitely many *different* types. It is this which prompts us to follow Husserl in saying that sometimes *imagination* of the token can found intuition of the type. Consider, for example, the assertion that each string of strokes *can* be extended by one more. This is the weakest expression of the idea that our 'language' is potentially infinite. But we cannot convince ourselves of it by perception or by the kind of mathematical intuition we have talked about so far, founded on actual perception. But if we imagine any string of strokes, it is immediately apparent that a new stroke can be added. One might imagine the string as a *Gestalt*, present all at once: then, since it is a figure with a surrounding ground, there is space for an additional stroke. However, this may not be the right way to look at the matter, since the imagination of an *arbitrary* string in this way will have to leave inexplicit its articulation into single strokes. Alternatively, we can think of the string as constructed step by step, so that the essential element is now succession in *time*, and what is then evident is that at any stage one can take another step.

Either way, one has to imagine *an arbitrary string of strokes*. We have a problem akin to that of Locke's general triangle. If one imagines a string in a specific way, one will imagine a string with a specific number of strokes, and therefore not a perfectly arbitrary string. There seems to be a choice between imagining *vaguely*, that is imagining a string of strokes without imagining its internal structure clearly enough so that one is imagining a string of n strokes for some particular n, or taking as paradigm a string (which now might be perceived rather than imagined) of a particular number of strokes, in which case one must be able to see the irrelevance of this internal structure, so that in fact it plays the same role as the vague imagining.

We naturally think of perception as at least sometimes uncorrupted by thinking, in that without conscious thinking one can take in some aspect of the environment and respond to it, and one can take a stance toward one's perceptions that is largely non-committal with respect to the judgements we would ordinarily be prepared to make. However that may be, it is clear that the kind of *Gedankenexperimente* I have been describing can be taken as intuitive verifications of such statements as that any string of strokes can be extended only if one carries them out on the basis of specific concepts, such as that of a string of strokes. If that were not so, they would not confer any generality.

Brouwer may have been trying to meet this difficulty, in a special case of this sort, with his concept of two-one-ness, according to which the activity of consciousness brings about 'the falling apart of a life-moment into two qualitatively distinct things', of which the moment then present retains the structure of the original, so that the resulting 'temporal two-ity' can be taken as a term of a new two-ity, giving rise to temporal three-ity.[20] Thus the process can always give rise to a new moment, which for Brouwer is the foundation for the infinity of the natural numbers. One has something similar in the figure–ground structure of perception, which was appealed to above. However, in all versions we think of whatever step it is as one that *can be iterated indefinitely*. In a sense, this is given by the fact that after the step of 'adding one more' one has essentially the same structure. But a concept such as that of a *string* of strokes involves the notion of such iteration. To spell that out, we are led into the circle of ideas surrounding mathematical induction.

[20] L. E. J. Brouwer, *Collected Works*, vol. 1, ed. A. Heyting (Amsterdam: North Holland, 1975), 417 (from 1929). Cf. p. 17 (from 1907), p. 480 (1948), and p. 510 (1952).

Although the view has been attributed to Brouwer that 'iteration' is the fundamental intuition of mathematics, my view is that the particular concept of intuition I am explicating runs out at this point, and it is only in a weaker and less clear sense that mathematical induction is a deliverance of intuition.

Although the concept of a string of strokes involves iteration, the proposition that every such string can be extended is not an inductive conclusion. A proof of it by induction would be circular. Such a proof would be called for only if we really needed the fact that every string of strokes can be obtained by iterated application of the operation of adding one more. In fact, I think the matter is thus: we have a structure of perception, a 'form of intuition' if you will, which has the essential feature of Brouwer's two-one-ness, that however the idea of 'adding one more' is interpreted, we still have an instance of the same structure. But to see the *possibility* of adding one more, it is only the general structure that we use, and not the specific fact that what we have before us was obtained by iterated additions of one more. This is shown by the fact that in the same sense in which a new stroke can be added to any string of strokes, it can be added to any bounded geometric configuration.

VI

It should be clear that we do not acquire in this way any reason to believe it *physically possible* to extend any string of strokes. At most, the structure of space and time is at stake here, and physical possibility requires something more, whatever makes the difference between the space of pure geometry and the physical universe, consisting at least of space containing matter. Actually we require less than the space of pure geometry, since even if we do hold to the spatiality of the strokes (which perhaps we can avoid), only very crude properties of space are appealed to, in particular not its metric properties.

We can call the possibility in question *mathematical possibility*; this expresses the fact that we are not thinking of the capabilities of the human organism, and it may even be extraneous to think of this 'construction' as an act of the *mind*. The latter construal agrees with the viewpoint of Kant and Brouwer. It is very tempting if we want to say that any string of strokes is *perceptible* or *imaginable*. (It is preferable to reserve these words for tokens, but then one can speak of the *intuitability* of the type.) The idea is that no matter how many times the operation of constructing one more stroke in imagination has been repeated, 'we' can still construct one more. However, I think there is really a hidden assumption that there is no constraint on what 'we' can perceive beyond the open temporality of these experiences and some very gross aspects of spatial structure. Kant and Brouwer thought these were contributions of our minds to the way we experience the world. Kant of course thought that we could not know these things a priori unless our minds had contributed them. I am not persuaded by this, and in any case I do not want my argument to rest on the notion of a priori knowledge. If we express the *content* of the proposition in a way as independent as possible of the description of the insight, then it is just that for an arbitrary string of strokes it is possible that there should be one that extends it by one stroke.

The nominalist seems to demand both more and less than we do. He may try to get on without even the potential infinity of a sequence like that of stroke-string-types, but then he will have to do without the infinity of the natural numbers. His position is really the embarrassing one that Russell found himself in about the axiom of infinity. He can treat it as a hypothesis to whose truth he is not committed, but then mathematics allows the possibility that where we have proved by ordinary mathematical means a proposition B, we are not entitled to reject its negation, since, where A is the relevant axiom of infinity, if A is false, both $A \to B$ and $A \to \sim B$ are true.

Alternatively, he may accept as an empirical hypothesis some proposition entailing the existence in space and time of a ω-sequence such as a sequence of tokens of stroke-strings, each one extending the previous one.[21] Since he is talking about physical existence, he is making a stronger claim than we do, which mathematics does not need. (He could be a traditional empiricist and discern such a sequence in some phenomenal field, but on empiricist grounds this seems very questionable, and it has the same mathematically irrelevant strength.) Such a hypothesis clearly has a theoretical character, and it might even be rejected if physics were to evolve in such a way that space-time came to be understood as both finite and discrete. Any reason we have for believing it depends on the historically given physics, constructed in tandem with an arithmetic with an infinity of numbers.

A third position which might be called nominalistic is one alluded to above: one continues to hold that in strict usage one should talk only of tokens; the relation *same type* is available but is understood as just a useful equivalence relation, not the foundation of identity of types; and one meets the problem of the potential infinity of types by a modal interpretation of quantifiers.[22] Earlier I gave some reasons for denying this view the title of

[21] This is, e.g., entailed by the position of Hartry Field's interesting *Science without Numbers* (Oxford: Blackwell, 1980). His main project is to interpret physics in an extension of synthetic geometry, in which the variables range over points and regions of space-time, which he asserts to be physical. A model of arithmetic can certainly be constructed in his theory.

[22] Chihara, *Ontology*, 191; Parsons, 'Ontology and Mathematics', 160–2. In the latter, the last two lines of p. 160 are ambiguous. One way of taking it would be to say that '$\exists x Fx$', where the variable 'ranges over natural numbers', is *true if and only if* we can construct a perceptible inscription which can be put into the empty place of an inscription of 'Fa' so that a truth results. Properly, this should be recast as a necessary statement about inscriptions. This truth condition has a substitutional

nominalist,[23] but although I still hold to them, I would now say that the question whether it *is* nominalist is in the end terminological. So long as we stay short of set theory and other impredicative mathematics, and the modality involved is mathematical possibility, the position is not importantly different from my own. However, the latter qualification is important. If the modal theory of tokens is understood as a theory of physical tokens, and the modality is physical, then I think the view faces the same difficulties as the actualist forms of nominalism.

VII

I now turn to the question whether our conception of an intuition of types faces serious objections because of the timelessness, acausality, or incompleteness of types as abstract entities. Stroke-string-types and other such expressions are minimally abstract, since they are types of tokens which are concrete. Our intuitions of them are founded on sense experience or on imaginings which imagine their objects as in space and time, even if not at any particular location. The timelessness of types is simply universality: since they can be instantiated anywhere, they are understood as located nowhere. Because the existence of a type depends on the possibility of a token, they cannot be understood as mereological sums. The *problem* about the timelessness of types is really epistemological: how can we know truths about types by a certain kind of perception of tokens, which are then valid for *any* tokens of the types involved. In my remarks above, I have done little more than try to make clear that we *do* have such knowledge. More explanation should be given, though some experience tends to show that explanations of such matters are always in the end question-begging. Observe, however, that the problem is not created by an ontology of types. On the nominalistic views I have mentioned, there is also a question about knowledge of the general truths about tokens that are the nominalistic versions of the truths about types.

It might be questioned whether types are acausal after all; for example, I might say, 'His words made me furious'. Suppose he said, 'You have no right to call yourself a philosopher'. But in fact we do not think of the *sentence* as making me furious (and not just because of the indexicality of this particular example). Nor do we attribute the effect to the proposition expressed, although that might be more plausible in this case. It is much more natural to

---

character, and then the resulting interpretation as a language of arithmetic is substitutional; inscriptions and construction thereof are talked of only in the metalanguage.

Other readings are possible that make the quantifier range over inscriptions.

[23] Parsons, 'Ontology and Mathematics', 162–4.

attribute the effect to the event of his saying the words, or his expressing that proposition, on the occasion on which he did. This preserves the acausality of the sentence, but its relation to causality is like its relation to space and time. Its tokens are caught up in the causal nexus, and indeed affect our senses. Once we see the relation between intuition of types and ordinary perception, I think this difficulty rather dissolves. However, an objector may be thinking of Benacerraf's dilemma (see n. 12). To deal with this requires a longer story than I can tell now.

About incompleteness, one might first think that the closeness to the concrete of such abstract objects as strings of strokes would make them *not* incomplete. For example, it would be simply false that ||| is identical with an object given in some other way, say the number 3. ||| has some properties 3 lacks, such as that it is composed of strokes. Another problem is cognitive relations, including *de re* propositional attitudes; if I see on a blackboard the formula '$\forall x\, (x \neq 0 \to \exists y\, (x = Sy))$', I do not see the number that corresponds to it under some arithmetization of the syntax of first-order arithmetic.

I suggest the following explanation. What is basic to the concept of type gives identity and difference relations only to other types in the same system of symbols. (Two inscriptions may be of the same type with respect to one symbolism and not with respect to another.) Since this is a distinctive feature of what types *are*, common sense tends to treat the types of a given symbolism as *sui generis*, so that none is identical with anything given in some other way. In one sense this resolves the incompleteness, since it determines all predicates (at least from the point of view of classical logic), but in a negative way: all atomic predicates except those from the structure and those expressing the basic facts about its instantiation, are false. But this inclination of common sense does not correspond to a feature of the nature of things, at least not to one that cannot be overridden when it comes to the regimentation of language.

However, this kind of consideration does show a significant disanalogy between this kind of mathematical intuition and ordinary perception. *What is intuited* depends on the concept brought to the situation by the subject. In some cases, such as natural language, the concepts involved may be innate or develop more or less spontaneously and unreflectively. In the more characteristically mathematical cases of geometric figures and the sort of artificial symbolism we have been discussing, this is not so. Therefore we do not have the scope that we have with ordinary perception for identifying the object of intuition independently of the subject's conceptual resources. If someone feels heat, and heat is the motion of molecules, then he feels something that is the motion of molecules. If we are using 'feel' in an object-relational way, he feels the motion of those molecules, even if he has no conception of molecules.

But no one could intuit a stroke-string-type unless he saw it *as* a type constructed from strokes, and this requires that he have the concept of stroke. If, in regimenting our theory, we identify ||| with the number 3, then perhaps we can say that he is intuiting 3, although he may have no idea that that is what he is doing. But we can only say that because he has *some* identifying concept. There is probably an ordinary concept of perception for which this holds as well, but it does not obviously hold for the most ordinary object-relational uses of 'see', 'hear', and perhaps 'feel'.[24]

VIII

Our investigation so far has reached a significant positive result. It is quite permissible to say that types of perceptible tokens are objects of intuition, where the concept of intuition involved is strongly analogous to that of perception. Moreover, we can represent some propositions about these objects as known intuitively.

This result is of very limited scope. Even though they form a model of arithmetic, from a mathematical point of view, strings of strokes are rather special objects. The perception-like character of what we call intuition of types may be thought to be due to the closeness to perception of the objects involved. Perhaps our concept of mathematical intuition will not carry us beyond elementary syntax and maybe traditional geometry. Are we prepared to say, for example, that the *natural numbers* are objects of intuition?

I have to deal with this question more briefly than I would like. Our discussion so far suggests a moderate position: intuition gives objects which form a model of arithmetic, and this model is as good as any, both for the foundations of arithmetic and for applications. But it may not be right to say that *the* natural numbers are objects of intuition, since intuition does not give a unique sequence to be 'the' natural numbers, and the concept of number does not rule out as the 'intended model' objects that are not objects of intuition.

However, we should try to come to terms with the higher-order aspects of the concepts of cardinal and ordinal number. I shall restrict myself to cardinal number. The formulation of a statement of number requires an operation on predicates, either a numerical quantifier like 'there are n x's such that Fx' or a term-forming operation like 'the number of x's such that Fx'. This point,

---

[24] In the case of natural language, we classify types according to the language, and not according to the conceptual apparatus of the perceiver. If I say to someone who has never heard English, 'Where is the American Embassy?', he hears that sentence of English, even though he does not recognize it as such, and is not able to recognize an utterance of the same sentence by someone with a different accent. This case is analogous to the role of natural kinds in the description of what someone sees.

however, imposes no constraint at all on what kind of objects the natural numbers are. We should resist the temptation to identify the numbers with the numerical quantifiers themselves (as 'second-level concepts' or the like) as well as the subtler temptation, to which Frege succumbed, to try to find an object that represents the numerical quantifier in an especially intrinsic way.

A more serious matter is that apparently the truth conditions for statements of number must incorporate the Fregean criterion for sameness of number: the number of F's is the same as the number of G's if and only if *there is a one-to-one correspondence of the F's and the G's.*

We might seek to accommodate this in a way which makes numbers objects of intuition by understanding numbers as a kind of generalized types, in which the tokens are numeral-tokens serving as counters.[25] The relation playing the role of sameness of type is that of 'representing the same number'. Since what is involved in this relation is what is involved in *counting*, observe that verifying by counting that there are n x's such that Fx involves exhibiting a one-to-one correspondence between the F's and a sequence of n 'counters'—standardly numerals. If the predicate 'F' is simple enough and the objects are objects of perception, 'there are n F's' can be verified by perception. We do not have to take it as *saying* that there *is* such a correspondence between the F's and the counters. But it does in some way imply it, and in order to establish the elements of number, we do have to reason about such correspondences.

For arithmetic, however, the correspondences we need are finite. I will assume that *finite sets* of objects of intuition are themselves objects of intuition. Space does not permit defending that assumption here. What I want to observe is that from it follows what is needed to justify the claim that the natural numbers, considered as 'numbering' objects of intuition, are objects of intuition. Two numerals a and b, perhaps from different notation systems, represent the same number if there is a one–one correspondence of the numerals up to a and those up to b. Our assumption implies that this does not involve reference except to objects of intuition.

However, this approach would suggest that arithmetic, as applied to objects *in general*, belongs to set theory. We could still say in this context that finite numbers are objects of intuition, on the ground that for the constitution of numbers as objects this full generality is not needed. However, the general principles of cardinal and ordinal number, applied to arbitrary sets, even arbitrary (possibly not hereditarily) finite sets, will not be intuitive knowledge unless sets in general are objects of intuition. I have not tried to argue that they are.

---

[25] Parsons, 'Frege's Theory of Number', 201. Cf. id., 'Ontology and Mathematics', 160.

Before we end this paper, we must say something about mathematical induction, which arises already for strings of strokes. I have not said much about our understanding of what an arbitrary string of strokes, or an arbitrary natural number, is. However, this understanding should surely yield the relevant induction principle. What bearing does this have on our remarks on intuition?

Let us concentrate on the more intuitive case of strings of strokes. In such a case, the conclusion of an inference by induction is a general statement about objects of intuition. It does not follow that it is therefore intuitive knowledge. There is a temptation to call our understanding of the general notion of a string of strokes an intuition, because it is clear and seems to make inductive inferences evident.[26] However, this would have to be a different, and in this context potentially confusing, sense of 'intuition', since what is involved is the understanding of a general term; this does not give any *object*.[27]

Because of the essential way in which this understanding is used, I am inclined to deny that even very simple inductive conclusions are intuitive knowledge. Gödel, however, in discussing a distinction between intuitive and abstract evidence, uses another criterion.[28] His line is drawn where one begins to refer to what he calls 'abstract objects', by which he means statements and proofs.[29] I do not deny that this is also an important distinction. Moreover, I do not claim to have shown that his terminology is inappropriate. More needs to be said about the epistemological aspects of the concept of intuition, even the very limited concept that I have developed, than I have said in this paper.[30]

[26] For the natural numbers, Dummett makes such a suggestion, in order to criticize it, in 'Platonism', in *Truth and Other Enigmas* (London: Duckworth, 1978), 202–14.

[27] Of course, there are strings of strokes which as a practical matter can never be intuited, and it is only by means of the general notion of a string of strokes that we conceive such objects. But the thought of 10^{100} strokes, however clear, is not an intuition of them.

[28] Gödel, 'Über eine noch nicht benützte Erweiterung des finiten Standpunktes', 280–2. In the translation referred to in n. 7, 'anschauliche Erkenntnis' (p. 281) is translated 'immediate concrete knowledge'.

[29] In my view, what is really essential is semantic reflection. See my 'Ontology and Mathematics', 165–7.

[30] This paper was first presented at a meeting of the Aristotelian Society on Monday, 17 March, 1980, at 6.30 p.m. At the time the author was a Fellow of the National Endowment for the Humanities and a Visiting Fellow of All Souls College, Oxford. He wishes to record his gratitude to both these institutions.

VI

PERCEPTION AND MATHEMATICAL INTUITION

PENELOPE MADDY

Set-theoretic realism is a view whose main tenets are that sets exist independently of human thought, and that set theory is the science of these entities.[1] The foremost advocate of this position, the late Professor Gödel, has stressed an analogy between mathematics and physical science.[2] According to Gödel, higher set theory bears a relation to the rest of mathematical knowledge and to practical mathematical dealings of everyday life which is analogous to the relation borne by theoretical physics to physical science in general and to common-sense knowledge of the world. Sense perception gives us knowledge of simple facts about physical objects, and a faculty of mathematical intuition gives us knowledge of sets, numbers,[3] and of some of the simpler axioms concerning them. In both cases, theories involving 'unobservable' entities or processes (that is, entities or processes beyond the range of sense perception or mathematical intuition) are formed in order to explain, predict, and systematize the elementary facts (of perception or intuition) and are judged by their success.

A view of this sort has several attractive features: (i) it allows a straightforward Tarskian semantics for set-theoretic discourse; (ii) it makes no mystery of how mathematical premises can combine with physical ones to yield

First published in *Philosophical Review*, 84/2 (1980): 163–96. Reproduced here by permission of the publisher and the author. © 1980 Cornell University Press.

[1] The term 'platonism' is often applied to views of this sort, but I will avoid it. To me 'realism' seems more appropriate, since sets, on the view I am concerned with, are taken to be individuals or particulars, not universals.

[2] Kurt Gödel, 'Russell's Mathematical Logic', in P. A. Schilpp (ed.), *The Philosophy of Bertrand Russell* (Evanston, Ill.: Northwestern University Press, 1944), 123–53, repr. in P. Benacerraf and H. Putnam (eds.), *Philosophy of Mathematics*, 1st edn. (Englewood Cliffs, NJ: Prentice-Hall, 1964), 211–32; 2nd edn. (Cambridge: Cambridge University Press, 1983), 447–69; and id., 'What is Cantor's Continuum Problem?', *American Mathematical Monthly*, 54 (1947): 515–25; repr. in Benacerraf and Putnam (eds.), *Philosophy of Mathematics*, 1st edn., 258–73; 2nd edn., 470–85.

[3] Numbers, at least in the version of realism I will sketch, are taken to be properties of sets, and are brought within the general set-theoretic epistemology in that way. The details of this move cannot be dealt with here. See my 'Sets and Numbers', *Noûs*, 15 (1981): 495–512.

testable consequences in physical science (that is, both sorts of premises are true in the same sense); (iii) it squares with the pre-philosophical views of most working mathematicians; and (iv) it allows set-theoretic practice to remain as it is; it does not demand reform. On these last two counts, it differs from the holistic form of realism advocated by Quine. On the holistic view, mathematics is a highly theoretical part of the web of knowledge, justified by its usefulness in physical science. But set theorists do not generally suppose that the truth or falsity of their axioms is so intimately linked with applicability in science, and their methods can be seen to reflect this.

Despite its obvious attractions, realism has met with considerable opposition from philosophers. Benacerraf[4] has argued that philosophies of mathematics can be divided into two groups: those inspired by ontological considerations and those inspired by epistemological considerations. Realism falls in the first group, dealing straightforwardly with questions of what mathematical objects exist and what mathematical statements mean, at the expense of questions of how we know mathematical facts. Benacerraf's central objection to Gödel is that even if the hypothetico-deductive model of the higher reaches of set theory were unproblematic,[5] this would not be very comforting until much more could be said about how mathematical intuition supplies us with the sort of knowledge on which the edifice is to be based. Specifically, an account of how the analogous faculty—sense perception—provides us with knowledge of physical objects begins with a causal interaction between the knower and the object (or fact) known, but no such interaction seems possible in the case of sets.

Since the invention of the Kripke/Putnam theory of reference,[6] this style of objection has been extended, for example, by Lear,[7] to include the realist's claim that set-theoretic statements are about sets. In physical science, we refer to things and kinds of things by virtue of standing at the end of a complex causal chain of usage leading back to dubbing. But, it seems that no sample of the kind 'set' could be vulnerable to such an initial baptism. So, the argument runs, not only are we unable to know facts about sets, but we are also unable to refer to them, so in fact, our theory of sets cannot be about them.

[4] P. Benacerraf, 'Mathematical Truth'; see Chapter I.

[5] Little is understood about the concepts of explanation and prediction in the context of mathematics, but some promising work has been done by M. Steiner (e.g., 'Mathematical Explanation', *Philosophical Studies*, 34 (1978): 135–51). Still, persuasive cases can be made for such axiom candidates as the axiom of measurable cardinals and the axiom of projective determinacy in terms of their explanatory scope and so on. I won't go into this part of the problem in this paper. For some discussion of these issues, see my 'Sets and Numbers'.

[6] S. Kripke, 'Naming and Necessity', in D. Davidson and G. Harman (eds.), *Semantics of Natural Language* (Dordrecht: Reidel, 1972), 253–355; H. Putnam, 'The Meaning of "Meaning"', in K. Gunderson (ed.), *Language, Mind and Knowledge*, Minnesota Studies in the Philosophy of Science, 7 (Minneapolis: University of Minnesota Press, 1975).

[7] J. Lear, 'Sets and Semantics', *Journal of Philosophy*, 74/2 (Feb. 1977): 86–102.

Taking Gödel's remarks as a starting-point, I will try to sketch an account of perception and intuition which will skirt these objections, and provide a basis for set-theoretic realism. I will not attempt to prove, or even argue for, the independent existence of sets, but rather, on the realistic assumption that they do so exist, I will try to show how we can refer to and know about them. Since accusations to the contrary are most often based on causal theories of reference and knowledge, I will use these theories as starting-points. I will not attempt to defend the causal theories (I don't pretend to know which are the correct theories of knowledge and reference), but I will try to show that they are reconcilable with set-theoretic realism. My success should be judged by the plausibility of the proposed epistemology, and by the extent to which it preserves the virtues listed above and squares with other reasonable tenets and motivations for set-theoretic realism.

I. THE OBJECTIONS OF BENACERRAF AND LEAR

Let's begin with the problem of reference. The part of the causal theory that is relevant here is the theory of reference to kinds. Of course, the theory of kinds on which this depends has not been completely worked out, but I think it is clear that some common nouns, like 'gold' and 'water' are more amenable to this treatment than some others, like 'bachelor' and 'home run', which seem better suited to a traditional theory according to which terms refer to whatever satisfies their conditions of definition. As one theorist puts it, in these latter cases:

> We do not have some kind of thing in mind, name it, and then seek to discover what it is we have named as we do in the case of 'gold' or 'tiger'. Rather, we have a certain specification or description in mind, and define anything that satisfies the description as having a right to the name.[8]

I think it is clear that the realist would take 'set' to be more like 'gold' than like 'bachelor' in terms of this contrast. Sets exist, and we inquire into their nature.

According to the causal theory, successful reference to a natural kind is accomplished by means of a chain of communication from the referrer back to an initial baptism. (Of course, the baptist refers without such a chain, but most of us are rarely in that role.) One member of this chain acquires the word by means of a causal interaction with the previous link; that is, I learn a word, in part, by hearing it, reading it, or some such sensory experience,

---

[8] S. Schwartz (ed.), *Naming, Necessity and Natural Kinds* (Ithaca, NY: Cornell University Press, 1977), 38.

caused, in part, by my predecessor in the chain.[9] The chain of communication for sets is no more problematic than for any other word. I learn it from my teachers, and intend my usage to refer to what theirs refers to.

The trouble is supposed to be with the initial baptism. This is an imaginary event in which the baptist isolates some samples of the kind, and picks out the kind of which these samples are members. According to Kripke, these baptisms are of at least two sorts: by description and by ostension. As an example of the first sort, we might imagine that the kind 'gold' was first picked out as the stuff instantiated by all or most of the samples in Fort Knox. Here there need be no causal connection between the baptist and the sample. This is perfectly acceptable so long as the description is well-founded,[10] that is, so long as the singular and general terms occurring in the description already refer. In our example, 'Fort Knox' must already refer to the appropriate building by means (presumably) of a causal chain back to another baptism.

In the case of sets, such a baptism by description doesn't seem possible. How could samples of the kind be picked out without assuming the kind is already picked out? (For example, suppose the baptist says, 'Set is the kind of which the set of my hands is a sample'.) Baptism by ostension is clearly what Lear had in mind when he wrote: 'There is no standard set with which one stands in the necessary causal relation to make it vulnerable to the appropriate dubbing'.[11] I will question this claim.

In an ostensive baptism of gold, a baptist would stand before some samples, look at them, and declare that these and all things like them are gold. Let us imagine an analogous initial baptism of sets. We imagine our baptist in his study saying things like: 'All the books on this shelf, taken together, regardless of order, form a set', and 'The globe, the inkwell, and the pages in this notebook, taken together, in no particular order, form a set'. By this process, the baptist picks out samples of a kind. The word 'set' refers to the kind of which these samples are members. One important feature of this treatment is that we can refer to sets without knowing much about them, just as we first referred to gold without knowing that it has atomic number 79, or how it differs from iron pyrites. And, just as gold on other planets, and gold that doesn't look like gold, are included within the scope of the kind 'gold', pure sets and infinite sets are included within the kind 'set'; all that matters is that they are the same kind of thing.

The obvious objection for Lear to this picture of set-theoretic reference is that, while the gold-dubber causally interacts with some samples, the

---

[9] For simplicity, and because they are not relevant to my central concerns, I have left out of this account many of the complexities of the causal theory of reference.
[10] See M. Jubien, 'Ontology and Mathematical Truth', *Noûs*, 11/2 (May 1977): 136.
[11] Lear, 'Sets and Semantics', 88.

set-dubber causally interacts only with the members of some samples. Here the realist might argue:[12] the extent of the causal interactions of both the set- and the gold-dubber is something like light bouncing off certain objects and bringing about some retinal changes. In the case of the gold-dubber, strictly speaking, the interaction is actually with the front side of a time slice of the sample. In other words, the thing whose kind we count the person as having dubbed is not the thing with which the dubber has actually causally interacted; the interaction is only with a fleeting aspect of that temporally extended object. Similarly, the set-dubber has only aspects of the sets within his causal grasp. But, if the interaction of the gold-dubber with an aspect is enough to allow the gold-dubber to pick out a sample, why shouldn't the set-dubber's interaction with an aspect do the same thing? The realist could argue that the relation of element to set is no more objectionable than the relation of fleeting aspect to temporally extended object.[13] If this argument can be filled in, it seems the realist can adopt a causal theory of reference, after all.

Of course, Benacerraf's epistemological objection to realism runs parallel to the above. On a theory of knowledge of the sort suggested by Goldman and Harman,[14] in the simplest perceptual cases, there must be an appropriate causal connection between the knower and the fact known. And, once again, sets seem unable to enter into causal relations. And, still again, in cases in which the causal theorist would admit that knowledge of a physical object is acquired, it is still only an aspect of the object in question which participates in the causal interaction.

Gödel responds to the challenge of the causal theorists by insisting that mathematical intuition is to play a role in the development of set theory analogous to that of sense perception in the development of physical science.[15] In particular, it is mathematical intuition that inspires us to form mathematical theories and convinces us of the truth of their axioms. To fill in the analogy, he considers the details of perceptual experience. According to Gödel, our physical concepts are formed on the basis of what is immediately given,

---

[12] That an argument of this sort might be available to the realist was first suggested to me by John P. Burgess.

[13] Of course, the set-dubber's interaction is also with mere fleeting aspects of the elements of his sets, so the relation between what he interacts with and what kind he dubs is the composition of the aspect/object and the element/set relations. This added complexity can be eliminated by imagining that the set-dubber uses sets of aspects, rather than sets of objects, as samples. A more reasonable course is to assume that if both the relations in question are legitimate, then so is their composition.

[14] A. Goldman, 'A Causal Theory of Knowing', *Journal of Philosophy*, 64/12 (June 1967), 357–72; G. Harman, *Thought* (Princeton: Princeton University Press, 1973).

[15] Gödel, 'Cantor's Continuum Problem?', 271–2. Quotations in this paragraph and the next are from this location.

but the immediately given is not to be taken as some form of unstructured data. He writes:

> That something besides the sensations actually is immediately given follows ... from the fact that our ideas referring to physical objects contain constituents qualitatively different from sensations or mere combinations of sensations, e.g., the idea of object itself.

Some structure, over and above mere stimulations, is immediately given, because our perceptual beliefs are about objects. He does not specify any more of these 'abstract elements contained in our empirical ideas', but he does emphasize that they are not contributed by the mind, because 'we cannot create any qualitatively new elements, but only reproduce and combine those that are already given'.

The question, then, is where these abstract elements of perceptual experience originate, if they do not enter by way of the sense-organs, and are not contributed by the mind. To account for their presence, Gödel postulates 'another kind of relationship between ourselves and reality', apart from the causal effect of physical objects on the sense-organs. Finally, he tells us that the immediately given of mathematical intuition, from which we form our concept of set and learn the truth of some of the axioms concerning it, is 'closely related' to these abstract elements. Presumably, the faculty of mathematical intuition itself is 'closely related' to this new sort of relationship between ourselves and reality which is responsible for the presence of these abstract elements in our perceptual experience.

This suggestion is hardly a complete solution, both because it has never been conclusively established that the mind can contribute nothing beyond reproductions and combinations of what is presented to it from outside, and because so little is said about this new relationship between ourselves and reality. At first, it might seem hardly more valuable than the suggestion given earlier to the effect that the set-theoretic realist must argue for some sort of similarity between the aspect/object relation and the element/set relation. But Gödel's approach does represent a definite advance, in that it suggests that the nature of mathematical intuition can be better understood in light of an investigation of the origin and role of the abstract elements of perceptual experience.

The central question is: what legitimates the gap between what is causally interacted with and what is known about, or what kind is dubbed, in the case the causal theorist accepts? I think the key here is that it is not enough simply for the baptist's or knower's retinas to be stimulated by light bouncing off the front surface of an object; what is required is that the baptist or knower perceive the object. This is made clear by examples of people, blind from birth, whose sense-organs are restored to perfect operating condition, but who cannot be said to perceive the objects around them.[16] The realist's point, then, is that stimulation of the sense-organs by light bounced off an object is not the same as perception of that object, as Gödel has pointed out. The realist's hope is that an account of what makes a pattern of sensory stimulation into a perception of a physical object could be modified into an account of how it is possible to interact with a set of physical objects in the way required by the causal theories of reference and knowledge.

To this end, I will argue below that people often perceive sets of physical objects. I will do this by extending a particular theory of perception of physical objects. The theory of object perception itself is naturally open to objection, but I will not defend it here, since this is done elsewhere by its original proponents, and an extension of the sort I am interested in can probably be added to most any theory of object perception. An obvious exception are sense-data theories according to which it is not even possible, strictly speaking, to perceive physical objects. Such theories seem to me false, and I refer the reader elsewhere in the literature for reasons.[17] Finally, though, let me repeat that I do not suppose that any arguments I might give for the theory of set perception and the theory of intuition sketched below are conclusive. First, I doubt that it is possible to give conclusive philosophical arguments for a view of this sort, since many of the questions involved are empirical. But the main obstacle is that I have not and will not attempt to prove the existence of sets, and a theory according to which they are perceptible certainly presupposes or entails this. As a result, philosophers of conceptualistic or nominalistic leanings will be able easily to provide alternative readings of the evidence, and to avoid accepting what is sketched below. My limited goal here is not to convince them, but to sketch a reasonable epistemology for the realist. So, what is said below should be judged by its plausibility from the realist's point of view, and by the extent to which it squares with other tenets of realism in the philosophy of set theory.

II. PERCEIVING OBJECTS

The following account of what it is for a person to perceive (visually) an object at a given location under normal conditions is derived from Pitcher's.[18]

[16] Cf. D. Hebb, *The Organization of Behavior: A Neurophysiological Theory* (New York: Wiley and Sons, 1949), ch. 2.

[17] e.g., G. Pitcher, *A Theory of Perception* (Princeton: Princeton University Press, 1971), ch. 1, and the references cited in P. Machamer, 'Recent Work on Perception', *American Philosophical Quarterly*, 7/1 (1970), sect. 2.

[18] Pitcher, *Theory of Perception*, ch. 2. Some irrelevant subtleties in Pitcher's account have been omitted. As is customary in such contexts, I will restrict my attention to sight, assuming that other senses can be dealt with similarly.

P perceives an x at l if and only if

(i) there is an x at l;
(ii) P acquires perceptual beliefs about x, in particular, that there is an x at l;
(iii) the x at l is involved in the generation of this perceptual belief state in an appropriate causal way (in the kind of way, for example, my hand is responsible for my perceptual belief that there is a hand before me when I look at my hand in good light).

It should be noted that this account is intended to capture a strict sense of perception according to which, for example, a hunter does not perceive a pheasant hen sitting motionless in her nest if she blends too perfectly with her surroundings for him to distinguish her from them. In this case, although the hunter may acquire some perceptual beliefs which are about the hen in a loose sense, he does not acquire any which are about her in a strict sense—for example, the perceptual belief that there is a pheasant hen in such-and-such a location.[19] Clause (iii) is familiar from Grice,[20] and guarantees that if P perceives an x at l, in this strict sense, then P knows that there is an x at l.

Several terms in clause (ii) require clarification. First, let us assume that beliefs are psychological states. Whether or not these states are dispositional, and whether or not dispositions are to be analysed in purely behaviouristic terms, are questions which need not be raised here, so long as it is admitted that behavioural evidence can be used to support claims that persons hold certain beliefs. Further, we assume that beliefs are neither necessarily conscious nor necessarily linguistic.[21] Perceptual beliefs, in particular, are nonconscious, and probably partly non-linguistic. Perceptual belief states are also extremely rich; that is, for example, for P to acquire the (visual) perceptual belief that there is a tree outside the window, he must also acquire a great variety of other perceptual beliefs, depending on the occasion, such as, that the tree is roughly so big, so far away, that it is in leaf, swaying in the breeze, and so on.[22] And perceptual beliefs are not inferred from other beliefs.[23] The components of a perceptual state—that is, the members of a complex set of

[19] Ibid. 78–9.
[20] H. P. Grice, 'The Causal Theory of Perception', *Proceedings of the Aristotelian Society*, suppl. vol. 35 (1961): 121–52.
[21] Cf. D. Armstrong, *Belief, Truth and Knowledge* (Cambridge: Cambridge University Press, 1973), chs. 2 and 3.
[22] Pitcher, *Theory of Perception*, 87–9.
[23] There is some advantage (see Harman, *Thought*, ch. 11; R. Gregory, *Eye and Brain* (New York: McGraw-Hill, 1966) and id., *The Intelligent Eye* (New York: McGraw-Hill, 1970)) to the view that perceptual beliefs are inferred from stimulations. Harman calls these inferences 'automatic' to distinguish them from ordinary inferences of beliefs from other beliefs. I think it is clear that stimulations should not be thought of as beliefs, since they are not behaviour-guiding until they become perceptions. I will reserve 'inference' for inference of beliefs from other beliefs, and call perceptual beliefs 'non-inferential'.

perceptual beliefs which are all acquired on one occasion—often influence one another. A perceptual belief about the identity of an object, for example, can influence perceptual beliefs about its shape and size, and (obviously) vice versa. The beliefs which make up a perceptual state arise in a body, not in an inferential sequence. Thus any belief acquired on a given occasion which influences and is influenced by perceptual beliefs acquired on that occasion is to be considered a perceptual belief, a part of the overall perceptual state.[24]

A philosophical theory of perception of the sort sketched above is defended by Pitcher; it bears a strong resemblance to that of Armstrong, and has affinities with various psychological theories of perception as information acquisition.[25] Many possible objections to a view of this sort, some of them fairly obvious, are met by Pitcher in his final version of the view, which is much more subtle than the crude first approximation presented here. Goldman[26] presents some additional, more difficult objections. He cites the behaviour-guiding character of some perceptual states, their cognitive content, their classificatory function, and the fact that they are often influenced by higher cognitions as evidence that many perceptual states have belief content. This is not, he adds, enough to show that all perceptual states are simply sets of beliefs. Some perceptual states might lack belief content, and others, though they possess belief content, might not be exhausted by it. This is surely correct, but we are interested here in a strict sense of 'P perceives an x at l', and I think some belief content—in particular, the perceptual belief that there is an x at l—is clearly required for that perceptual state. In other words, the perceptual states in which we are interested do all have belief content, and it will make no difference to us whether or not this belief content exhausts the perceptual state. So the controversy between Goldman and the belief theorists, Pitcher and Armstrong, need not detain us.

Difficult questions which are relevant here include the following: what perceptual beliefs is a given perceiver capable of acquiring on a given occasion, and how can we tell which ones the perceiver actually does acquire on that occasion?[27] Clearly, P cannot acquire a perceptual belief that there is a DC-10 overhead if P doesn't know what a DC-10 is, or, in more philosophical terminology, if P lacks the concept of a DC-10. I will assume, with Armstrong,[28] that to have a concept is just to have the capacity for beliefs of a certain sort, and that having a capacity, in turn, is being in a certain psychological state. Judging whether or not someone is in a certain psychological

---

[24] Pitcher, *Theory of Perception*, 103, 108.
[25] Ibid.; D. Armstrong, *Perception and the Physical World* (Atlantic Heights, NJ: Humanities Press, 1961); Gregory, *Eye and Brain*; id., *Intelligent Eye*.
[26] A. Goldman, 'Perceptual Objects', *Synthese*, 35 (1977), sect. 6.
[27] Cf. Pitcher, *Theory of Perception*, 93–4.
[28] Armstrong, *Belief, Truth and Knowledge*, ch. 5, sect. 1.

PERCEPTION AND MATHEMATICAL INTUITION

state, be it a belief state or the state of having a certain concept or capacity, is usually done on behavioural grounds. If psychological states are not analysable in exclusively behavioural terms, and we are not assuming that they are, behavioural evidence cannot be considered conclusive, though it is often the best evidence available. In certain situations, introspective evidence might also be admissible. And, assuming there is a correspondence, not necessarily identity, between psychological states and brain states, neurophysiological evidence might also be adduced. (Of course, substantial scientific progress will be needed before this is a real possibility, but it is not ruled out a priori.)

The Gödelian question from which this inquiry arose was: how does the idea of a physical object come to be present in our experience? This can now be rephrased as: what makes us capable of acquiring perceptual beliefs about physical objects? or, how do we come to have the concept of a physical object? Psychologists and neurophysiologists have produced some interesting behavioural and neurophysiological results and theories in an attempt to provide an answer to these questions, and, in general, to the question of how and when various conceptual elements enter human perceptual states. I will review a small portion of this work.

There is considerable experimental evidence that the ability to perceive a primitive distinction between a figure and its background is inborn in humans and many laboratory animals.[29] The structure of the retina is probably responsible for the presence of this conceptual information in the human perceptual state, as such a connection has been demonstrated in the case of the frog. McCulloch and his co-workers have isolated various structures in the frog's retina which send impulses to the frog's brain only under certain complex conditions (independent of the level of general illumination)—for example, in the presence of sharp boundaries between relatively light and relatively dark patches, or dark areas with curved edges, or movement of such curved edges. In fact, one fibre

> responds best when a dark object, smaller than a receptive field, enters that field, stops, and moves about intermittently thereafter. The response is not affected if the lighting changes or if the background (say a picture of grass and flowers) is moving, and is not there if only the background, moving or still, is in the field. Could one better describe a system for detecting an accessible bug?[30]

As might be expected, the researchers came to think of these fibres as 'bug-detectors', and the frog's behaviour certainly suggests that this mechanism enables it to acquire perceptual beliefs about nearby bugs. Similar mechanisms in humans are probably responsible for perceptual beliefs

[29] Hebb, *Organization of Behavior*, 19–21.
[30] W. McCulloch *et al.*, 'What the Frog's Eye Tells the Frog's Brain', in *Embodiments of Mind* (Cambridge: Cambridge University Press, 1965), 254.

concerning figure and background, and perhaps some concerning distance and size.[31]

Beyond this fairly simple level, however, the evidence is that the capacity to acquire perceptual beliefs of the familiar sort is not present at birth.[32] Psychologists talk of a phenomenon called 'identity' in perception. A figure is seen with identity if it is immediately seen as similar to some other figures but dissimilar to others (that is, as falling in some categories but not in others), and it is easily recalled, recognized, or named. For example, when I see a triangular figure, I immediately see it to be more similar to other triangles than to squares; I can recall it, recognize it, and call it and similar figures 'triangles'. In our terminology, I have acquired the perceptual belief that there is a triangle in front of me. Experiments on newly sighted human patients who have been blind from birth, and on chimpanzees raised in total darkness, demonstrate that the capacity to acquire such a belief (the concept of a triangle) is acquired after considerable sensory experience. For example, Hebb reports:

> Investigators (of vision following operation for congenital cataract) are unanimous in reporting that the perception of a square, circle, or triangle, or of sphere or cube, is very poor. To see one of these as a whole object, with distinctive characteristics immediately evident, is not possible for a long period. The most intelligent and best-motivated patient has to seek corners painstakingly even to distinguish a triangle from a circle.... A patient was trained to discriminate square from triangle over a period of 13 days, and had learned so little in this time 'that he could not report their form without counting corners one after another.... And yet it seems that the recognition process was beginning already to be automatic, so that some day the judgement "square" could be given with simple vision, which would then easily lead to the belief that form was always simultaneously given'.[33]

Similar results are obtained with the chimpanzees.

Given that such a simple capacity as the ability to see a triangle as more like another triangle than like a square is the product of considerable sensory experience, it is to be expected that so complex a talent as that of seeing a series of different patterns as aspects of one thing—that is, as a series of views of one physical object—is not present at birth. This expectation is substantiated by experiments of Piaget and his colleagues.[34] The child's ability to acquire perceptual beliefs about physical objects, as judged from behaviour, develops between the ages of about one month and two years. At the beginning

---

[31] e.g., T. Bower, 'The Visual World of Infants', *Scientific American*, 217 (1968): 80–92.

[32] This has little to do with the philosophical controversy over innate ideas, because even defenders of that view admit that something sensory is needed to 'draw out' or 'awaken' innate ideas.

[33] Hebb, *Organization of Behavior*, 28, 32.

[34] J. Piaget, *The Construction of Reality in the Child*, trans. M. Cook (New York: Basic Books, 1954); J. Phillips, *The Origins of Intellect: Piaget's Theory*, 2nd edn. (San Francisco: Freeman, 1975), ch. 2.

of this period, objects exist for the child when they are in the child's field of vision, and cease to exist afterward.[35] At its end, the child clearly possesses the concept of an independently existing physical object, and is fully able to acquire perceptual beliefs about physical objects.

Supposing then, as it seems we should, that the ability to perceive triangles or physical objects—that is, the ability to acquire perceptual beliefs about them—is itself acquired, what can be said about how this is accomplished? The most promising theory is a neurophysiological one presented by Hebb.[36] He sketches an account of the changes that take place in the brain when someone acquires the concept of a triangle; that is, when someone develops the ability to see triangles with identity, to gain perceptual beliefs about triangles, to perceive them. The upshot of his view is that repeated eye fixations on the various parts of triangles results (by the growth of synaptic knobs) in the growth of what he calls a 'cell assembly'.

Once the subject has acquired this complete assembly, looking at any triangle will cause it to reverberate for half a second or more, a tremendous advance over the life span of the stimulation from a visual pattern for which no assembly has been formed. This longer trace should persist long enough to allow the organic structural changes required for long-term memory to take place. In other words, with the completed cell assembly, the subject is able to see triangles with identity, to acquire perceptual beliefs about them. Thus, the cell assembly is a triangle-detector in much the same sense as the fibre isolated by McCulloch is a bug-detector.

The ability to perceive physical objects is not unlike the ability to perceive triangles, though it is more complex. The trick is to see a series of patterns as constituting views of a single thing. It seems likely[37] that what is involved is the development of some complicated phase sequence of cell assemblies. Crudely put, the theory is that human beings develop neural object-detectors which are responsible for their ability to acquire perceptual beliefs about physical objects. In Gödel's terminology, the presence of the idea of a physical object in our physical experience is due to our object-detectors. Gödel has suggested that the presence of this abstract element in our experience is

---

[35] This wording may be a bit too strong. I am grateful to the referee when this material was first published in *Philosophical Review* for making me aware of recent work which suggests that it might be more accurate to say that for the younger child, objects don't exist independently of their location or trajectory. Still, the main point remains: a period of development is needed before the child possesses our concept of physical object. For more up-to-date discussions, see T. Bower, 'The Object in the World of the Infant', *Scientific American*, 210 (Oct. 1971): 31–8; T. Bower and J. Paterson, 'The Separation of Place, Movement, and Object in the World of the Infant', *Journal of Experimental Child Psychology*, 15 (1973): 161–8; and T. Bower, *Development in Infancy* (San Francisco: Freeman, 1974).

[36] Hebb, *Organization of Behavior*, chs. 4 and 5.

[37] J. Bruner, 'On Perceptual Readiness', *Psychological Review*, 64 (1957): 237.

'due to another kind of relationship between ourselves and reality', that is, a relationship other than 'the action of certain things upon our sense organs'.[38] The action of a given physical object on our sense-organs or, more precisely, the causal interaction of an aspect of that object with our retinas is responsible only for our sensations, the pattern of our sensory stimulations. The object-detector is responsible for the idea of the object itself, and the presence in us of the object-detector is the result of a much more complex interaction between us and our environment than that which produces the sensations, just as Gödel suspects. Part of what is responsible for P's ability to acquire perceptual beliefs about a certain object on a given occasion is the structure of P's brain at birth—a result of evolutionary pressures of the environment on P's ancestors—and part is the sum of those early interactions between P and objects which resulted in P's object-detector. This evolutionary pressure of the environment, plus our youthful interactions with it, make up Gödel's 'other relationship'. Note that while it is complex, it is still causal.

Now recall that what was required by the causal theories over and above a causal interaction of the baptist or knower with an aspect of the object was that the baptist or knower perceive the object. And P will perceive an x at l if P acquires true perceptual beliefs about it, and the x at l is appropriately causally involved in the generation of P's perceptual belief state. Finally, I have argued that the x at l is appropriately causally involved if an aspect of the x at l stimulates P's object-detector. Thus, the object-detector is what legitimizes the gap between what is known about, or what kind is baptized, and what is actually causally interacted with.

III. PERCEIVING SETS

What I want to suggest now is simply that we do acquire perceptual beliefs about sets of physical objects, and that our ability to do this develops in much the same way as that in which our ability to perceive physical objects develops, as described in the previous section. Consider the following case: P needs two eggs for a certain recipe, reaches into the refrigerator for the egg carton, opens it, and sees three eggs there. This belief (that there are three eggs before P) is perceptual, because it is an integral part of the body of beliefs making up (or perhaps partly constituting) P's perceptual state. Other perceptual beliefs acquired on this occasion probably include details about the size and colour of the eggs, the fact that two eggs can be selected from among the three in various ways, the locations of the particular eggs in the nearly

[38] Gödel, 'Cantor's Continuum Problem', 1st edn., 272; 2nd edn., 484.

empty carton, and so on. The numerical beliefs are clearly part of this complex of perceptual beliefs, because they can influence the others as well as being influenced by them. (For example, the welcome fact that there are enough eggs for the recipe can make the eggs themselves look larger.) So, the various numerical beliefs acquired on this occasion are perceptual, and I further claim that they are beliefs about a set; that is, I claim P acquires the perceptual beliefs that there is a set of eggs before P, that it is three-membered, and that it has various two-membered subsets.

The most obvious objection to this claim is that sets do not have location, so P cannot perceive a set before him in the egg carton. Here I must agree that many sets, the empty set or the set of real numbers, for example, cannot be said to have location, but I disagree in the case of sets of physical objects. It seems perfectly reasonable to suppose that such sets have location in time—for example, that the singleton containing a given object comes into and goes out of existence with that object. In the same way, a set of physical objects has spatial location in so far as its elements do. The set of eggs, then, is located in the egg carton—that is, exactly where the physical aggregate made up of the eggs is located.[39]

A more difficult question is why the numerical perceptual beliefs in question should not be considered to be beliefs about the physical aggregate, not the set. These beliefs are beliefs that something or other has a number property, and Frege[40] has soundly defeated the view that a physical aggregate alone can have such a property. Frege's own solution[41] is that such beliefs are about concepts; but it seems no less plausible to suppose that they are actually about extensions of concepts, or, in other words, sets. Another popular candidate for the object of such beliefs is the physical aggregate coupled in some way with a property (in our example, the physical aggregate made up of the eggs together with the property of being an egg). Clearly, this view enjoys no ontological advantage over the one I have suggested, involving, as it does, properties instead of sets. Furthermore, I find it hard to be sure what the difference is between believing that 'three' applies to a particular

[39] The arguments of Benacerraf and Lear considered earlier involve an inference from the fact that sets are abstract objects to the claim that we cannot causally interact with sets. Abstract objects are supposed not to exist in space and time, which presumably provides at least part of the support for this inference. I have now denied that abstract objects cannot exist in space and time, and suggested that sets of physical objects do so exist. (Note that Jubien, 'Ontology and Mathematical Truth', 146–7, seems to agree. These are what he calls 'saturated sets'.) On the basis of this assumption, I will not exactly deny that sets cannot participate in causal interactions (causal interactions are still, strictly speaking, with aspects of objects), but I will suggest that such sets can play a role in the generation of our perceptual beliefs about them which is analogous to that played by physical objects in the generation of our perceptual beliefs about them.

[40] G. Frege, *The Foundations of Arithmetic*, trans. J. L. Austin Oxford: Blackwell, 1959; repr. Evanston, Ill.: (Northwestern University Press, 1968), sect. 23.

[41] Ibid., sect. 46.

physical aggregate under the property 'egg' and believing that a particular set of eggs is three-membered. What is the set over and above the physical aggregate individuated in a certain way? If there is no difference between these, then it would be impossible to acquire a perceptual belief about the one, without, at the same time, acquiring a perceptual belief about the other. Perhaps, on some views, the difference is that if one egg is moved to a different slot in the egg carton, we are still confronted with the same set (although its location is different), but we might (depending on our definitions) be confronted with a different aggregate. If this is the difference, it seems to be more evidence for the claim that the belief that there are three eggs in the carton is actually about the set of eggs, and not about the physical aggregate, because P surely believes that moving one egg (barring mishap) will not affect the fact that there are three. This dispute about the object of belief probably cannot be finally resolved without a careful comparison of the merits of various philosophies of mathematics in which the competing answers might be embedded, but fortunately, this need not be done here. From the point of view of set-theoretic realism—that is, from the point of view adopted here—the supposition that these perceptual beliefs are about the set of eggs in the carton is clearly the simplest and most reasonable, so we will make it.

Given, then, that these perceptual beliefs are about sets, how do we come by the capability to acquire such beliefs? Once again, the behavioural evidence collected by Piaget and his colleagues suggests that this capability develops in stages similar to those marked in the development of the ability to perceive physical objects, though at a later age, between about seven and eleven years.[42] Before the beginning of this period, a child may be able to classify objects into groups in a consistent way (say squares with squares, triangles with triangles), but the child does not grasp the inclusion relation. For the younger child, the set ceases to exist when its subsets are attended to, while for the older child, the set remains stable and contains various other sets as subsets.

A similar confusion is observed in connection with a set's number properties. The younger child imagines that the number of elements in a set changes when it is rearranged, especially when its elements are moved closer together or farther apart. In older children, by contrast, once a one-to-one

[42] J. Piaget and A. Szeminska, *The Child's Conception of Number*, trans. C. Gattegno and F. Hodgson (Atlantic Heights, NJ: Humanities Press, 1952); Phillips, *Origins of Intellect*, ch. 4. Once again, the referee has brought to my attention some more recent work which suggests this time that these abilities are acquired much earlier, say, between 2½ and 5 years of age. Of course, for my argument, the only crucial point is that some of this material is learned in much the same way as the analogous material about physical objects. For the more recent work, see E. Rosch *et al.*, 'Basic Objects in Natural Categories', *Cognitive Psychology*, 8 (1976): 382–439, and R. Gelman, 'How Young Children Reason about Small Numbers', in N. Castellan *et al.* (eds.), *Cognitive Theory* (New York: Erlbaum, 1977).

correspondence between two sets has been established, the belief in their equinumerosity cannot be shaken; indeed, the very question seems silly to them. Once the concept is in place and a set is perceived, the thought that it should change its number properties when its elements are moved about (barring mishap) is preposterous.

It should be noted that the child's development of the set concept is not a linguistic achievement. Of course, children are rarely taught the word 'set', but they are taught number-words, and it might be thought that their early errors are primarily verbal, and that it is verbal instruction that brings about their correction. The evidence, however, is against this assumption.[43] One must expect that the set concept could be developed, just as the object concept, in the complete absence of language.

How is the set concept—in particular, the ability to acquire perceptual beliefs about sets—itself acquired? It has been indicated that behavioural evidence suggests the set concept is developed over a period of time like the object concept. These studies also suggest that the determining factor in these developments is repeated exposure to and manipulation of the sort of things in question. The development of the object concept is brought about by children's experiences with various physical objects in their environment, and the set concept by experiences with sets of physical objects—for example, by forming one-to-one correspondences between them, by regrouping them to form subsets, and so on. Hebb's theory of the formation of the neural triangle-detector made essential use of the behavioural evidence that development of the ability to see triangles with identity requires repeated fixations on corners of triangular figures, eye movements from one corner to another, and even, in some cases, active seeking out of corners. It has already been theorized that an object-detector develops in a similar way, as a result of various experiences with physical objects in the environment. Given the evidence that the set concept requires a similar developmental period and repeated experience with sets in the environment parallel to the required experiences with triangles and physical objects, it seems reasonable to assume that these interactions with sets of physical objects bring about structural changes in the brain by some complex process resembling that suggested by Hebb, and that the resulting neural 'set-detector' is what enables adults to perceive sets.

On this account, then, when P looks into the egg carton, (i) there is a set of eggs in the carton; (ii) P acquires some perceptual beliefs about this set of eggs; and (iii) the set of eggs in the carton is appropriately causally responsible for P's perceptual belief state. The involvement of the set of eggs in the generation of P's belief state is the same as that of my hand in the generation

---

[43] Cf. Phillips, *Origins of Intellect*, 145.

of my belief that there is a hand before me when I look at it in good light; namely, an aspect of the thing interacts causally with the retinas, stimulating the appropriate detector. P perceives the set of eggs before him; P knows there is a set of eggs before him; indeed, P knows this set of eggs is three-membered and contains various two-element subsets, because these facts are appropriately causally responsible for P's belief in them. As in the case of knowledge of physical objects, it is the presence of the appropriate detector which legitimizes the gap between what is causally interacted with, and what is known about.

It should be noted that on this account a given causal interaction of an aspect of an object with P's retinas is enough to satisfy the causal requirement for P's perceiving, or acquiring knowledge about, various different things—for example, a time slice, an object, or a set. The thing perceived, or the pieces of knowledge acquired, on a given occasion are determined by the belief content of P's perceptual state, that is, by which perceptual beliefs P actually acquires on that occasion. This, in turn, depends on which detectors are actually stimulated. Though P is capable of perceiving sets, he may, on a given occasion, perceive only the objects themselves, because he is not interested in how many there are, or how they can be classified. In such cases, the set-detector is not stimulated, probably as a result of neural gating mechanisms which correspond to the degree of attention or inattention.[44]

It is now clear that any baptist with the requisite detector could have picked out samples of the kind 'set' in the way suggested earlier. This is sufficient for dubbing the kind 'set' if, in fact, sets do form a kind. I have indicated that, from the point of view of the set-theoretic realist, the treatment of sets as forming a kind is much more likely to be correct than a more traditional theory according to which a set is anything which satisfies certain conditions. Using the account of perception just given, it is possible to give further support to the view that sets form a kind by noting a perceptual similarity relation associated with that kind.[45] Our perceiver P is likely to think a dozen eggs is more like an encyclopedia than like the sky. The reason is that the first two subjects immediately suggest classification and numerability, and attention to such details is the alteration of the neural gating mechanisms which brings the set-detector to a state of increased sensitivity and makes it more likely to fire. In other words, P is likely to acquire perceptual beliefs about the set of eggs (that there are twelve) and the set of books (that it is made up of two subsets, positioned on adjacent shelves)—in short, to perceive sets—in the first two cases. This is not so in the third; hence

[44] Bruner, 'On Perceptual Readiness', 241–4.
[45] For the connection between kinds and similarity, see W. Quine, 'Natural Kinds', in *Ontological Relativity and Other Essays* (New York: Columbia University Press, 1969), 114–38.

the perceived similarities and dissimilarities. The kind, then, consists of those things similar to the samples. I cannot argue conclusively that sets form a kind, since the theory of kinds on which the causal theory rests has not been fully worked out by its advocates. Given what has been said so far in support of the claim that they do, however, and given that causal theorists arguing against set-theoretic realism tend to deny that there is an appropriate connection between baptist and sample—not that sets form a kind—I will assume that a complete theory of kinds should include the kind 'set'. This kind can be dubbed by picking out samples, as described above. Particular sets and less inclusive kinds can then be picked out by description; for example, 'the set with no elements' for the empty set, or 'those sets whose transitive closures contain no physical objects' for the kind of pure sets.[46]

IV. INTUITION

Three outstanding difficulties for the set-theoretic realist were mentioned at the outset. What has been said so far indicates that at least two of these can be overcome, namely, that of how reference to the kind 'set' is established, and that of how simple facts about particular sets of physical objects can be known. But, the set-theoretic epistemology sketched above requires that more than simple facts about particular sets and their interrelations be knowable in some 'quasi-perceptual' or 'intuitive' way. Specifically, at least some of the basic axioms of set theory must be included within this domain of 'intuitive evidence'. We are left with the problem of how these basic general truths can be known. In this section, I will try to show how this third obstacle can be overcome.

So far, I have sketched a theory according to which various concepts are acquired by means of a complex interaction between a human being and the world over a period of time. To have a concept is to have the ability to acquire certain beliefs, in particular, in the cases we are concerned with, the ability to acquire perceptual beliefs about particular things of the kind concerned on appropriate occasions. It has been theorized that a complex neural structure is the source of the ability to acquire these perceptual beliefs—thus, one perceives a triangle, physical object, or set when one's triangle-, physical object-, or set-detector is stimulated. What I want to point out is that the acquisition of some very general beliefs about things of that kind; indeed, the structure of the detector itself determines some very general beliefs about things of that kind; indeed, the structure of the detector itself determines

---

[46] Cf. Jubien, 'Ontology and Mathematical Truth'. These descriptions are now 'well-founded' in his sense.

some very general beliefs about things of the kind it detects. For example, three-sidedness is, in a sense, 'built into' the triangle-detector in the form of mechanisms stimulating eye movement from one corner to another, just as three-angledness is built in the form of the detector's three distinct cell assemblies for the corners themselves. Crudely put, then, the very form of one's triangle-detector guarantees that one will believe any triangle to be three-sided. This is, in effect, a general belief about triangles.

Similarly, I suggest that the structure of one's object-detector gives one some very general beliefs about physical objects, among them probably such beliefs as that physical objects can look different from different points of view, or that they do not cease to be when one ceases to see them. Of course, it is deceptive to describe these beliefs in this way, since they are not linguistic. A child of 3 is perfectly capable of perceiving physical objects, and thus has some of these general beliefs, but lacks the linguistic equipment to express them as I have. In short, such beliefs can be had by those who lack the linguistic terms, but not by those who lack the concept. They consist of various beliefs of the form that any thing of the kind of those things which stimulate the object-detector has certain properties. I will call these 'intuitive beliefs'. Like the others, the set-detector embodies intuitive beliefs about things of the kind of those which stimulate it. Among these are probably beliefs that might be expressed as 'sets have number properties', 'sets (other than singletons) have many proper subsets', 'any property determines a set of things which have that property', 'the number property of a set is not changed (barring mishap) by moving its elements'.

As has been stressed, it is possible, in fact usual, to acquire these concepts and intuitive beliefs without acquiring linguistic forms with which to express them. For this reason, I have been somewhat inaccurate even in referring to the concepts involved in our examples as the concepts of triangle, physical object, and set. Obviously, what we ordinarily think of as the geometer's, physicist's, or set theorist's concepts of triangle, physical object, or set, or even what might be described as the everyday meanings of the terms 'triangle', 'physical object', and 'set', are much more sophisticated than the pre-linguistic concepts described above. But consider what happens when these 'kind' words are introduced into the language or, better, into a particular speaker's idiolect. Some samples of the kind are specified, some triangular figures displayed, some varied physical objects pointed out, some sets indicated,[47] and if the teaching is successful, the subject associates the word with the appropriate detector. Without any more scientific training than this,

---

[47] A set, rather than a physical aggregate, can be indicated by making sure that attention is being paid to number properties. This can be done, e.g., verbally or by emphasizing one-to-one correspondences.

a subject with the requisite linguistic tools is likely to assent most readily to such assertions as:

(1) Physical objects exist in space and time.
(2) At any moment, a given physical object is in a certain place and moving at a certain speed.

and

(1′) Given any two objects, there is a set whose elements are just those two objects.
(2′) Any things can be collected into a set.

One might say, with Gödel, that such assertions 'force themselves upon us as being true'.[48] My suggestion is that the reason they do this is that they are fairly successful linguistic formulations of various pre-linguistic intuitive beliefs. The more accurately the linguistic form reflects the pre-linguistic belief, the more striking is the phenomenon of its forcing itself upon us. Attempts to formulate intuitive beliefs in linguistic terms I will call 'intuitions' or 'intuitive principles'.

From this account at least two things follow. First, intuitions can be false. Of course, they can be inaccurate formulations of intuitive beliefs, and false for that reason, but it seems they can also be accurate formulations of incorrect intuitive beliefs. I say this because it seems possible that we could be badly mistaken in the concepts we form and the intuitive beliefs that go with them. That is, it seems possible that stimulation by aspects of certain things might cause us to form a detector which embodied features very different from those which the things involved actually have. For example, I suppose that physical objects might cease to exist when no one perceives them, or that sets might fail to have number properties, though it is hard for us to imagine such things. Second, it follows that intuitions are more likely to be accurate formulations of intuitive beliefs if they are widely shared. Thus it is legitimate to suspect the claim of a single scientist that a certain principle is intuitive if few others share this opinion. However, it should be noted that scientists often use the words 'intuitive' and 'intuition' in senses different from the ones I am trying to capture here. For example, a scientist sometimes says that his 'intuitions' favour one theory over another. If most anyone can see that the favoured theory is simply more compelling than the other, this is probably a case of intuition in the sense I am using here. If, however, the scientist involved actually favours that theory because he knows more about the subject than others who disagree or have no opinion, and sees or suspects that it will turn out better for various theoretical reasons, 'intuition' is being used in a different sense, one which I will avoid here. In such a case, I will say that the favoured theory is supported by theoretical evidence, not by intuition.

---

[48] Gödel, 'Cantor's Continuum Problem', 1st edn., 271; 2nd edn., 483–4.

My suggestion is that intuitions of the sort described here form a basis both for the meaning of the associated linguistic term and for the scientific theory of things of the appropriate kind. In physics, though intuitions clearly play a role, they are rarely made explicit, except when they are overthrown by theoretical considerations, as, for example, (2) above. Intuitions can be false, so no matter how obvious they seem, they must be confirmed like any theory, and like any theory, they can be overthrown. Their status as intuitions, the fact that they force themselves upon us, is some evidence in their favour to begin with, but sufficient disconfirming evidence can outweigh this initial advantage.

From the writings of Zermelo on his original axioms and the axiom of choice, the work of Fraenkel on replacement and choice, and the remarks of numerous authors on the iterative conception, historical evidence could be adduced for my claim that intuitions form a basis (but do not exhaust) the scientific theory of sets, that they can be confirmed or disconfirmed like any theory, and that their status as intuitions is evidence in their favour. As an account of how we come to believe various intuitive principles (for example, extensionality or pairing), the theory summarized by this claim seems reasonable, but for an account of how we come to know these principles, something more must be said. The problem is that evidence that a principle is intuitive—that is, evidence for the supposition that it is an accurate formulation of an intuitive belief—is taken as evidence for its truth. If true (or correct)[49] intuitive beliefs could be counted as intuitive knowledge, then this would be acceptable; the strength of the 'forced-upon-us' phenomenon would determine the strength of the intuitive evidence for the truth of a given intuitive principle, evidence to be weighed along with other confirming and disconfirming evidence. So, the question is whether or not the notion of intuitive knowledge can be reconciled with the causal theory.

The essential steps in this direction have been made by our causal theorist Goldman himself in his discussion of innate knowledge.[50] Recall that Goldman, in an effort to overcome Gettier problems, added to the justified true belief account of knowledge the requirement that there be an appropriate causal connection between the fact known and the knower. Detractors of the idea of innate knowledge have argued that innate beliefs (if there are such) cannot be justified, and thus, cannot be knowledge. Goldman cites examples

---

[49] There is a difficulty here, in that intuitive beliefs are non-linguistic, and hence, it is unclear whether or not they are propositional. If not, it is probably inappropriate to classify them as true or false. Still, they can be classified as correct or incorrect, as tending toward success or failure in their behaviour-guiding function. (See Goldman, 'Perceptual Objects', 276.) I will continue to use the word 'true'.

[50] See A. Goldman, 'Innate Knowledge', in S. Stich (ed.), *Innate Ideas* (Berkeley: University of California Press, 1975), 111–20, for the details of the following argument.

PERCEPTION AND MATHEMATICAL INTUITION 135

to show that justification above and beyond the satisfaction of the causal requirement is not necessary for knowledge. If the causal connection between fact believed and believer is appropriate, that is, if it is 'an instance of a kind of process which *generally* leads to true beliefs of the sort in question',[51] then the belief constitutes knowledge. In the case of innate beliefs, he continues, the fact believed is causally responsible for the belief by evolutionary adaptation. This sort of causal connection is appropriate because (presumably) it generally leads to true beliefs. Thus, innate true beliefs constitute knowledge.

Fortunately, in order to apply this work of Goldman, it is not necessary for us to determine whether or not the intuitive beliefs we are concerned with are innate. Certainly, human children are born with a propensity for forming the concepts of physical object and set (indeed such rudimentary abilities as those to distinguish figure from ground and to perceive simple groupings are already present), and interaction with objects and sets in the environment causes the child to do so. Assuming for the moment that the intuitive beliefs acquired along the way are true, these facts are causally responsible for the child's beliefs, partly by evolutionary adaptation, and partly by more commonplace causal interactions. This causal connection is (presumably) of a sort that generally leads to true beliefs, and thus, the child's intuitive beliefs constitute knowledge, as required.

In summary, I have sketched a view according to which normal adults possess intuitive, non-linguistic knowledge of general facts about sets, and intuitive principles like the simpler axioms and the iterative conception are justified by their accuracy in formulating this intuitive knowledge, and by their theoretical merits. This account of our knowledge of the simple axioms of set theory, taken together with the account of the process by which more theoretical axioms are confirmed, forms a view of set-theoretic knowledge consistent with the ontological and semantic principles of set-theoretic realism in general, and with Gödel's few remarks on the subject in particular.[52]

V. CHIHARA'S OBJECTIONS

I will conclude by considering a few of the criticisms directed at Gödel's view by Chihara, its most hostile opponent. Many of Chihara's objections

[51] Ibid. 116.
[52] It should be remarked in passing that the sort of realism presented here suggests the following answers to the traditional questions in the philosophy of mathematics: (i) set-theoretic truths are not analytic; they are true by virtue of facts about independently existing sets; (ii) intuitive beliefs are a priori because once the concepts are in place, no further experience is needed to support them, and no further experience will count against them (though theoretical evidence might count against their linguistic expressions); (iii) it seems that if an object exists, its singleton must also, but the current notion of necessity is inadequate for dealing with this issue.

concern the arguments by which Gödel purports to establish what Chihara calls 'the equisupportive claim', that is, the claim that the assumption that sets exist 'is quite as legitimate as the assumption of physical bodies, and there is quite as much reason to believe in their existence'.[53] I am not concerned here with whether or not Gödel's science/mathematics analogy establishes this claim, but I am concerned with getting the analogy straight, and I think Chihara fails to do this. I will indicate where I think he goes wrong.

On the interpretation of Gödel's view which has been offered here, we can acquire perceptual beliefs about the existence of physical objects and of sets of these, and if the physical object or set concerned is appropriately involved in the generation of the belief state, these beliefs amount to knowledge. I have suggested that the physical mechanisms by means of which we acquire these beliefs are similar in the two cases. In response to such reasonings, Chihara remarks:

> At best, the above argument only shows that the same sort of justification can be given for the existence of mathematical objects as for the existence of physical objects. Thus, suppose that the most satisfactory theory we can now come up with to explain the data gathered by scientists working with some new accelerator is a quark theory. Then, at one level of abstraction, the same sort of justification could be given for the existence of quarks as has been given for the existence of positrons. But surely we can't conclude that we would have as much reason to believe in quarks as we have to believe in positrons.[54]

Here Chihara has switched from the realm of perceptual justification to that of theoretical justification. This suggests that he is thinking of all mathematical existence assumptions as highly theoretical, that he is not taking seriously Gödel's emphasis on a perceptual faculty. This passage can, however, be read metaphorically, that is, simply as claiming that though Gödel's justification for believing in sets is of the same sort as the usual justification for believing in physical objects, it is not of the same strength, just as the theoretical case for quarks is not of the same strength as the theoretical case for positrons. But it seems to me that this is not at all obvious unless one (mistakenly) supposes that the justification for believing in sets is purely theoretical. It is probably true that perceptual set-detectors cannot develop before perceptual object-detectors, but once they are in place, I see no reason to suppose that one is more dependable than the other.

The error of substituting theoretical justification for perceptual justification is clearly present when Chihara claims:

> Gödel can be seen to be addressing himself to the following sort of question: Given that some people have such and such beliefs and such and such experiences, is it

[53] Gödel, 'Russell's Mathematical Logic', 1st edn., 220; 2nd edn., 456–7; C. Chihara, 'On a Gödelian Thesis Regarding the Existence of Mathematical Objects', unpublished, p. 1.
[54] Chihara, 'On a Gödelian Thesis', 4.

reasonable for them to infer *P*? But are we supposed to have been in a comparable position at some time before we came to believe in the existence of physical objects?[55]

Of course we were never in such a position with respect to physical objects. Ever since the human eye and brain developed to the point that we were able to perceive objects, we have believed (and presumably, known) that they exist without inferring it. Similarly, we have never been in such a position with respect to sets. Ever since the slightly later period of human prehistory when we acquired the ability to count and to perceive sets, we have also believed in their existence. Naturally, this period is incomparable to that in which the existence of molecules was a matter of scientific debate, so the fact that Gödel presents no theoretical argument for the existence of sets along the lines of that once presented for the existence of molecules should come as no surprise. Instead, Gödel's case depends on the everyday experience of perceiving sets of physical objects, as, for example, I perceive the set of my fingers when I set out to count them.

A similar confusion arises in connection with Gödel's discussion of the significance of the failure of Russell's no-class theory, a confusion encouraged by a recalcitrant passage from Gödel:

[Sets] are in the same sense necessary to obtain a satisfactory theory of mathematics as physical bodies are necessary for a satisfactory theory of our sense perceptions and in both cases it is impossible to interpret the propositions one wants to assert about these entities as propositions about the 'data', i.e. in the latter case, actually occurring sense perceptions.[56]

This passage surely suggests that the failure of the no-class theory is to be seen as analogous to the failure of phenomenalism, though it can be interpreted otherwise.[57] Three pages later, however, Gödel writes:

The whole scheme of the no-class theory is of great interest as one of the few examples, carried out in detail, of the tendency to eliminate assumptions about the existence of objects outside the 'data' and to replace them by constructions on the basis of these data.... All this is only a verification of the view defended above that logic and mathematics (just as physics) are built up on axioms of real content which cannot be 'explained away'.[58]

This passage I take as suggesting that the failure of the no-class theory is analogous to the failure of operationalism in physics.

---

[55] Ibid. 14.

[56] Gödel, 'Russell's Mathematical Logic', 1st edn., 220; 2nd edn., 456–7.

[57] A satisfactory theory of our sense perceptions would presumably include an account of how a physical object brings about my belief that there is a physical object before me by impinging on my sense-organs. And the actual existence of the object is required by such a satisfactory theory. This is not to say that the object is a theoretical construction out of something more primitive. I take 'actually occurring sense perceptions' to be perceptual beliefs about objects, not some primitive, unanalysed data. As Gödel claims, a statement about a physical object cannot be translated exhaustively into a statement about the perceptual beliefs of various observers.

[58] Gödel, 'Russell's Mathematical Logic', 1st edn., 223–4; 2nd edn., 460.

Whichever of these was Gödel's intention, it is clear that the second form of the analogy is the correct one on the version of set-theoretic realism I have offered here. A phenomenalist might claim that 'I see a book before me' is equivalent in some sense to a statement about actual and possible sensations. The impossibility of such a translation is widely recognized. The analogous claim for the nominalist or no-class theorist would be something to the effect that 'I see a set of books before me' is equivalent to a statement about actual or possible sensations, but so far as I know, no one has suggested this. What is sometimes claimed is that there being such a set is no more than there being an open sentence of a certain sort, or better yet, a sequence of marks of a certain sort. But, for my belief that there is a set of books before me to be in fact a belief about a sequence of marks, those marks must be appropriately involved in the generation of my belief state, and they clearly are not. From this point of view, nominalism seems even more implausible than phenomenalism.

The better form of the analogy is between the no-class theory and operationalism. The idea is that classical mathematics, especially analysis, cannot be done without existence assumptions beyond those of nominalistic systems like no-class theory, just as physics cannot be done without treating theoretical entities as something more than operationally definable fictions. Chihara's response to this form of the analogy[59] is to suggest that there is no reason to suppose that classical mathematics is true, unless it is its applications in science, and that as much mathematics as is needed in science can be done within a nominalistically acceptable system. Whether or not this can be done, Gödel certainly does not believe that the only reason for believing classical mathematics to be true is its applicability; mathematical perception and intuition give us additional reason to believe it true. Of course, Gödel does not present a theory of these faculties, so Chihara is to be forgiven his scepticism. I hope the theory presented here has increased the plausibility of the view that there are mathematical experiences (perceptions) and intuitions, and that these lend support to the theoretical parts of set theory beyond what is available from the successful applications of mathematics in science.

Chihara raises an objection of a different sort when he charges that the independence of the continuum hypothesis from the axioms of Zermelo-Fraenkel presents a problem for the set-theoretic realist. He cites Mostowski's remark that:

Probably we shall have in the future essentially different intuitive notions of sets just as we have different notions of space, and will base our discussion of sets on axioms which correspond to the kind of sets which we want to study.... everything in the

---

[59] Chihara, 'On a Gödelian Thesis', 5, and id., *Ontology and the Vicious Circle Principle* (Ithaca, NY: Cornell University Press, 1973), ch. 5.

recent work on foundations of set theory points toward the situation which I just described.[60]

From this, Chihara concludes that 'For Gödel ... the proliferation of set theories poses the thorny problem of determining which of the many set theories is the one that most truly describes the real world of sets'.[61]

Of course, the analogy between the status of the parallel postulate and that of the continuum hypothesis is not exact.[62] Gödel points out[63] that when the parallel postulate was shown to be independent, mathematical evidence was no longer relevant to the question of its truth or falsity. Non-Euclidean geometries were developed so that the question of which geometry best suits the physical world could be decided by physical evidence. Naturally, we retain different mathematical concepts of space. In the case of the continuum hypothesis, the intended interpretation is not physical, but mathematical, so though different set theories may be developed, mathematical evidence will still be relevant to the question of which one is correct. If this were decided, we would probably not retain different mathematical concepts of set. Furthermore, for each logician who, like Mostowski, expects such a proliferation of set theories, there is probably one who does not. For example, Cohen writes: 'In the case of the Continuum Hypothesis, this tendency may possibly, *though unlikely*, lead to a splitting of set theory depending on how one evaluates the power of the continuum.'[64] Also included in this group are those who look forward to a solution of the continuum problem by the addition of new axioms.[65]

I have tried to suggest that such a proliferation of set theories is less likely than Chihara makes it sound, but even if one were to occur, I do not think Gödel's position would be seriously affected, Chihara continues:

> It may be suggested that Gödel might evade this problem by allowing set theory to bifurcate into two theories, both of which treat entities that resemble the 'sets' of our earlier theories, but in one of which, the continuum hypothesis holds, whereas in the other, the hypothesis does not hold. Unfortunately, this maneuver is hardly satisfactory. Gödel has argued that we have good reason for thinking there are sets. But could he go on to argue that we have equally good reasons for postulating the existence of

---

[60] A. Mostowski, 'Recent Results in Set Theory', in I. Lakatos (ed.), *Problems in the Philosophy of Mathematics* (Amsterdam: North Holland, 1972), 94; Chihara, *Ontology*, 64–5.

[61] Chihara, *Ontology*, 65.

[62] Cf. G. Kreisel, 'Informal Rigour and Completeness Proofs', in Lakatos (ed.), *Problems*, 150.

[63] Gödel, 'Cantor's Continuum Problem', 271.

[64] P. Cohen, 'Comments on the Foundations of Set Theory', in D. Scott (ed.), *Axiomatic Set Theory* (AMS, 1971), 12, emphasis mine.

[65] e.g., J. Shoenfield, *Mathematical Logic* (Reading, Mass.: Addison-Wesley, 1967), 314; P. Cohen, *Set Theory and the Continuum Hypothesis* (Benjamin, 1966), 150–2, and id., 'Comments', 12; D. Martin and R. Solovay, 'Internal Cohen Extensions', *Annals of Mathematical Logic*, 2/2 (1970): 143; and, of course, Gödel, 'Cantor's Continuum Problem', 1st edn., 266–7; 2nd edn., 476–7.

two new kinds of entities? Surely his argument would support belief in, at best, one of the new kind of 'sets'.[66]

This last seems to me incorrect. If set theory were to bifurcate, we would say there were two universes, which satisfied $ZFC + 2^{\aleph_0} = \aleph_1$, and $ZFC + 2^{\aleph_0} > \aleph_1$, respectively. Of course, there are many logicians who distinguish between two universes of this description even now, namely, L and V. This suggests that the imagined bifurcation might not involve two disjoint universes of different kinds of entities; rather, one might just leave out some of the sets of the other. Of course, this causes no trouble for Gödel's view.

Apparently, though, this is not what Chihara has in mind. He imagines that we might someday have reason to believe that there are A-sets and B-sets, and that the universes of A-sets and B-sets satisfy $ZFC + 2^{\aleph_0} = \aleph_1$ and $ZFC + 2^{\aleph_0} > \aleph_1$, respectively. Gödel's perceptions and intuitions, Chihara claims, can only involve one of these. Why should this be true? It might be that all the perceivable sets are both A- and B-sets, that all the intuitive beliefs are true of both universes, and that only some theoretical considerations suggest that there are, in fact, two universes, corresponding perhaps to different power set operations that coincide at the lower levels, but diverge later on. This situation is exactly the L versus V situation, and causes no trouble. Chihara must mean that A-sets and B-sets are essentially different kinds of entities. But even on this assumption, it is possible that we have perceived both kinds of sets without noticing their differences, as is not uncommon in the physical sciences. If further development of the theory indicates that two different sorts of entities have been confused (like gold and iron pyrites, water and heavy water, sets and classes), it is just as natural on Gödel's view as on any other to distinguish the theory of one from the theory of the other.

Chihara also discusses Gödel's remark that 'we do have something like a perception also of the objects of set theory, as is seen from the fact that the axioms force themselves upon us as being true'.[67] I have described above the sense in which I think this remark should be understood: in the process of acquiring the ability to perceive sets, we also acquire very general intuitive beliefs about them, and the most simple set-theoretic axioms are linguistic versions of these, so that, when we hear them, they seem obvious; they are 'forced upon us'. Chihara argues that when the beginning student of set theory is taught the axioms and finds them obvious in this way, the student has simply been introduced to the idea of the universe of sets, and recognizes the axioms as 'true to the concept', not as objectively true. I think this is incorrect.

---

[66] Chihara, *Ontology*, 65.
[67] Gödel, 'Cantor's Continuum Theory', 1st edn., 271; 2nd edn., 484; Chihara, *Ontology*, 79.

PERCEPTION AND MATHEMATICAL INTUITION

The same sort of discussion precedes the student's recognition of various intuitive physical truths at the beginning of formal training in physics, but Chihara is not tempted to deny these their objective truth. Both the student of physics and our student of set theory have had perceptual contact with some of the more familiar objects of their study (medium-sized physical objects and sets of these, respectively), and when these obvious intuitive truths are introduced, they do not take them to be true to the concept of physical object, or of set, but to be objectively true of those things they are already familiar with, namely, objects and sets.

In conclusion, I think Chihara's objections to set-theoretic realism either do not apply to, or are not conclusive against, the version offered here. Though the science/mathematics analogy, as specified here, does not prove the existence of sets, I think it does provide an epistemological picture of the sort set-theoretic realism requires. In particular, I have argued that if sets do exist, we can know about them and refer to them in ways that should be acceptable to the causal theorists.[68]

---

[68] This paper developed out of chapter 3 of my doctoral dissertation, *Set Theoretic Realism* (Princeton: Princeton University Press, 1979). I am deeply grateful to my adviser, John P. Burgess, for his help and encouragement during all phases of this project.

VII

TRUTH AND PROOF: THE PLATONISM OF MATHEMATICS

W. W. TAIT

1

What is the relation in mathematics between truth and proof?

An arithmetical proposition A, for example, is about a certain structure, the system of natural numbers. It refers to numbers and relations among them. If it is true, it is so in virtue of a certain fact about this structure. And this fact may obtain even if we do not or (for example, because of its relative complexity) cannot know that it does. This is a typical expression of what has come to be called the *Platonist* (or *platonist* or *realist*) point of view towards mathematics.

On the other hand, we learn mathematics by learning how to do things— for example, to count, compute, solve equations, and, more generally, to prove. Moreover, we learn that the ultimate warrant for a mathematical proposition is a proof of it. In other words, we are justified in asserting A— and therefore, in any ordinary sense, the truth of A—precisely when we have a proof of it.

Thus, we seem to have two criteria for the truth of A: it is true if (indeed, if and only if) it holds in the system of numbers, and it is true if we can prove it. But what has what we have learned or agreed to count as a proof got to do with what obtains in the system of numbers? I shall call this the *truth/proof problem*. It underlies many contemporary attacks on platonism.

The argument against platonism begins with the observation that the first criterion, holding in the system of numbers, is inapplicable because we have no direct apprehension of this structure. Sometimes this argument is augmented by the thesis that 'apprehension of' would involve causal interaction with the elements of the structure, and, since numbers are 'abstract' (i.e., not in space-time), no such interaction is possible. In any case, the argument continues: it follows that a proof cannot be a warrant—or even incomplete

First published in *Synthese*, 69 (1986): 341–70. Reproduced here with permission.

evidence—for holding in the structure. For no kind of evidence is available that the canons of proof apply to the structure. Thus, if proof is a warrant for *A*, then *A* cannot be about the system of numbers. If, on the other hand, proof is not a warrant, then we have no mathematical knowledge at all. An even stronger argument, to the effect that we cannot even meaningfully *refer* to numbers, is based on the thesis that reference also involves causal interaction.

Because of these considerations, many writers have felt that mathematics is in need of a foundation in the revisionist sense that we must so construe the meaning of mathematical propositions as to eliminate the apparent reference to mathematical objects and structures. Some of these writers see platonism itself as a foundation, that is, as a theory of what mathematics is about—but one which, no matter how naïvely plausible, is refuted by the truth/proof problem.

There are many interesting problems which might reasonably be called problems in the foundations of mathematics; but I shall argue here that among them is *not* the need for a foundation in this revisionist sense. The truth/proof problem, which seems to demand such a revision, will resolve itself once we are clear about what truth and proof in mathematics mean and what is involved in the notion of a proposition holding in a structure. These notions seem to me to be surrounded in the literature by a good deal of confusion which gets attached to platonism. Free of this confusion, platonism will appear, not as a substantive philosophy or foundation of mathematics, but as a *truism*.

2

Many who reject platonism on the grounds of the truth/proof problem take discourse about sensible objects to be, not only unproblematic, but a paradigm case of the apparent content of a proposition being its real content. Thus

(1)　　　　　　　　There is a prime number greater than 10

is not really about the system of numbers, as we might naïvely read it, because our warrant for it is a proof—and what has that to do with the system of numbers? On the other hand,

(2)　　　　　　　　There is a chair in the room

really is about the sensible world—about chairs and rooms—because we verify it by looking about the room and seeing a chair. Thus, Dummett 1967 begins: 'Platonism, as a philosophy of mathematics, is founded on a simile: the comparison between the apprehension of mathematical truth to [sic] the perception of physical objects, and thus of mathematical reality to the physical universe.' He then argues that there is no analogue to observation in the case of mathematics, and so the simile is misconceived. And Benacerraf (1973) writes: 'One of its [i.e., the "standard" Platonistic account's] primary advantages is that the truth definitions for individual mathematical theories thus construed will have the same recursion clauses as those employed for their less lofty empirical cousins' (p. 669; p. 21 this volume).

The 'standard' account is that, for example, (1) has the 'logico-grammatical form'

There is an F which bears the relation G to b,

and he takes it as unproblematic that (2) has this form. But, 'For a typical "standard" account (at least in the case of number theory or set theory) will depict truth conditions in terms of conditions on objects whose nature, as normally conceived, places them beyond the reach of the better understood means of human cognition (e.g., sense perception and the like)' (pp. 667-8; p. 20 this volume).

I shall have more to say bearing on these passages in the course of this essay; but for now I intend them only as instances of the view that, whereas the naïve (platonistic) construal of (1) is problematic, the naïve reading of (2) is acceptable, indeed as a paradigm case.[1]

3

Why does the experience that I describe as 'seeing a chair in the room' warrant the assertion of (2) any more than a proof warrants the assertion of (1)?

---

[1] Many other contemporary authors could of course have been cited for essentially the same point. I shall focus primarily on Benacerraf 1973 (Ch. I) and Dummett 1967, 1975 (Ch. IV), in citing the literature, because these seem to me to represent most clearly and fully the two most important formulations of difficulties with platonism. Benacerraf's paper is frequently cited as grounds for revisionist foundations of mathematics—e.g., in Field 1980, 1982; Kitcher 1978, 1983; and Steiner 1975. It consists in arguing that, in the context of mathematics, there is an apparent conflict between our best theory of truth, which is Tarski's, and our best theory of knowledge, which is causal, because we do not causally interact with mathematical objects. As I understand him, Benacerraf himself, unlike many who cite him, is not calling for a revision of our conception of mathematics but only for a resolution of the apparent conflict. Dummett's critique of platonism rests on a conception of meaning which he argues is incompatible with platonism and, indeed, leads to the intuitionistic conception of mathematics; and so it is revisionist. My purpose, however, is not to review these papers. I cite them only because I wish to undermine conceptions which I myself cannot coherently formulate. On the other hand, I shall, I believe, resolve the difficulties that they find with platonism in the course of this paper.

I am not referring here to the possibility of perceptual error or illusion: the nearest analogue to that in the case of (1) would perhaps be error in proof or ambiguity of symbols. Rather, I am asking a traditional sceptical question: what have my experiences to do with physical objects and their relationships *at all*?

For, in the case of (2) *also*, I am applying the canons of verification that I have been trained to apply. Among other things, this training involved learning to say and react to sentences such as 'I see a chair', 'There is no chair in the room', etc., under suitable circumstances. It is true that, unlike the case of (1), these circumstances involve sensory experience. But (2) is about physical objects, not my sensations.

One may feel that the crucial difference between (1) and (2) is this: in the former case, proving is inextricably bound up with what I have been trained to do; whereas in the latter case, the role of training is confined to language learning, and this consists simply in learning to put the right (conventional) names to things. And, after that, training plays no further role: I simply read the true proposition off the fact as I observe it.

This view of (2) is in essentials the so-called Augustinian view of language, which, in my opinion, is thoroughly undermined by Wittgenstein's *Philosophical Investigations*, §§1–32. My learning to put names to things *consists* in my learning to use and respond, verbally and otherwise, to expressions involving these names. For example, how is it that I am naming the chair as opposed to naming its shape, colour, surface, undetached chair part, temporal slice, etc.? The answer is that it is the way we *use* the word 'chair' that determines this. And the point is not merely that the act of naming is ambiguous as to which among several categories it refers. Ambiguity itself presupposes language: to understand words like 'shape', 'colour', etc., is to have a mastery of a language. My point—or rather, Wittgenstein's point—is that *nothing* is established by the act which we call an act of naming, without a background of language or, at least, without further training in how the name is to be used.[2] We do not read the grammatical structure of propositions about sensible objects off the sensible world, nor do we read true propositions about sensibles off pre-linguistic 'facts'. Rather, we master language, and, *in* language, we apprehend the structure of the sensible world and facts.

---

[2] The issue here is not 'inscrutability of reference'. That idea makes sense in connection with translating one language into another. But in what sense is *our* reference to the chair or to the number 2 inscrutable? When Wittgenstein (1953) writes 'What is supposed to show what [the words] signify, if not the kind of use they have' (§10), his point is not that there is a well-defined universe of things (perhaps described in the language of God) and that a word succeeds in referring to one of these things rather than another because of the kind of use it has. Rather, it is that we call a word 'referring' because of the kind of use that it has. And we ask the question 'To what does the word "X" refer?' in language, and it can only be answered there, by pointing perhaps or by saying 'X refers to Y', where 'Y' is 'X' or some other term.

To apprehend the fact that A is simply to apprehend that A. And this apprehension presupposes language mastery.

So, if A is a proposition about the sensible world of rooms and chairs, then it is true if and only if it holds in that world. But we sometimes count what we experience as verification for A. And why should these two things, what holds in the world of rooms and chairs and what we experience, have anything to do with each other? Note that it is not sufficient to point out that verification is not conclusive in the way that the existence of a proof is, since the question is why verification should have *anything* to do with what holds in the world of rooms and chairs.

Thus, I see nothing special to mathematics about the truth/proof problem. We have described a truth/verification problem which is its analogue in the case of the sensible world. Moreover, the latter is not really a new argument, but, in essentials, has been a standard part of the sceptics armoury. It is perhaps this analogy that Gödel had in mind when he wrote (1964, p. 470) that 'the question of the objective existence of the objects of mathematical intuition . . . is the exact replica of the question of the objective existence of the outer world'. At any rate, I know of no argument against the existence of mathematical objects which does not have a replica in the case of sensible objects. For example, some writers argue against Platonism that, if there is a system of numbers, then why shouldn't there be more than one of them, all indistinguishable—how would we distinguish them? And why should our theorems refer to one such system rather than another? Answer: why shouldn't there be more than one physical world, all indistinguishable from one another and such that my 'seeing a chair' is a seeing a chair in all of them? Why should (2) be about one of these worlds rather than another? If you answer that it is about the world you inhabit, then I shall ask: which you?, etc. Sceptics about mathematical objects should be sceptics about physical objects too.

Of course, scepticism about either is misplaced, and both the truth/proof and the truth/verification problems are consequences of confusion, and are not real problems.

Perhaps this becomes more evident when we note that, in both cases, the problem purports to challenge our canons of warrant (i.e., proof and verification, respectively); but carried to its logical conclusion, it also challenges our canons of meaningfulness. Why should the structure of reality be what is presupposed by the grammatical structure of our language as we have learned it? For example, the meaningfulness of a sentence involving '+' presupposes the truth of the sentence which expresses that '+' is well defined in the numbers. So scepticism about truth will already imply scepticism about meaning.

4

For both Benacerraf and Dummett in the above cited papers, what is special about discourse about physical objects is the possibility of sense perception, and the difficulty that they raise for platonism is based on the absence in the case of mathematical objects of any such 'better understood means of human cognition'.

Some writers, for example, Gödel (1964), Parsons (1979–80; see Ch. V), and Maddy (1980; see Ch. VI), attempt to meet this difficulty by arguing that there is perception or something like it in the case of mathematical objects. But, of course, from the point of view of the truth/proof problem, the issue is not whether we perceive mathematical objects, but whether our canons of proof obtain their meaning and validity from such perceptions. And the answer to this seems to me to be clearly no. We perceive sets, for example, only when we have mastered the concept 'set', that is, have learned how to use the word 'set'. For example, it does not seem reasonable to suppose of people, before the concept of set was distinguished, that when they perceived a heap of pebbles, they also perceived a set—a different object—and simply spoke ambiguously. And, in whatever sense we may perceive numbers, it is hard to see how that can provide a foundation for the use of induction to define numerical functions, for example. The canons of proof are like canons of grammar; they are norms in our language governing the use of words like 'set', 'number', etc. What we call 'mathematical intuition', it seems to me, is not a *criterion* for correct usage. Rather, having mastered that usage, we develop feelings, schematic pictures, etc., which guide us. Of course, such intuitions may play a causal role in leading us to correct arguments and even to new mathematical ideas; but that is a different matter. In any case, the appropriate response to the anti-platonists is not to argue that there is something like perception in the case of mathematics. Rather, it is to point out that, even in the supposedly paradigm case of sensible objects, perception does not play the role that they claim for it. That this is so is manifest from the truth/verification problem.[3]

---

[3] Gödel 1964 wrote: 'But, despite their remoteness from sense experience, we do have something like a perception also of the objects of set theory, as is seen from the fact that the axioms force themselves upon us as being true' (in Benacerraf and Putnam 1983: 483–4). Many authors regard Gödel as an archetypal platonist and this passage as a bold statement of what every platonist must hold if he is to account for mathematical knowledge. In the words of Benacerraf (1973) (who, incidentally, inadvertently left out the words 'something like' in quoting the above passage): '[Gödel] sees, I think, that something must be said to bridge the chasm, created by his realistic and platonistic interpretation of mathematical propositions, between the entities that form the subject matter of mathematics and the human knower. . . .he postulates a special faculty through which we "interact" with these objects' (p. 675; p. 26 this volume).

5

Platonism is often identified with a certain 'account' of truth in mathematics, namely Tarski's. That this is so for Benacerraf (1973) is clear from the first of the above quotes from that paper and

> I take it that we have only one such account [of truth], Tarski's, and that its essential feature is to define truth in terms of reference (or satisfaction) on the basis of a particular kind of syntactico-semantical analysis of the language, and thus that any putative analysis of mathematical truth must be an analysis of a concept which is a truth concept at least in Tarski's sense (p. 667; p. 19 this volume).

It is difficult to understand how Tarski's 'account' of truth can have any significant bearing on any issue in the philosophy of mathematics. For it consists of a definition *in* mathematics of the concept of truth for a model in a formal language L, where the concept both of a formal language and of its models are mathematical notions. For example, $\exists$ in L is interpreted in terms of the mathematical 'there exists'. But Benacerraf is concerned with mathematical truth, not with truth of a formal sentence in a model. How can

---

But I don't think that this is a fair reading of Gödel's remark. To understand what he means by 'something like perception', one should look at his argument for it: 'the axioms force themselves upon us as being true.' One should also look at the paragraph immediately following the quoted one:

> It should be noted that mathematical intuition need not be conceived of as a faculty giving an *immediate* knowledge of the objects concerned. Rather it seems that, as in the case of physical experience, we *form* our ideas also of those objects on the basis of something else which *is* immediately given. Only this something else here is not, or not primarily, the sensations. That something besides the sensations actually is immediately given follows (independently of mathematics) from the fact that even our ideas referring to physical objects contain constituents qualitatively different from sensations or mere combinations of sensations, e.g., the idea of object itself, whereas, on the other hand, by our thinking we cannot create any qualitatively new elements, but only reproduce and combine those that are given. Evidently the 'given' underlying mathematics is closely related to the abstract elements contained in our empirical ideas. It by no means follows, however, that the data of the second kind, because they cannot be associated with actions of certain things upon our sense organs, are something purely subjective, as Kant asserted. Rather they, too, may represent an aspect of objective reality, but, as opposed to the sensations, their presence in us may be due to another kind of relationship between ourselves and reality. (Benacerraf and Putnam 1983: 484.)

If anything is being 'postulated' here, it is this other kind of relationship, not a faculty. This relationship is to account for the objective validity, not only of the 'something like a perception' of mathematical objects, but also of our ideas referring to physical objects. For it concerns the 'given' underlying mathematics, which is closely related to the abstract elements contained in our empirical ideas—e.g., the elements giving rise to our idea of an object (cf. *Theaetetus*, 184d–186). That Gödel intends this relationship to be necessary for the objective validity of empirical as well as mathematical knowledge is indicated by the first sentence of the next paragraph, which I have already partially quoted, indicating that the question of the objective existence of mathematical objects is the exact replica of that concerning the objective existence of the outer world.

But he writes that the former question 'is not decisive for the problem under discussion here. The mere psychological fact of the existence of an intuition which is sufficiently clear to produce the axioms of set theory and an open series of extensions of them suffices to give meaning to the truth or falsity of propositions like Cantor's continuum hypothesis.' The point seems clear: the 'something

THE PLATONISM OF MATHEMATICS 149

Tarski's account apply here? What is the locus of the definition; that is, what is the metalanguage? Not the language of mathematics, of course, since that is the language whose meaning they wish to explain.

Benacerraf's remark that truth is defined 'in terms of reference (or satisfaction)' is at first sight puzzling, since truth is a special case of satisfaction. But by 'satisfaction' he undoubtedly means *valuation*, that is, assigning values to variables. But it is misleading to speak here of reference. The model assigns values to the constants of L; but this, like the notion of valuation, is expressed in terms of the notion of function, and the concept of reference does not enter in. It is the more misleading when Benacerraf goes on to advocate a causal theory of reference.

An enlightening way to look at Tarski's truth definition is in terms of the notion of an interpretation: with each formula ϕ of L, we define by induction on ϕ a formula $I(\phi)$ (in the same variables) of the metalanguage, that is, of some part of the ordinary language of mathematics in which we defined the model. The truth definition now is just the 'material condition for truth': a sentence ϕ of L is true iff $I(\phi)$.

Dummett 1975 writes: 'On a platonistic interpretation of a mathematical theory, the central notion is that of truth: a grasp of the meaning of a sentence belonging to the language of the theory consists of a knowledge of what it is for that sentence to be true' (p. 223). But when he speaks of what it is for ϕ to be true or 'of what the condition is which has to obtain for [ϕ] to be true', to what condition can he be referring here other than the condition that $I(\phi)$? But now, in what consists a grasp of the meaning of $I(\phi)$? (Wittgenstein 1953, §198: 'every interpretation, together with what is being interpreted, hangs in the air.') Dummett is aware of this infinite regress, but he uses it as an argument against classical reasoning in mathematics, which he identifies with platonism. But of course the infinite regress disappears when we note that platonism does not consist in an interpretation of mathematical theories. We do indeed

like a perception', viz., mathematical intuition, is not what bestows objective validity on our theorems, any more than the perceptions of the brain-in-the-vat bestow objective validity on its assertions about the physical world. Yet, the brain-in-the-vat will have grounds for asserting (2); and, in the same way, mathematical intuition yields grounds for asserting (1). Thus, the 'something like a perception' is not the 'another kind of relationship between ourselves and reality' to which Gödel refers.

I do not entirely agree with Gödel here. What is objective about the existence of mathematical or empirical objects is that we speak in a common language about them—and this includes our agreement about what counts as warrant for what we say. And this view guides my estimation, stated above, of the role of mathematical intuition *vis-à-vis* grounds for asserting mathematical propositions. I cannot make the distinction Gödel seems to want to make between subjective validity, founded on our intuition, and objective validity. But it is worthwhile to point out that Gödel's 'something like a perception' is not a 'special faculty through which we interact with [mathematical] objects'. Indeed, he was far less naïve about the role of ordinary sense perception in empirical knowledge than the many writers who have focused on the passage in question as the Achilles heel of platonism.

interpret theories in mathematics, as when we construct inner models of geometries or set theory or when we construct examples of groups, etc., with certain properties. But we do this in the language of mathematics, and our 'grasp' of this consists in our ability to use it. Dummett agrees with this (p. 217); but because he takes platonism to be an interpretation, he believes that this conclusion is an argument against platonism.

Benacerraf and Dummett seem to me to be typical of those who adopt a particular picture of platonism. The picture seems to be that mathematical practice takes place in an object language. But this practice needs to be explained. In other words, the object language has to be interpreted. The platonist's way to interpret it is by Tarski's truth definition, which interprets it as being about a model—a model-in-the-sky—which somehow exists independently of our mathematical practice and serves to adjudicate its correctness. So there are two layers of mathematics: the layer of ordinary mathematical practice in which we prove propositions such as (1) and the layer of the model at which (1) asserts the 'real existence' of a number.

This is the picture that seems to lie behind the distinction in Chihara (1973: 61–75) between the 'mythological' platonist and the 'ontological' platonist. The former simply does mathematics while refraining from commitment to the interpretation. The latter accepts the interpretation, and so is committed to the 'real' existence of a prime number greater than 10 and to the 'real' existence of 10.

But one cannot explain what this interpretation is supposed to be. An interpretation in the ordinary sense is a translation. Into what language are we supposed to be translating the language of ordinary mathematics?

The platonist, on this picture, is the realist that Wittgenstein (1953) criticizes at §402, along with the idealist and the solipsist—and he might have added the nominalist in the contemporary sense—when he says that the latter 'attack the normal form of expression as if they were attacking a statement' and that the realist defends it as though he 'were stating facts recognized by every reasonable human being'. Needless to say, it is not this version of platonism that I am defending or that I even understand. Thus, I should not be understood to be taking part in any realism/anti-realism dispute, since I do not understand the ground on which such disputes take place. As a mathematical statement, the assertion that numbers exist is a triviality. What does it mean to regard it as a statement outside of mathematics?[4]

---

[4] The 'external question' of the existence of numbers would seem to presuppose a univocal and non-question-begging notion of existence against which to measure mathematical existence. But what is it? Quine (1953, n. 1), indeed, attempts an argument to the effect that the desire to distinguish mathematical from space-time existence on the grounds that the latter, but not the former,

6

It is ironic that Dummett should think that platonism is founded on a comparison between mathematical reality and the physical universe and that Benacerraf should think that it is motivated by the desire to have the same account of truth for mathematics as for its less lofty empirical cousins. Plato, who was, as far as we know, the first platonist, was entirely motivated by his recognition of the fact that the exact empirical sciences of his day—geometry, arithmetic, astronomy, and music theory, for example—were not literally true of the sensible world in the semantical sense, and, indeed, did not literally apply to it. He did not have our distinction between mathematics and empirical science, nor the idea of mathematical objects. Thus, in no sphere did he think that scientific truth was truth in the semantical sense. And Tarski's truth definition, while it concerns the semantical notion of truth, is a piece of mathematics, concerning the mathematical notion of a model of a formal language.

The fact is that *we can* regard numerical propositions, say, as being about a well-defined structure—the system of natural numbers. This would be misleading only if it led us to think that our propositional knowledge of this structure derives from some sort of non-propositional cognition of it or of its elements. In the case of sensibles, on the other hand, there is no such well-defined structure. For example, if my desk remains the same object after I scratch it as before—and clearly we must agree to this for a sufficiently light scratch—then transitivity of identity fails for sensible objects, since a finite number of such scratches will reduce the desk to a splinter which we would not identify with it. Nor can we avoid this conclusion by such resources as speaking of the desk-at-an-instant; for this is no longer a sensible object. Moreover, the predicates we apply to sensibles—for example, of shape, colour, or size—are inherently vague. Thus, the canons of exact reasoning, as

---

involves empirical investigation is unfounded. His argument is that showing that there is no ratio between the number of centaurs and the number of unicorns involves empirical investigation. But the mathematical fact here is that 0 has no reciprocal; and that needs no empirical investigation.

I think that Carnap (1956) is right that 'external questions' of existence have no prima-facie sense. But his attempt to make an absolute distinction between theoretically meaningful questions and those without theoretical meaning on the basis of his notion of a linguistic framework fails. For example, his framework for number theory is a formal system. But correct and sufficiently expressive formal systems for number theory are incomplete, and, moreover, do not express all the properties of numbers. In later writings, Carnap attempted to solve the problem of incompleteness by allowing the system to contain the infinitary ω-rule. But now the internal question 'Does there exist a number n such that $\phi(n)$' can only mean 'Does there exist an infinitary deduction of $\exists x \phi(x)$?' But this is an external question, and may be mathematically non-trivial. But anyway, linguistic frameworks are constructed in our everyday language; and it is hard to see how, lacking a precise notion of theoretical meaningfulness for it, we can convincingly determine when we have a 'good' framework and when we do not.

embodied say in some system of deductive reasoning, do not apply to the domain of sensible objects.

And when we idealize the domain of sensibles so that it takes on the character of a well-defined structure and logic applies, then the other part of the picture of empirical knowledge painted by Dummett and Benacerraf becomes manifestly problematic. For example, if reference to my desk is replaced by reference to a space-time region and reference to colours, shapes, and sizes by reference to magnitudes, then the relevance of sense perception becomes less direct. It can no longer be understood on the model of observation to observation sentence, and, at least judging by the literature on the subject of theory confirmation, it is *not* one of the 'better understood means of human cognition'. The relevant perceptions are of measurement: and the measuring devices and measurements perceived are not elements of the idealized domain, but are sensible objects like my desk. The role of sense perception in confirming or applying mathematical models of the phenomena is very complex. Yet it is only when we are thinking of such a model, and not of the sensible world itself, that the picture of the universe as a well-defined structure applies. Thus, when Benacerraf ignores the vagueness of the terms 'large' and 'older than', he is not merely setting aside a complication. He is raising the 'less lofty empirical cousins' to an altogether loftier state, where they too may suffer their share of the attacks on platonism.[5]

7

Benacerraf seems to believe that the 'better understood means of human cognition' all involve causal interaction between the knower and the known. He writes:

[5] In the nominalism of Field (1980, 1982), the mathematical model is identified with the physical world. Thus, space-time regions become nominalistically acceptable objects, and mathematics is involved only in so far as such objects as numbers, sets, and function are. Regions are real because we causally interact with them, or at least can do so with some of them. This idea is developed in the 1980 book to show how to free Newton's theory of gravitation of mathematics, to make it a nominalistic theory. Of course, there is a difficulty in that, for a wide range of phenomena, Newton's theory is inadequate and, if we replace it by Einstein's theory, e.g., the 'nominalization' has yet to be demonstrated. Moreover, Einstein's theory does not account for other ranges of phenomena, and it is open whether it is compatible with an account of them. Finally, even if we had a reasonable universal physics, i.e., an account of all known forces, we should still have to ask (at least if we took Field's position) whether it was true. Well, let us suppose that we have such a 'true' universal physics, which is a space-time physics. Won't causation be a relation between space-time points or regions? But, unless some Supreme Court decisions—made with greater precision than not only is it accustomed to, but than it is *in principle* capable of—are begged. I am not a space-time region, and so do not causally interact with such things. The world of chairs and rooms and us is different from the world of mathematical physics. The latter is called an *idealization* of the former; and this only means that we can use the mathematical theory in a certain way.

THE PLATONISM OF MATHEMATICS

I favor a causal account of knowledge in which for X to know that S is true requires some causal relation to obtain between X and the referent of the names, predicates, and quantifiers of S. I believe in addition in a causal theory of *reference*, thus making the link to my saying knowingly that S *doubly* causal (p. 22 this volume).

His problem, then, is that on the platonist view we would not be able to refer to mathematical objects, much less know anything about them, since we do not causally interact with them. (Of course, one may feel that the same problem arises for the referents of the predicates 'large' and 'older than', to take Benacerraf's examples.) His argument for 'some such view' is that we would argue that X does not know that S by arguing that he lacks the necessary causal interactions with the grounds of truth of S—for example: he wasn't there. Of course, this argument is plausible only if S is an empirical proposition (and Benacerraf's example is empirical), and so it would seem to be a complete *non sequitur* in the case of mathematical knowledge. In the latter case, we might rather argue that X does not know that S by arguing that he hasn't the competence to produce a proof of S.

However, Benacerraf thinks that whatever account we give of mathematical knowledge, it should be extendable to embrace empirical knowledge as well. And if that is so, then indeed the correct account of knowledge of sensibles had better be extendable to mathematics; and so there is no *non sequitur*. But his argument that our account of mathematical knowledge should be extendable to empirical knowledge is that to 'think otherwise would be, among other things, to ignore the interdependence of our knowledge in different areas'. But this seems to me to be a very weak argument. Consider a case of interdependence: a mathematical prediction of the motion of a physical object. First, we read the appropriate equations off the data—that is, we choose the appropriate idealization of the phenomenon. Second, we solve the equations. Third, we interpret the solution empirically. When Benacerraf speaks of mathematical knowledge in his paper, the relevant kind of knowledge is knowledge that S, where S is a mathematical proposition. But *that* kind of knowledge is involved only at the second step, and it involves nothing empirical.[6] The first and third steps involve only knowing how to apply mathematics

---

[6] Putnam (1979) also seems confused on this point. He writes that Wittgenstein may have had in mind the following 'move':

> One might hold that it is a presupposition of, say, '2 + 2 = 4', that we shall never *meet* a situation that we would *count* as a counterexample (this is an empirical fact): and one might claim that the appearance of a 'factual' element in the statement '2 + 2 = 4' arises from *confusing* the mathematical assertion (which has *no* factual content, it is claimed) with the empirical assertion first mentioned.

The 'empirical fact' and 'empirical assertion first mentioned', I assume, is that we shall never meet a situation that we would count as a counter-example. But this is, for Wittgenstein and in fact, no more an empirical fact than that we shall never meet, in a game of chess, a situation which we would

to the phenomena. But I don't see why an account of this kind of knowing requires that, if empirical knowledge involves causal interaction, then so does mathematical knowledge. The fact is that we do know how to apply mathematics, and we do not causally interact with mathematical objects. Why doesn't this fact simply refute a theory of knowing how that implies otherwise?

We may wish to explain why it is that idealization of the phenomena works. We may also wish to explain why language and inductive inference work. But these seem to me to be scientific 'why's', to be answered by an account of how we process information and of how this means of processing information (and so, creatures like us) evolved.

Although it is unnecessary for the purpose at hand, let me comment briefly (and certainly insufficiently) on the double causal interaction that Benacerraf thinks must be involved in knowledge about objects. It seems to me that when we speak of mathematical knowledge in the ordinary way, we are referring to the ability to state definitions and theorems, to compute, to prove propositions, etc.: in general it is a matter of knowing *how*, of competence. Anyway, it is this kind of knowledge that we test students for. The ideas of propositional knowledge (knowledge *that*) and knowledge in the sense of acquaintance with (knowledge *of*) also seem to me ultimately to reduce to the idea of knowledge how; and this is so, not only in the (relatively simple) domain of mathematics, but in general. This is of course very different from the Cartesian notion of knowledge, since knowledge in this sense presupposes a communal practice against which competence is to be measured and so

---

count as one in which the king is captured. Of course, neither of these assertions is a prediction about our future behaviour or an assertion about our past behaviour; they are each part of a description of a certain game. It is indeed an empirical fact that we play the game—that we do mathematics and play chess—but that is another matter. Putnam goes on:

> *This* move, however, depends heavily on overlooking or denying the circumstance that an empirical fact can have a partly mathematical *explanation*. Thus, let T be an actual (physically instantiated) Turing machine so programmed that if it is started scanning the input '111', it never halts. Suppose that we start T scanning the input '111', let T run for two weeks, and then turn it off. In the course of the two-week run, T did not halt. Is it not the case that the *explanation* of the fact that T did not halt is simply the *mathematical* fact that a Turing machine with that program never halts on the input, *together with* the empirical fact that T instantiates that program (and continued to do so throughout the two weeks)?

The answer is simply: yes. But what has this to do with the fact that the mathematical proposition '2 + 2 = 4' or 'Turing machine t with input "111" never halts' is not the sort of proposition for which the idea of empirical counter-examples makes sense? This example is no different from our explanation of the motion of a physical object. We model the behaviour of T with t. If it is a good model (and this idea defies precise analysis), then the fact that t doesn't halt (in the mathematical sense) should lead us to believe that T doesn't (in the physical sense) halt. But what has this to do with the conceivability of an empirical counter-example to the statement that t doesn't (in the mathematical sense) halt? The sense in which it is claimed that '2 + 2 = 4' has no 'factual content' is not intended to imply that it has no empirical applications.

THE PLATONISM OF MATHEMATICS 155

cannot serve as an external foundation for a critique of that practice. Critique must come from within, measuring our practice against the purpose of that practice. Also, knowledge in this sense is not a matter of all or nothing: we recognize degrees of knowing. (For example, when is giving a proof *really* giving a proof—with understanding? Compare this with Wittgenstein's discussion of reading.) We may indeed obtain a causal account of knowledge in this sense; but it does not seem at all plausible that such an account, even in the case of knowledge about physical objects, will involve causal interaction with those objects. (The appearance of plausibility here arises from the possibility that I might unwarrantedly believe a true proposition. But one may reasonably doubt that there is a sense in which my belief is unwarranted that would not show up in what I am disposed to *do*.) As for the view that reference involves causal interaction, the motive for this seems to me to confuse the question of how we come to use a word in the way we do with the question of how it is in fact used. (Cf. Wittgenstein 1953, §10.)

8

Platonism is taken to be an account of mathematics which says, for example, that number theory is about a certain model. And then it is challenged to tell us what that model is. One asks: how do we get to know this model? Or: how do we know when we speak together that we are speaking about the same model?, etc. It is as if we have a formal system and are told that there is an intended model for it. But no one can tell us what this model is, and so we do not even know why the formal system is grammatically correct, much less valid.

Thus, Dummett (1967: 210) writes:

To say that we cannot communicate our intuition of the natural numbers unequivocally by means of a formal system would be tolerable only if we had some other means to communicate it.... We cannot know that other people understand the notion of all properties (of some set of individuals) as we do, and hence have the same model of the natural numbers as we do.

and 'we arrive at the dilemma that we are unable to be certain whether what someone else refers to as the standard model is really isomorphic to the standard model we have in mind' (pp. 210–11).

What is my intuition *of* the numbers? I can only be said to have intuitions *about* them—and then, only when I have some minimum understanding of number theory. And this understanding is not an 'intuition' (although there may be accompanying feelings and pictures); it is a competence. What does it mean to 'have a model' or to 'have one in mind'? And what does it mean for us to have the same one? This can only mean that we do the same number theory—the one which is part of our common language.[7] And I can ask: how do we know that you have the same physical universe that I have? Dummett seems to believe that we must *explain* our ability to communicate mathematics and that platonism is inadequate because it fails to do this. But no explanation is necessary, unless one is calling for a general empirical account of human communication. Mathematics *presupposes* the fact of communication—the fact of our common disposition to use and react to symbols in specific ways. If we lacked such common dispositions, we could not be said to have mathematics any more than, if we lacked legs, could we be said to walk.

Every reasonably schooled child understands the language of arithmetic. It is the schizophrenic parent of the child who, motivated by an inappropriate picture of meaning and knowledge, develops 'ontological qualms'. The picture is read into platonism and then, because it is inappropriate, platonism, that is, our ordinary conception of mathematics, is rejected. The fact that the picture is *generally* inappropriate is simply ignored. We owe to Chihara a clear illustration of the schizophrenia, namely, in his mythological platonist.

9

Strangely, Dummett understands that the notion of a model is a mathematical notion and that we construct or describe models *in* mathematics (1967: 213–14 and 1963). He is ascribing to platonism an idea that he must find incomprehensible. Why? Part of the answer at least may be found in the terms in which he argues in the 1967 paper that Frege's context principle undermines realism:

---

[7] Consider systems $\mathfrak{A} = \langle A, a, f \rangle$, where A is a type of object, a is an object of type $A(a : A)$, and f is a function from A to A ($f : A \to A$). Dedekind (1888) characterized the system $\mathfrak{N} = \langle N, 0, S \rangle$ of numbers as such a system in which $0 \neq Sn$ for all $n : N$, $Sm = Sn \to m = n$ and, if X is any set of numbers containing 0 and closed under S, then it is the set of all numbers. There is no question of *identifying* the system of numbers: it is, as Dedekind puts it (§73), a 'free creation of the human mind'. We have created it in the sense that we have specified once and for all its grammar and logic. Moreover, given any other system $\mathfrak{A}$ satisfying this characterization, the proof that $\mathfrak{A}$ is isomorphic to $\mathfrak{N}$ is a triviality, and we shall not disagree about that. We might indeed disagree about the principles used to construct some set P of numbers or some system $\mathfrak{A}$; but that is a different matter and, anyway, if we leave aside those who wish to use only constructive principles, there is as a matter of fact, there is no such disagreement (cf. §15). Moreover, the possibility of this kind of disagreement exists even in constructive mathematics, which Dummett (1975; Ch. IV this volume) is advocating. In that case, Dedekind's characterization should be replaced by the classically equivalent one essentially given by Lawvere (1964), namely, that $\mathfrak{N}$ has the property of *unique iteration*: given any system $\mathfrak{A}$, the equations $g0 = a$ and $gS = fg$ define a unique function $g : N \to A$. But we may still disagree about when a system $\mathfrak{A}$ has been legitimately introduced.

THE PLATONISM OF MATHEMATICS

When we scrutinize the doctrines of the arch-Platonist Frege, the substance of the existential affirmation finally appears to dissolve. For him mathematical objects are as genuine objects as the sun and moon: but when we ask what these objects are, we are told that they are the references of mathematical terms, and 'only in the context of a sentence does a name have a reference'. In other words, if an expression functions as a singular term in sentences for which we have provided a clear sense, i.e. for which we have legitimately stipulated determinate truth conditions, then that expression *is* a term (proper name) and accordingly has a reference: and to know those truth conditions *is* to know what its reference is, since 'we must not ask after the reference of a name in isolation'. So, then, to assert that there are, e.g., natural numbers turns out to be to assert no more than that we have correctly supplied the sentences of number theory with determinate truth conditions; and now the bold thesis that there are abstract objects as good as concrete ones appears to evaluate to a tame assertion that few would want to dispute.

I, for one, would dispute the 'tame assertion' that we have 'correctly supplied the sentences of number theory with determinate truth conditions', unless, of course, we are speaking about some formal system of number theory and we have explained their meaning in ordinary mathematical terms. We interpret formal systems; but in what language do we interpret ordinary mathematics to give it 'determinate truth conditions'?

There are many difficulties and complexities in connection with Frege's context principle; he applies it in Frege (1884) to justify his definition of the numbers, and he applies it in Frege (1893) to justify the introduction of course-of-values in terms of which the numbers are defined. One complication is that he is proposing an extension of the ordinary mathematical discourse of his time—a new norm for mathematics—and another is that his extension is inconsistent. Also, he formulates his argument (1884, §60) that numbers are objects against the background of his object/function ontology. Also, because he was concerned with mathematics, he did not concern himself with the problem of terms such as 'Homer' which function grammatically like terms but may not denote. Moreover, and possibly for the same reason, he was concerned only with the role of names in the context of declarative sentences and not in other kinds of linguistic expressions. Finally, his formulation of the principle leaves open the question of the meaning of sentences. But one nowhere finds him saying that mathematical objects are the references of mathematical terms in answer to the question of what they are. Rather, he is giving a criterion for the meaningfulness of terms, and he suggests in §60 that the criterion extends beyond mathematics. And he then says that there is nothing more to the question of the self-subsistence of numbers than the role that number-words play in propositions. I take this to mean that to say that a term refers is to say that it *is* a meaningful term. There is certainly no implication here that every mathematical object is the reference of a term.

One should note that, anyway, Wittgenstein's reformulation of the context principle, replacing the context of a proposition by the context of a language (cf. 1953, §10 and the discussion in note 2), must, for Dummett, also tame the same bold thesis. But, if so, Wittgenstein's argument also tames the bold thesis that there are physical objects. The issue between the realist and the idealist of §402 is a non-issue too. So if abstract objects are not 'as good as' Dummett conceives concrete objects to be, then neither are concrete objects. Dummett (1975; Ch. IV this volume) wishes to accept Wittgenstein's critique of language to the extent of accepting the formula that meaning is determined by use as a rough guide to the analysis of mathematical language. But I think that the above passage shows that he does not accept the full consequences of the critique. However, that is already shown by the fact, noted by Lear (1982), that he adopts the above formula to argue for a revisionist view of mathematics contrary to Wittgenstein (1953, §124).

10

On the basis of the preceding discussion, I think that we can now begin to resolve the truth/proof problem. This problem arises because there seem to be two, possibly conflicting, criteria for the truth of a mathematical proposition: that it hold in the relevant structure and that we have a proof of it.

The first step of the resolution is to see that the first criterion is not a criterion at all. The appearance that it is arises from the myth of the model-in-the-sky, of which we must—but do not seem to—have some sort of non-propositional grasp, with reference to which our mathematical propositions derive their meaning and to which we appeal to determine their truth. The fact is that there are no such models; there are only models, that is, structures that we construct *in* mathematics. Our grasp of such a model *presupposes* that we understand the relevant mathematical propositions and can determine the truth of at least some of them—for example, those whose truth is presupposed in the very definition of the model. Thus, rather than saying that holding in the model is a criterion for truth, we would better put it the other way around: being true is a criterion for holding in the model.

The myth of the model tends to get attached to platonism (or at least to 'epistemological' platonism in the sense of Steiner (1975)) because the view that mathematics is about things like the system of numbers is compared with the view that propositions about sensible things are about the physical world; and here there is a tendency to believe that there *is* such a non-propositional grasp, namely, sense perception, which does endow meaning on what we say and to which we appeal to determine truth. But I hope that, if not what I have

THE PLATONISM OF MATHEMATICS

said, then Wittgenstein's critique of this view of discourse about sensibles will convince the reader that it is inadequate.

11

However, the first step of the resolution of the truth/proof problem may appear to have thrown out the baby with the bath water so far as platonism is concerned; and both Benacerraf and Dummett think that this is so. For, if we reject the myth of the model, then how *are* we to understand the notion of truth in mathematics? There might seem to be no alternative here to identifying it with the notion of provability. But then the independence of truth from the question of what we know or can know, which is the essence of platonism, would be lost.

Benacerraf takes the less dogmatic line, not that this is the *only* alternative, but that it is the only one that has been substantially considered. But he takes the notion of proof here to be that of deducibility in some formal system, and he argues for the obviously correct conclusion that this yields an inadequate notion of truth. Dummett (1975; Ch. IV this volume) takes the view that, in giving up the myth of the model, we are giving up the notion of truth and, with it, classical mathematics. He holds that the only viable alternative is to replace the notion 'A is true' by 'p is a proof of A', where the notion of proof here is the intuitionistic one.

Although I have argued in an earlier paper (Tait 1983) that Dummett is wrong here, and, indeed, that the intuitionistic conception is not entirely coherent, I nevertheless think that his response is, in a sense, in the right direction. Namely, I think that the intuitionists' view that a mathematical proposition A may be regarded as a type of object and that proving A amounts to constructing such an object is right. Of course, to say that we may regard A as a type of object does not mean that we normally regard theorem proving as a matter of constructing objects. Indeed, when we are interested in constructing an object, say a real number, characteristically we are concerned with constructing one with a particular property. As a proposition, 'real number' is trivial. In the case of propositions, we are generally concerned with finding *some* proof, and only rarely are we concerned with its properties. My point is rather that, independently of what we would say we are doing when we are theorem proving, what we are actually doing may be faithfully understood as constructing an object. The basic mathematical principles of proof that we use, for example, the laws of logic, mathematical induction, etc., are naturally understood as principles of construction.

However, the intuitionists also hold that the objects that a proposition A stands for, the objects of type A, *are* its proofs; and that I think is wrong. A proof of A is a *presentation* or *construction* of such an object: A *is true* when there is an object of type A and we prove A by constructing such an object.

Here then is the answer to one of our questions: why is proof the ultimate warrant for truth? The answer is of course that the only way to show that there is an object of type A is to present one. (To prove that there is an object of type A will mean nothing more than to prove A, and that means to exhibit an object of type A.)

Consider the equation $s = t$ between closed terms of elementary number theory. What does this equation mean? We may say that it expresses something about the system of numbers. That is certainly so, but it is also not to the point until we say what that something is, without simply repeating the equation in the same or other terms.

The intuitionists seem to me very convincing when they say that what the equation expresses is that there is a certain kind of computation, namely, one which reduces s and t to the same term. For not only do we initially learn the meaning of such terms and equations by learning how to compute, but we take the existence of such computations as the ultimate warrant for the equation. Thus, it seems entirely natural to construe the equation as standing for the existence of such a computation and to take the equation to be true precisely when there is one.

Dummett (1975; Ch. IV this volume) accepts this analysis of such equations, but Dummett (1967) feels that, in accepting it, one is rejecting the platonist point of view. His argument is that once we have accepted it, there is no reason to invoke the notion of truth in the sense of 'holding in the system of numbers' to account for the meaning of the equation. But, of course, we are not accounting for its meaning in this way and, indeed, could not do so without circularity. *That* it holds in the system of numbers—in other words, the *fact* about this system which it expresses—is that there is such a computation. And we *prove* the equation by producing one.

At least part of the reason why Dummett believes that the above analysis of equations amounts to a rejection of platonism is that he, along with the intuitionists, identifies the proofs of the equations, that is, the *presentations* of the computations, with the computations themselves; and when we do that, we can no longer account for the possibility of true but unprovable equations (Dummett 1967: 203). One might object that the platonist need not account for this possibility providing he can account for there being *some* true but unprovable propositions. But the identification of computation with proof is a special case of the intuitionistic identification of the object with its construction in general. I do not believe that this identification is ultimately intelligible; but one sees that, in accepting it, there is in general no possibility of true but unprovable propositions.

However, it seems to me that, even in the case of the above sort of equations, the intuitionists are wrong, and that one should not identify computations with proofs. For example, we easily prove $10^{10} = (10^5)^2$ as an instance of a more general theorem; but in the canonical notation 0, S0, SS0, ... for numbers, I shall be unable to explicitly compute 10^{10}, and even for terms with much shorter computations, the chance of my computing accurately is very small. Dummett (1977) makes the distinction here between 'canonical proofs', which in the present context are the explicitly presented computations, and the sort of proof one obtains from proofs of more general propositions, which are shorthand descriptions of canonical proofs. But when we know that the computation is longer than human beings, individually or collectively, are able to perform, we must ask the question: canonical proof for whom? To answer this by reference to an 'ideal computer' seems highly unsatisfactory. In the first place, proof is a *human* activity—and this would seem especially important to an intuitionist. But secondly, I am unable to see a significant difference between referring to an ideal computer who can compute $f(n)$ for *each* n and one who can compute it for *all* n and hence can decide whether $f(n) = 0$ for all n or not. I don't mean that there isn't a *formal* difference, but rather that it is hard to see why the one idealization is legitimate and the other not. Yet the intuitionists reject the latter one, which would lead to the law of excluded middle for arithmetic propositions.

Computations are mathematical objects, forming a mathematical system like the system of numbers. One may object to the use of the term 'computation' here, because of its association with computing as a human activity. But the term is also used in my sense, for example in the mathematical theory of computability. The ease with which one can confuse the two senses may contribute to the apparent plausibility of the intuitionistic identification of the computation with its presentation.

13

When we extend the conception of mathematical propositions as types of objects to propositions other than equations, the distinction between object of type A and proof of A becomes even more evident. For example, let ϕ be a function which associates with each object a of type A, expressed by $a : A$, a type ϕa. Then

$\forall x : A \, . \, \phi x$

is the type of all functions f defined on A such that $fx : \phi x$ for all $x : A$, and

$\exists x : A \, . \, \phi x$

is the type of all pairs (x, y) such that $x : A$ and $y : \phi x$. These definitions of the quantifiers are essentially forced on us by the propositions as types conception.[8] The remaining logical constants are definable from the quantifiers, the null type 0, and the two-element type 2, whose objects we denote by ⊤ and ⊥. Thus, if we identify the type B with the constant function $\phi \equiv B$, then implication and negation are defined by

$A \to B = \forall x : A \, . \, B \quad \neg A = A \to 0,$

and, if $\psi\top = A$ and $\psi\bot = B$, then

$A \wedge B = \forall x : 2 \, . \, \psi x \quad A \vee B = \exists x : 2 \, . \, \psi x$

Again, these definitions are essentially forced on us.[9]

---

[8] Suppose that we already have that, for any $x : A$, $\phi(x)$ is already identified with a type of object. Then $\exists x : A\phi(x)$ means that, for some $x : A$, $\phi(x)$, and so that, for some $x : A$, there is a $y : \phi(x)$, and so that there is a pair (x, y) of the required type. $\forall x : A\phi(x)$ means that $\phi(x)$ for all $x : A$, and so that, for each $x : A$, there is a $y : \phi(x)$. So we have 'reduced' the meaning of $\forall$ to 'for all x, there exists a y'. We avoid an infinite regress here only by taking the latter to mean that we have a function f which gives us $y = fx$ for each x. This is, as a matter of fact, the way in which we do reason. The appearance that it isn't arises from the fact that we often are thinking of the reasoning as taking place in a model in which no such f occurs. So $\forall x : A\phi(x)$ may be true in the model without there being the required f in the model. But that of course is different from saying that there is no such f.

Our analysis of the quantifiers yields the axiom of choice in the form

$\forall x : A \exists y : B\psi(x, y) \to \exists z : A \to B \forall x : A\psi(x, zx).$

For let f be of the antecedent type. Then for each $x : A$, fx is of the form (y, u), where u is of type $\psi(x, y)$. Let $z : A \to B$ be defined by $zx = y$ and let $v : \forall x : A\psi(x, zx)$ be defined by $vx = u$. Then (z, v) is of the type of the conclusion. The argument above for our analysis of the universal quantifier looks itself like an application of the axiom of choice:

$\forall x : A \exists y(y : \phi(x)) \to \exists f \forall x : A(fx : \phi(x)).$

But there are two respects in which it is different. First, it contains two variables, y and f, whose type is unspecified, and secondly, it involves 'propositions' of the form '$u : C$'. Concerning the first point, the notion of a mathematical object in general seems a problematic notion, and certainly is no part of the platonistic conception that I am discussing. Concerning the second, '$u : C$' is not a mathematical proposition in the sense that I am discussing. Otherwise, we would have to know what are the objects v of type $u : C$, the objects of type $v : (u : C)$, etc., leading to an infinite regress. The fact is that we have a type C only when we have agreement as to what counts as an object of type C. Thus, statements such as '$u : C$' are grammatical, statements. It is the wrong picture to think that there is a universe of 'mathematical objects', and then we must determine for one of them, u, what type it has. (This seems to me to be the view behind Jubien 1977, where its absurdity is well illustrated.) 0 is a mathematical object because it is a number. If I have introduced an object u of type C, then either $u : C$ or else I have been indulging in nonsense.

[9] Cf. Tait 1983. In the case of negation, note that $A \wedge B \to 0$ should be a requirement for a negation B of A. But when this holds, $B \to \neg A$; and so $\neg A$ is the weakest candidate for negation. One may feel, none the less, that negation presents a counter-example to the view of propositions as types; since, if 'A is true' is to mean that there is an object of type A, then 'A is false' ought to mean

In this way, the logical operations appear as operations for constructing types and the laws of logic as principles for constructing objects of given types. In this respect, there is no essential difference between constructing a number or set of numbers and proving a proposition. As Brouwer insisted should be the case, the logic of mathematics becomes part of mathematics and not a postulate about some transcendent model. However, Brouwer's view that the objects of mathematics be mental objects does not seem to me coherent. And the intuitionists' view that, for example, when we construct a number, we should be able to determine its place in the sequence 0, 1, 2, ... ignores the difficulty that we cannot in any case do this for sufficiently complex constructions. Anyway, it is a restriction on ordinary mathematical practice that is inessential to the conception of propositions as types. The law of excluded middle amounts to admitting objects of types $A \vee \neg A$ which we may not otherwise be able to construct; and this does indeed lead to the construction of numbers whose positions in the above sequence are not computable. But it is not essential to our conception that they should be.

An object of an $\forall$-type is a function, and I have argued elsewhere (cf. Tait 1983) that, even in the case of constructive mathematics, one must distinguish between a function and a presentation of it, by a rule of computation or otherwise. I shall assume that, in the case of non-constructive mathematics, no argument is needed for this and, therefore, that the distinction between objects of type A and proofs of A is clear.

14

Of course, we have not really specified the types 0 and 2 nor the operations $\forall$ and $\exists$ until we have specified the principles of construction or proof associated with them. A brief discussion of this occurs in Tait 1983 (though the treatment of equations there is inadequate), and a fuller treatment is in preparation. These principles underlie mathematical practice in the sense that arguments that cannot be reduced to them are as a matter of fact regarded as invalid.

Questions about the legitimacy of principles of construction or proof are not, in my opinion, questions of fact. For mathematics presupposes a common mathematical practice, and it is this that such principles codify. Without

that there is no such object, and this is not an existence statement. But if there is no object of type A, then there is an object of type $A \to 0$, namely, the null function. But, in any case, there is something deceptive about the discussion. What does 'not' mean when we say that it is not the case that there is an object of type A? For this too is a mathematical proposition, and, indeed, simply means $\neg A$. We should not think that there is a meaning of 'not' that somehow transcends mathematical practice.

agreement about these principles and their application, there are no mathematical 'facts' (cf. n. 8). Of course, many factors, including the requirement of logical consistency, would be involved in explaining why our mathematics takes the form it does; but the view that there is some underlying reality which is independent of our practice and which adjudicates its correctness seems to me ultimately unintelligible.[10]

In this respect, the controversy between constructivists and non-constructivists is similar to controversies about what is good or just between people of different moral or political outlook. In the latter case, one may ask what precisely *is* the issue. Why not simply use the terms 'just$_1$' and 'just$_2$'? It seems to me that the answer is that there is agreement about what I shall call the *normative* content of the term 'just' (or 'good'). Namely, to hold an action X to be just is to be disposed to act in certain ways. And I am not referring here entirely to linguistic acts such as affirming that one ought to do X. Rather, I have in mind Aristotle's practical syllogism: to hold that X is just is to be disposed to *do X*. If there were no agreement about this normative content of the term 'just', then there would be no point in disputing its *material* content, that is, the question of what acts are to count as just. But the latter sort of dispute seems to me not necessarily to involve matters of fact, inasmuch as there may not be a sufficient basis of ethical agreement to decide the issue.

In the same way, there is a normative content of the term 'valid'. To hold an inference to be valid is to be disposed to make the inference. *Because* we agree about this normative content, it is significant to argue about its material content, about what inferences are to count as valid. But, here too, there may be no matter of fact, only a matter of persuasion and adjustment of mathematical 'intuitions'. It is no accident that the dispute over the law of excluded middle often takes a moralistic tone. There are no non-circular arguments for this law, and, in spite of all efforts to show otherwise, there are no arguments against it which are not essentially to the effect that it leads to non-computable objects.

Constructivists do not deny any instance of the law of excluded middle, of course: that would lead to inconsistency. Rather, they refrain from its application. Thus, in principle, constructive mathematics may be viewed as a restriction within ordinary mathematics on the methods of proof or

---

[10] The question of the truth *of* mathematics, as opposed to truth *in* mathematics has historically been the concern of many philosophers. In some cases, e.g., Plato and Leibniz, this question has been distinguished from that of why mathematics applies to the phenomena, and in others, such as Aristotle and Kant, it has not. This latter question, of why mathematization of the phenomena works, has itself been a source of anti-platonism. But, as I have indicated in §7, the only kind of answer to that question would be in terms of cognitive science and an account of why it is that we have evolved.

construction.[11] Aside from this, it is a striking fact that there simply is no disagreement concerning the valid principles of mathematical reasoning. Of course, I have not mentioned all of the type-forming operations involved in mathematics; nor is it clear that one could do so. For example, set theory involves the types obtained by 'iterating' the operation of passing from a type A to $PA = A \to 2$ into the transfinite. This involves the idea of creating new types by inductive definitions. However, although there might be disagreement about what inductive definitions one ought to admit, there is none about the principles of proof to be associated with such a definition when it is admitted.[12]

15

The answer to the initial question of this paper, concerning the relation between truth and proof in mathematics, is that a proposition A is true when there is an object of type A and that a proof of it is the construction of such an object. *That* there is an object of type A is the 'fact' about, say, the system of numbers that A expresses. It is clear from this why proof is the ultimate warrant for truth.

The platonist view that truth is independent of what we know or can know is entirely correct on this view. In the first place, there may be propositions which we can in principle prove on the basis of existing mathematics, but whose proofs are too complex for us to process. Secondly, there may be

---

[11] Of course, if one is interested only in constructive mathematics, one may diverge from the classical development of, say, analysis, by choosing concepts more amenable to constructive treatment than the classical analogues. My point is only that the principles of construction and reasoning used in the development remain classically valid. Apparent counter-examples such as Brouwer's proof that every real-valued function on the continuum is continuous are a result of ambiguity, not of using classically invalid principles.

[12] There is another method of obtaining new types which derives from Dedekind (1888) and which we may refer to as 'Dedekind abstraction'. For example, in set theory we construct the system $\langle \omega, \phi, \sigma \rangle$ of finite von Neumann ordinals, where $\sigma x = x U \{x\}$. We may now abstract from the particular nature of these ordinals to obtain the system $\mathfrak{N}$ of natural numbers. In other words, we introduce $\mathfrak{N}$ together with an isomorphism between the two systems. In the same way we can introduce the continuum, e.g., by Dedekind abstraction from the system of Dedekind cuts. In this way, the arbitrariness of this or that particular 'construction' of the numbers or the continuum, noted in connection with the numbers in Benacerraf 1965, is eliminated. It is, incidentally, remarkable that some authors such as Kitcher (1983) have taken Benacerraf's observation to be an argument against identifying the natural numbers with sets, but have been content to identify the real numbers with sets, although there are again various ways to do that. Kitcher 1978 contains an amazing argument based on Benacerraf's observation, to the effect that platonism is false: on grounds of economy, all 'abstract' objects should be sets. Numbers are abstract. But there is no canonical representation of the numbers as sets. Therefore, the view that there are such things as numbers is false. (A person makes up a budget and, on grounds of economy, fails to budget for food. But we need to eat. So the notion of a budget is incoherent.)

propositions which are not provable on the basis of what we now accept, but are provable by means that we *would* accept. When I speak here of new means of proof, I do not of course mean the acceptance of new logical principles concerning 0, 2, ∀, ∃, inductive definitions, etc., but rather the introduction of further types to which we can apply these principles. For example, by the introduction of new types we may construct numerical functions, that is, 'proofs' of $N \to N$, which we cannot otherwise construct.

It is, incidentally, this open-endedness of mathematics with respect to the introduction of new types of objects that refutes the formalistic conception of mathematics, even if we leave aside the fact that mathematical concepts such as the number concept have a wider meaning than that given by their role in mathematics itself. The formalists seem to me right—in any case, we have not one example to refute them—that the above type-forming operations are completely determined in mathematics by the principles of inference we as a matter of fact associate with them. The incompleteness of formal systems such as elementary number theory is best seen as an incompleteness with respect to what can be expressed in the system rather than with respect to the rules of inference. For example, Gödel's undecidable proposition for elementary arithmetic can indeed be proved by induction; but the induction must be applied to a property not expressed in the system itself.[13]

REFERENCES

Benacerraf, P. (1965), 'What Numbers Could Not Be', *Philosophical Review*, 74: 47–73.
—— (1973), 'Mathematical Truth', *Journal of Philosophy*, 70: 661–79; repr. in this volume as Chapter I.
——, and H. Putnam (1983) (eds.), *Philosophy of Mathematics: Selected Readings*, 2nd edn. (Cambridge: Cambridge University Press).
Carnap, R. (1956), 'Empiricism, Semantics, and Ontology', in *Meaning and Necessity*, 2nd edn. (Chicago: University of Chicago Press). Repr. in Benacerraf and Putnam 1984.
Chihara, C. (1973), *Ontology and the Vicious Circle Principle* (Ithaca, NY: Cornell University Press).
Dedekind, R. (1888), *Was sind und was sollen die Zahlend* (Brunswick: Vieweg).
Dummett, M. (1963), 'The Philosophical Significance of Gödel's Theorem', *Ratio*, 5: 140–55, repr. in Dummett 1978; page references are to the latter.
—— (1967), 'Platonism', first published in Dummett 1978: 202–14.

---

[13] Earlier versions of this paper were presented to the Philosophy Department of the University of Wisconsin at Madison in the winter of 1984, at the Tarski Memorial Conference at Ohio State in the spring of 1984, and at the Pacific Division meeting of the APA in the spring of 1985. I received many valuable comments on all these occasions and, in particular, from Paul Benacerraf and Clifton McIntosh, who commented on my paper at the APA meeting. I should also like to thank Michael Friedman for his comments on an earlier version and for our many discussions of its subject-matter.

—— (1975), 'The Philosophical Basis of Intuitionistic Logic', in H. E. Rose and J. C. Shepherson (eds.), *Logic Colloquium '73* (Amsterdam: North Holland), 5–40, repr. in Dummett 1978; page references are to the latter. Also repr. in this volume as Chapter IV.
—— (1977), *Elements of Intuitionism* (Oxford: Clarendon Press).
—— (1978), *Truth and Other Enigmas* (Cambridge, Mass.: Harvard University Press).
Field, H. (1980), *Science without Numbers* (Princeton: Princeton University Press).
—— (1982), 'Realism and Anti-Realism about Mathematics', *Philosophical Topics*, 13: 45–69; repr. in Hartry Field, *Realism, Mathematics and Modality* (Oxford: Blackwell, 1989).
Frege, G. (1884), *Die Grundlagen der Arithmetik* (Breslau: Verlag von Wilhelm Koebner).
—— (1893), *Grundgesetze der Arithmetik*, vol. 1 (Jena: Pohle).
Gödel, K. (1947), 'What Is Cantor's Continuum Problem?', *American Mathematical Monthly*, 54: 515–25. A revised and supplemented version appears in Benacerraf and Putnam 1984; page references are to the latter. 'Gödel 1964' refers to the supplement in the later version (which first appeared in the 1st edn. of Benacerraf and Putnam in 1964).
Jubien, M. (1977), 'Ontology and Mathematical Truth', *Noûs*, 11: 133–50.
Kitcher, P. (1978), 'The Plight of the Platonist', *Noûs*, 119–36.
—— (1983), *The Nature of Mathematical Knowledge* (New York: Oxford University Press).
Lawvere, W. (1964), 'An Elementary Theory of the Category of Sets', *Proceedings of the National Academy of Science*, 52: 1506–11.
Lear, J. (1982), 'Leaving the World Alone', *Journal of Philosophy*, 19: 382–403.
Maddy, P. (1980), 'Perception and Mathematical Intuition', *Philosophical Review*, 89: 163–96; repr. in this volume as Chapter VI.
Parsons, C. (1979–80), 'Mathematical Intuition', *Proceedings of the Aristotelian Society*, 145–68; repr. in this volume as Chapter V.
Putnam, H. (1979), 'Analyticity and Aprioricity: Beyond Wittgenstein and Quine', *Midwest Studies in Philosophy*, 4 (*Studies in Metaphysics*).
Quine, W. V. (1953), 'On What There Is', in *From A Logical Point of View* (Cambridge, Mass.: Harvard University Press).
Steiner, M. (1975), *Mathematical Knowledge* (Ithaca, NY: Cornell University Press).
Tait, W. (1983), 'Against Intuitionism: Constructive Mathematics Is a Part of Classical Mathematics', *Journal of Philosophy*, 80: 173–95.
Wittgenstein, L. (1953), *Philosophical Investigations* (Oxford: Blackwell; New York: Macmillan).

VIII

MATHEMATICS WITHOUT FOUNDATIONS

HILARY PUTNAM

Philosophers and logicians have been so busy trying to provide mathematics with a 'foundation' in the past half-century that only rarely have a few timid voices dared to voice the suggestion that it does not need one. I wish here to urge with some seriousness the view of the timid voices. I don't think mathematics is unclear; I don't think mathematics has a crisis in its foundations; indeed, I do not believe mathematics either has or needs 'foundations'. The much-touted problems in the philosophy of mathematics seem to me, without exception, to be problems internal to the thought of various system-builders. The systems are doubtless interesting as intellectual exercises; debate between the systems and research within the systems doubtless will and should continue; but I would like to convince you (of course I won't, but one can always hope) that the various systems of mathematical philosophy, without exception, need not be taken seriously.

By way of comparison, it may be salutary to consider the various 'crises' that philosophy has pretended to discover in the past. It is impressive to remember that at the turn of the century there was a large measure of agreement among philosophers—far more than there is now—on certain fundamentals. Virtually all philosophers were idealists of one sort or another. But even the non-idealists were in a large measure of agreement with the idealists. It was generally agreed that any property of material objects—say, *redness* or *length*—could be ascribed to the object, if at all, only as a power to produce certain sorts of sensory experiences. When the man on the street thinks of a material object, according to this traditional view, he really thinks of a subjective object, not a real 'external' object. If there are external objects, we cannot really imagine what they are like; we know and can conceive only their powers. Either there are no external objects at all (Berkeley)—that is, no objects 'external' to minds and their ideas—or there are, but they are *Dinge an sich*. In sum, then, philosophy flattered itself to have discovered not just a crisis, but a fundamental mistake, not in some special science, but in our most common-sense convictions about material objects. To put it crudely, philosophy thought itself to have shown that no one has ever really perceived a material object and that, if material objects exist at all (which was thought to be highly problematical), then no one *could* perceive, or even imagine, one.

Anyone maintaining at the turn of the century that the notions 'red' and 'hard' (or, more abstractly, 'material object') were reasonably clear notions; that redness and hardness are *non*-dispositional properties of material objects; that we see red things and see *that* they are red; and that *of course* we can imagine red objects, know what a red object is, etc., would have seemed unutterably foolish. After all, the most brilliant philosophers in the world all found difficulties with these notions. Clearly, the man is just too stupid to see the difficulties. Yet today this 'stupid' view is the view of many sophisticated philosophers, and the increasingly prevalent opinion is that it was the arguments purporting to show a contradiction in the view, and not the view itself, that were profoundly wrong. Moral: not everything that passes—in philosophy anyway—as a difficulty with a concept is one. And second moral: the fact that philosophers all agree that a notion is 'unclear' doesn't mean that it *is* unclear.

More recently there was a large measure of agreement among philosophers of science—far more than there is now—that, in some sense, talk about theoretical entities and physical magnitudes is 'highly derived talk' which, in the last analysis, reduces to talk about observables. Just a few years ago we were being told that 'electron' is a 'partially interpreted' term, whereas 'red' is 'completely interpreted'. Today it is becoming increasingly clear that 'electron' is a term that has complete 'meaning' in every sense in which 'red' has 'meaning'; that the 'purpose' of talk about electrons is not simply to make successful predictions in observation language any more than the 'purpose' of talk about red things is to make true deductions about electrons; and that the whole question about how we 'introduce' theoretical terms was a mare's nest. I refrain from drawing another moral.

Today there is a large measure of agreement among philosophers of mathematics that the concept of a 'set' is unclear. I hope the above short review of some history of philosophy will indicate why I am less than overawed by this agreement. When philosophy discovers something wrong with science, sometimes science has to be changed—Russell's paradox comes to mind, as does Berkeley's attack on the actual infinitesimal—but more often it is philosophy that has to be changed. I do not think that the difficulties that philosophy finds with classical mathematics today are genuine difficulties; and I think that the philosophical interpretations of mathematics that we are being offered on every hand are wrong, and that 'philosophical interpretation' is just what mathematics doesn't need. And I include my own past efforts in this direction.

I do not, however, mean to disparage the value of philosophical inquiry. If philosophy got itself into difficulties with the concept of a material object, it also got itself out; and the result is some modest but significant increase in our clarity about perception and knowledge. It is this sort of clarity about mathematical truth, mathematical 'objects', and mathematical necessity that I should like to see us attain; but I do not think the famous 'isms' in the philosophy of mathematics represent the road to that clarity. Let us therefore make a fresh start.

A SKETCH OF MY VIEW

I think that the least mystifying way for me to discuss this topic is as follows: first to give a very cursory and superficial sketch of my own views, so that you will at least be able to guess at the positive position that underlies my criticism of others, and then to survey the alleged difficulties in set theory. Of course, any philosopher hates ever to say briefly, let alone superficially, what his own view on any topic is (although he is delighted to give such a statement to the view of any philosopher with whom he disagrees), because a superficial statement may make his view seem naïve or even downright stupid. But such a statement is a great help to others, at least in getting an initial orientation, and for that reason I shall accept the risk involved.

In my view, the chief characteristic of mathematical propositions is the very wide variety of equivalent formulations that they possess. I don't mean this in the trivial sense of cardinality: of course, every proposition possesses infinitely many equivalent formulations; what I mean is rather that in mathematics the number of ways of expressing what is in some sense the same fact (if the proposition is true) while apparently not talking about the same objects is especially striking.

The same situation does sometimes arise in empirical science, that is, the situation that what is in some sense the same fact can be expressed in two strikingly different ways, the most famous example being wave–particle duality in quantum mechanics. Reichenbach coined the happy expression 'equivalent descriptions' for this situation. The description of the world as a system of particles, not in the classical sense but in the peculiar quantum-mechanical sense, may be associated with a different picture than the description of the world as a system of waves, again not in the classical sense but in the quantum-mechanical sense; but the two theories are thoroughly inter-translatable, and should be viewed as having the same physical content. The

MATHEMATICS WITHOUT FOUNDATIONS

same fact can be expressed either by saying that the electron is a wave with a definite wavelength λ or by saying that the electron is a particle with a sharp momentum p and an indeterminate position. What 'same fact' comes to here is, I admit, obscure. Obviously what is *not* being claimed is *synonymy* of *sentences*. It would be absurd to claim that the *sentence* 'There is an electron-wave with the wavelength λ' is *synonymous* with the *sentence* 'There is a particle electron with the momentum h/λ and a totally indeterminate position'. What is rather being claimed is this: that the two theories are compatible, not incompatible, given the way in which the theoretical primitives of each theory are now being understood; that indeed, they are not merely compatible but equivalent: the primitive terms of each admit of definition by means of the primitive terms of the other theory, and then each theory is a deductive consequence of the other. Moreover, there is no particular advantage to taking one of the two theories as fundamental and regarding the other one as *derived*. The two theories are, so to speak, on the same explanatory level. Any fact that can be explained by means of one can equally well be explained by means of the other. And in view of the systematic equivalence of statements in the one theory with statements in the other theory, there is no longer any point to regarding the formulation of a given fact in terms of the notions of one theory as more fundamental than (or even as *significantly* different from) the formulation of the fact in terms of the notions of the other theory. In short, what has happened is that the systematic equivalences between the sentences of the two theories have become so well known that they *function* virtually as synonymies in the actual practice of science.

Of course, the fact that two theories can be related in this way is not by itself either surprising or important. It would not be worth remarking that two theories are related in this way if the pictures associated with the two theories were not apparently incompatible or at least very different. In mathematics, the different equivalent formulations of a given mathematical proposition do not call to mind apparently *incompatible* pictures as do the different equivalent formulations of the quantum theory, but they do sometimes call to mind radically different pictures, and I think that the way in which a given philosopher of mathematics proceeds is often determined by which of these pictures he has in mind, and this in turn is often determined by which of the equivalent formulations of the mathematical propositions with which he deals he takes as primary.

Of the many possible 'equivalent descriptions' of the realm of mathematical facts, there are two which seem to me to have especial importance. I shall refer to these, somewhat misleadingly, I admit, by the titles 'mathematics as modal logic' and 'mathematics as set theory'. The second, I take it, needs no explanation. Everyone is today familiar with the conception of mathematics as the description of a 'universe' of 'mathematical objects'—and, in particular, with the conception of mathematics as describing relations among *sets*. However, the picture would not be significantly different if one said 'sets and numbers'—that numbers can themselves be 'identified' with sets seems today a matter of minor importance; the important thing about the picture is that mathematics describes 'objects'. The other conception is less familiar, and I shall say a few words about it.

Consider the assertion that there is a counter-example to Fermat's 'last theorem'; that is, that there is an nth power which is the sum of two nth powers, $2 < n$, all three numbers positive. Abbreviate the standard formula that expresses this statement in first-order arithmetic as '~ *Fermat*'. If ~ *Fermat* is provable, then, in fact, ~ *Fermat* is provable already from a certain easily specified finite subset of the theorems of first-order arithmetic. (NB, this is owing to the fact that it takes only one counter-example to refute a generalization. So the portion of first-order arithmetic in which we can prove all true statements of the form $x^n + y^n \neq z^n$, x, y, z, n *constant* integers, is certainly strong enough to *disprove* Fermat's last theorem if the last theorem be false, notwithstanding the fact that *all* of first-order arithmetic may be too weak to *prove* Fermat's last theorem if the last theorem be true. And the portion of first-order arithmetic just alluded to is known to be finitely axiomatizable.) Let '*AX*' abbreviate the conjunction of the axioms of the finitely axiomatizable subtheory of first-order arithmetic just alluded to. Then Fermat's last theorem is *false* just in case '*AX* ⊃ ~ *Fermat*' is valid, that is, just in case

(1) $\qquad\qquad\qquad \Box(AX \supset \sim Fermat).$

Since the truth of (1), in case (1) *is* true, does not depend upon the meaning of the arithmetical primitives, let us suppose these to be replaced by 'dummy letters' (predicate letters). To fix our ideas, imagine that the primitives in terms of which AX and ~ *Fermat* are written are the two three-term relations 'x is the sum of y and z' and 'x is the product of y and z' (exponentiation is known to be first-order-definable from these, and so, of course, are *zero* and *successor*). Let AX(S, T) and ~ FERMAT(S, T) be like AX and ~ *Fermat* except for containing the 'dummy' triadic predicate letters S, T, where AX and ~ *Fermat* contain the constant predicates 'x is the sum of y and z' and 'x is the product of y and z'. Then (1) is essentially a truth of pure modal logic (if it is true), since the constant predicates occur 'inessentially'; and this can be brought out by replacing (1) by the abstract schema:

(2) $\qquad\qquad\qquad \Box\,[\text{AX}(S, T) \supset \sim \text{FERMAT}(S, T)]$

—and this is a schema of pure first-order modal logic.

Now then, the mathematical content of the assertion (2) is certainly the same as that of the assertion that *there exist numbers x, y, z, n* ($2 < n$, $x, y, z \neq 0$) such that $x^n + y^n = z^n$. Even if the expressions involved are not synonymous, the mathematical equivalence is so obvious that they might as well be synonymous, as far as the mathematician is concerned. Yet the pictures in the mind called up by these two ways of formulating what one might as well consider to be the same mathematical assertion can be quite different. When one speaks of the 'existence of numbers', one gets the picture of mathematics as describing eternal objects; while (2) simply says that AX(S, T) entails FERMAT(S, T), no matter how one may interpret the predicate letters 'S' and 'T', and this scarcely seems to be about 'objects' at all. Of course, one can strain after objects if one wants. One can, for example, interpret the dummy letters 'S' and 'T' as quantifiers over 'the totality of all properties', if one wishes. But this is hardly necessary, since one can find a particular substitution instance of (2), even in a nominalistic language (apart from the '$\Box$') which is equivalent to (2) (just choose predicates S^* and T^* to put for S and T such that it is not mathematically impossible that the objects in their field should form an ω-sequence, and such that, if the objects in their field did form an ω-sequence, S^* would be isomorphic to addition of integers, and T^* to multiplication, in the obvious sense). Or one can interpret '$\Box$' as a predicate of statements, rather than as a statement connective, in which case what (2) asserts is that a certain object, namely, the statement 'AX(S, T) $\supset$ ~ FERMAT(S, T)' has a certain property ('being necessary'). But still, the only 'object' this commits us to is the statement 'AX(S, T) $\supset$ ~ FERMAT(S, T)', and one has to be pretty compulsive about one's nominalistic cleanliness to scruple about *this*. In short, if one fastens on the first picture (the 'object' picture), then mathematics is wholly extensional, but presupposes a vast totality of eternal objects; while if one fastens on the second picture (the 'modal' picture), then mathematics has *no* special objects of its own, but simply tells us what follows from what. If 'Platonism' has appeared to be *the* issue in the philosophy of mathematics of recent years, I suggest that it is because we have been too much in the grip of the first picture.

So far I have only indicated how one very special mathematical proposition can be treated as a statement involving modalities, but not special objects. I believe that, by making a more complex and iterated use of modal notions, one can analyse the notion of *a standard model for set theory*, and thus extend the objects–modalities duality that I am discussing to the whole of classical mathematics. I shall not show this now; but, needless to say, I would not deal at such length with this one special example if I did not believe it to represent, in some sense, the general situation. For the moment, I shall ask you to accept it on faith that this extension to the general case can be carried out.

What follows, I believe, is that each of these two ways of looking at mathematics can be used to clarify the other. If one is puzzled by the modalities (and I am concerned here with necessity in Quine's narrower sense of logical validity, excluding necessities that depend on alleged synonymy relations in natural languages), then one can be helped by the set-theoretic notion of a *model* (necessity = truth in all models; possibility = truth in some model). On the other hand, if one is puzzled by the question recently raised by Benacerraf: how numbers can be 'objects' if they have *no* properties except order in a particular ω-sequence, then, I believe, one can be helped by the answer: call them 'objects', if you like (they *are* objects, in the sense of being things one can quantify over); but remember that these objects have the special property that each fact about them is, in an equivalent formulation, simply a fact about *any* ω-sequence. 'Numbers exist'; but all this comes to, for mathematics anyway, is that (1) ω-sequences are *possible* (mathematically speaking); and (2) there are *necessary* truths of the form 'if α is an ω-sequence, then . . .' (whether any *concrete* example of an ω-sequence exists or not). Similarly, there is not, from a mathematical point of view, any significant difference between the assertion that *there exists a set of integers* satisfying an arithmetical condition and the assertion that *it is possible to select* integers so as to satisfy the condition. Sets, if you will forgive me for parodying John Stuart Mill, are permanent possibilities of selection.

THE QUESTION OF DECIDABILITY

The sense that there is a 'crisis in the foundations' of mathematics has many sources. Morris Kline cites the development of non-Euclidean geometry (which shook the idea that the axioms of a mathematical discipline must be *truths*), the lack of a consistency proof for mathematics, and the lack of a universally acceptable solution to the antinomies. In addition to these, one might mention Gödel's theorem (Kline does mention it, in fact, in connection with the consistency problem). For Gödel's theorem suggests that the truth or falsity of some mathematical statements might be impossible in principle to ascertain, and this has led some to wonder if we even know what we mean by 'truth' and 'falsity' in such a context.

Now, the example of non-Euclidean geometry does show, I believe, that our notions of what is 'self-evident' have to be subject to revision, not just in the light of new observations, but in the light of new *theories*. The intuitive evidence for the proposition that two lines cannot be a constant distance apart for half their length (i.e., in one half-plane) and then start to approach each other (as geodesics can in general relativity, e.g., light rays which come in from infinity parallel and then approach each other as they pass on opposite sides of the sun) is as great as the intuitive evidence for the axioms of number theory. I believe that under certain circumstances revisions in the axioms of arithmetic, or even of propositional calculus (e.g., the adoption of a modular logic as a way out of the difficulties in quantum mechanics), is fully conceivable. The philosophical ploy which consists in saying 'then terms would have changed meaning' is uninteresting—except as a question in the philosophy of linguistics, of course—unless one can show that in their 'old meaning' the sentences of the theory in question can still (after the transition to non-Euclidean geometry, or non-Archimedean arithmetic, or modular logic) be admitted to have formerly expressed propositions that are clear and true. If in some sense there are 'Euclidean straight lines' in our space, then the transition to, say, Riemannian geometry *could* (not necessarily *should*) be regarded as a mere 'change of meaning'. But (1) there are *no* curves in space (if the world is Riemannian) that satisfy Euclid's theorems about straight lines; and (2) even if the world is Lobatchevskian, there are no *unique* such curves—to choose any particular remetricization which leads to Euclidean geometry and say '*This* is what "distance", "straight line", etc., *used* to mean' would be arbitrary. In short, the price one pays for the adoption of non-Euclidean geometry is to deny that there are *any* propositions which might *plausibly* have been in the minds of the people who believed in Euclidean geometry and which are simultaneously clear and true. Similarly, if one accepts the interpretation of quantum mechanics that is based on modular logic, then one has to deny that there has been a change in the meaning of the relevant sentences, or else deny that there are any unique propositions which might have been in the minds of those who formerly used those sentences and which were both clear and true. You can't have your conceptual revolution and minimize it too!

Yet all this does not, I think, mean that there is a crisis in the foundations of mathematics. It does not even mean that mathematics becomes an empirical science in the ordinary sense of that term. For the chief characteristic of empirical science is that for each theory there are usually alternatives in the field, or at least alternatives struggling to be born. As long as the major parts of classical logic and number theory and analysis have no alternatives in the field—alternatives which require a change in the axioms and which effect the simplicity of total science, including empirical science, so that a choice has to be made—the situation will be what it has always been. We will be justified in accepting classical propositional calculus or Peano number theory not because the relevant statements are 'unrevisable in principle', but because a great deal of science presupposes these statements, and because no real alternative is in the field. Mathematics, on this view, does become 'empirical' in the sense that one is allowed to try to *put* alternatives into the field. Mathematics can be wrong, and not just in the sense that the proofs might be fallacious or that the axioms might not (if we reflected more deeply) be really self-evident. Mathematics (or, rather, some mathematical theory) might be wrong in the sense that the 'self-evident' axioms might be false, and the axioms that are true might not be 'evident' at all. But this does not make the pursuit of truth impossible in mathematics, any more than it has in empirical science; nor does it mean that we should not trust our intuitions when we have nothing better to go on. After all, a mathematical theory that has become the basis of a successful and powerful scientific system, including many important empirical applications, is not being accepted *merely* because it is 'intuitive', and if someone objects to it, we have the right to say 'Propose something better!' What this does do, rather, is make the 'foundational' view of mathematical knowledge as suspect as the 'foundational' view of empirical knowledge (if one cares to retain the 'mathematical–empirical' distinction at all).

Again, I cannot weep bitter tears about the lack of a consistency proof for classical mathematics. Even if such a proof were possible, it would only be a development within mathematics and not a foundation for mathematics. Not only would it be possible to raise philosophical questions about the branch of mathematics that was used for the consistency proof; but, in any case, science demands much more of a mathematical theory than that it should merely be *consistent*, as the example of the various alternative systems of geometry already dramatizes.

The question of the significance of the antinomies, and of what to do about the existence of several different approaches to overcoming them, is far more difficult. I propose to defer this question for a moment and to consider first the significance of Gödel's theorem and, more generally, of the existence of mathematically undecidable propositions.

Strictly speaking, all Gödel's theorem shows is that, in any particular consistent axiomatizable extension of certain finitely axiomatizable subtheories of Peano arithmetic, there are propositions of number theory that can neither be proved nor disproved. (I think it is fair to call this 'Gödel's theorem', even though this statement of it incorporates strengthenings due to Rosser and Tarski, Mostowski, and Robinson.) It does not follow that any proposition of number theory is, in some sense, absolutely undecidable. However, it may well be the case that some proposition of elementary number theory is neither provable nor refutable in any system whose axioms rational beings will ever have any good reason to accept. This has caused some to doubt whether every mathematical proposition, or even every proposition of the elementary theory of numbers, can be thought of as having a truth value.

A similar consideration is raised by Paul Cohen's recent work in set theory, when that work is taken together with Gödel's classical relative consistency proof of the axiom $V = L$ (which implies the axiom of choice and the generalized continuum hypothesis). Together, these results of Gödel and Cohen establish the full independence of the continuum hypothesis (for example) from the other axioms of set theory, assuming those other axioms to be consistent. A striking feature of both proofs is their invariance under small (or even moderately large) perturbations of the axioms. It appears quite possible today that no decisive consideration will ever appear (such as a set-theoretic axiom we have 'overlooked') which will reveal that a system in which the continuum hypothesis is provable is the correct one, and that no consideration will ever appear which will reveal that a system in which the continuum hypothesis is refutable is the correct one. In short, the truth value of the continuum hypothesis—assuming it has a truth value—may be undiscoverable by rational beings, or at least by the 'rational beings' that actually do exist, or ever will exist. Then, what reason is there to think that it has a truth value?

This 'argument' is sometimes taken to show that the notion of a set is unclear. For, since the argument 'shows' (sic!) that the continuum hypothesis has no truth value and the continuum hypothesis involves the concept of a set, the only plausible explanation of the truth-value failure is some unclarity in the notion of a set. (It would be an interesting exercise to find *all* the faults in this particular bit of reasoning. It is horrible, isn't it?)

The first point to notice is that the existence of propositions whose truth value we have no way of discovering is not at all peculiar to mathematics. Consider the assertion that there are infinitely many binary stars (considering the entire space-time universe, i.e., counting binary stars past, present, and future). It is not at all clear that we can discover the truth value of this assertion. Sometimes it is argued that such an assertion is 'verifiable (or at least confirmable) in principle', because it may *follow from a theory*. It is true that in one case we can discover the truth value of this proposition. Namely, if either it or its negation is derivable from laws of nature that we can confirm, then its truth value can be discovered. But it could just happen that there are infinitely many binary stars, without this being required by any law. Moreover, the distribution might be quite irregular, so that ordinary statistical inference could not discover it. Indeed, at some point I cease to understand the question 'Is it always possible *in principle* to discover the truth value of this proposition?—for the methods of inquiry permitted ('inductive' methods) are just too ill-defined a set. But I suspect that, given any *formalizable* inductive logic, one could describe a logically possible world in which (1) there were infinitely many binary stars; and (2) one could never discover this fact using that inductive logic. (Of course, the argument that the proposition is 'confirmable in principle' because it could follow from a theory does not even purport to show that in every possible world the truth or falsity of this statement could be induced from a finite amount of observational material using some inductive method; rather, it shows that in *some* possible world the truth of this statement (or its falsity) could be induced from a finite amount of observational material.) Yet I, for one, see no reason—not even a prima-facie one—to suspect that this proposition does not have a truth value. Why *should* all truths, even all empirical truths, be discoverable by probabilistic automata (which is what I suspect we are) using a finite amount of observational material? Why does the fact that the truth value of a proposition may be undiscoverable by us suggest to some philosophers—indeed, why does it count as a *proof* for some philosophers—that the proposition in question doesn't *have* a truth value? Surely, some kind of idealistic metaphysics must be lurking in the underbrush!

What is even more startling is that philosophers who would agree with me with respect to propositions about material objects should feel differently about propositions of mathematics. (Perhaps this is due to the pernicious tendency to think of mathematics solely in terms of the mathematical objects picture. If one doesn't understand the nature of these objects—that is, that they don't have a 'nature', that talk about them is equivalent to talk about what is impossible—then talk about them may seem like a form of theology, and if one is anti-theological, that may be a reason for rejecting mathematics as a make-believe.) Surely, the *mere* fact that we may never know whether the continuum hypothesis is true or false is by itself just *no* reason to think that it doesn't have a truth value!

'But what does it *mean* to say that the continuum hypothesis is true?' someone will ask. It means that if S is a set of real numbers, and S is not finite and not denumerably infinite, then S can be put in one-to-one correspondence with the unit interval. Or, equivalently, it means that the sentence I have just written holds in any standard model for fourth-order number theory (actually, it can be expressed in third-order number theory). 'But what is a *standard model*?' It is one with the properties that (1) the 'integers' of the model form an ω-sequence under the < of the model—that is, it is not *possible* to select positive 'integers' $a_1, a_2, a_3, \ldots$ from the model so that, for all i, $a_{i+1} < a_i$—and (2) the model is maximal with this property—that is, it is not *possible* to add more 'sets' of 'integers' or 'sets of sets' of 'integers' or 'sets of sets of sets' of 'integers' to the model. (This last explanation contains the germ of the idea which is used in expressing the notion of a 'standard model' in modal-logical, as opposed to set-theoretic, language.)

I think that one can see what is going on more clearly if we imagine, for a moment, that physics has discovered that the physical universe is finite in both space and time and that all physical magnitudes are discrete (finiteness 'in the small'). That this is a possibility we must take into account was already emphasized by Hilbert in his famous article on the infinite—it may well be, Hilbert pointed out, that we cannot argue for the consistency of any theory whose models are all infinite by arguing that physical space, or physical time, or anything else physical, provides a model for the theory, since physics is increasingly tending to replace infinities and continuities by finites and discretes.

If the whole physical universe is thoroughly finite, both in the large and in the small, then the statement '$10^{100} + 1$ is a prime number' may be one whose truth value we can never know. For, if the statement is true (and even intuitionist mathematicians regard this decidable statement as possessing a truth value), then to verify that it is true by using any sieve method might well be physically impossible. And, if the shortest proof from axioms that rational beings will ever have any reason to accept is too long to be physically written out, then it might be physically impossible for being to whom only those things are 'evident' that are in fact 'evident' (or ever will be 'evident' or that we will ever in fact have good reason to believe) to know that the statement is true.

Now, although many people doubt that the continuum hypothesis has a truth value, everyone believes that the statement '$10^{100} + 1$ is a prime number' has a truth value. Why? 'Because the statement is decidable.' But what does that mean, 'the statement is decidable'? It means that it is *possible* to try out all the pairs of possible factors and see if any of them 'work'. It means that it is *possible* to decide the statement. Thus, the man who asserts that this statement is decidable, is simply making an assertion of mathematical possibility. Moreover, he believes that just one of the two statements

(3) If all pairs n, m ($n, m < 10^{100} + 1$) were 'tried' by actually computing the product nm, then in some case the product would be found to equal $10^{100} + 1$.

(4) If all pairs n, m ... (same as in (3)), then in no case would the product be found to equal $10^{100} + 1$.

expresses a *necessary* truth, although it may be *physically* impossible to discover which one. Yet this same mathematician or philosopher, who is quite happy in this context with the notion of mathematical possibility (and who does not ask for any nominalistic reduction) and who treats mathematical necessity as well-defined in this case, for a reason which is essentially circular, regards it as 'platonistic' to suppose that the continuum hypothesis has a truth value.[1] I realize that this is an *ad hominem* argument, but still—if there

---

[1] Incidentally, it may also be 'platonism' to treat statements of physical possibility or counterfactual conditionals as well-defined. For (1) 'physical possibility' is *compatibility* with the laws of nature. But the relation of compatibility is interdefinable with the modal notions of possibility and necessity, and, of course, the laws of nature themselves require many mathematical notions for their

is such an intellectual sin as 'platonism' (and it is remarkably unclear what this supposed sin consists of), why is it not already to commit it, if one supposes that '$10^{100} + 1$ is a prime number' has a truth value, even if no nominalistic reduction of this statement can be offered? (When one is defending a common-sense position, very often the only argument is *ad hominem*—for one has to keep throwing the burden of the argument back to the other side, by asking to be told *precisely* what is 'unclear' about the notions being attacked, or why a 'reduction' of the kind being demanded is necessary, or why a 'foundation' for the science in question is needed.)

In passing, I should like to remark that the following two principles, which many people seem to accept, can be shown to be inconsistent, by applying the Gödel theorem:

(I) That, even if some arithmetical (or set-theoretical) statements have no truth value, still, to say of any arithmetical (or set-theoretical) statement that it has (or lacks) a truth value is itself always either true or false (i.e., the statement either has a truth value or it doesn't).

(II) All and only the decidable statements have a truth value.

For the statement that a mathematical statement S is decidable may itself be undecidable. Then, by (II), it has no truth value to say 'S is decidable'. But, by (I), it has a truth value to say 'S has a truth value' (in fact, *falsity*, since if S has a truth value, then S is decidable, by (II), and, if S is decidable, then 'S is decidable' is also decidable). Since it is false (by the previous parenthetical remark) to say 'S has a truth value', and since we accept the equivalence of 'S has a truth value' and 'S is decidable', then it must also be *false* to say 'S is decidable'. But it has no truth value to say 'S is decidable'. Contradiction.

THE SIGNIFICANCE OF THE ANTINOMIES

The most difficult question in the philosophy of mathematics is, perhaps, the question raised by the antinomies and by the plurality of conflicting set theories. Part of the paradox is this: the antinomies do not at all seem to affect the notion 'set of material objects' or the notion 'set of integers' or the notion 'set of sets of integers', etc. Yet they *do* seem to affect the notion '*all* sets'. How are we to understand this situation?

One way out might be this: to conclude that we understand the notion 'set' in some contexts (e.g., 'set of integers', 'set of sets of integers'), but to

---

statement. (2) A counterfactual conditional is true just in case the consequent *follows* from the antecedent, together with certain other statements that hold both in the actual and in the hypothetical world under consideration. And, of course, no nominalistic reduction has ever succeeded, either for the notion of physical possibility or for the subjunctive conditional.

conclude that we do not understand it in the context 'all sets'. But we do seem to understand *some* statements about all sets, for example, 'For every set x and every set y, there is a set z which is the union of x and y'. So must we really abandon hope of making sense of the locution 'all sets'?

It is at this point that I urge we attend to the objects–modalities duality that I pointed out a few pages ago. The notion of a set has been used by a number of authors to clarify the notions of mathematical possibility and necessity. For example, if we identify the notion of a 'possible world' with the notion of a model (or, more correctly, with the notion of a structure of the appropriate type), then the rationale of the modal system S5 can easily be explained (as, for instance, by Carnap in *Meaning and Necessity*), and this explanation can be extended to the case of quantified modal logic by methods due to Kripke, Hintikka, and others. Here, however, I wish to go in the reverse direction, and assuming that the notions of mathematical possibility and necessity are clear (and there is no paradox associated with the notion of necessity as long as we take the '$\Box$' as a statement connective (in the degenerate sense of 'unary connective') and not—in spite of Quine's urging—as a predicate of sentences), I wish to employ these notions to try to give a clear sense to talk about 'all sets'.

My purpose is not to start a *new* school in the foundations of mathematics (say, 'modalism'). Even if in some contexts the modal logic picture is more helpful than the mathematical objects picture, in other contexts the reverse is the case. Sometimes we have a clearer notion of what 'possible' means than of what 'set' means; in other cases the reverse is true; and in many, many cases both notions seem as clear as notions ever get in science. Looking at things from the standpoint of many different 'equivalent descriptions', considering what is suggested by *all* the pictures is both a healthy antidote to foundationalism and of real heuristic value in the study of first-order scientific questions.

Now, the natural way to interpret set-theoretic statements in the modal-logical language is to interpret them as statements of what would necessarily be the case if there were standard models for the set theories in question. Since the models for von Neumann–Bernays set theory and its strengthenings (e.g., the system recently proposed by Bernays) are also models for Zermelo set theory, let me concentrate on Zermelo set theory. In order to 'concretize' the notion of a model, let us think of a model as a graph. The 'sets' of the model will then be pencil points (or some higher-dimensional analogue of pencil points, in the case of models of large cardinality), and the relation of membership will be indicated by 'arrows'. (I assume that there is nothing inconceivable about the idea of a physical space of arbitrarily high cardinality; so models of this kind need not necessarily be denumerable, and may even be standard.) Such a model will be called a 'concrete model' (or a 'standard concrete model' if it be standard) for Zermelo set theory. The model will be called standard if (1) there are no infinite-descending 'arrow' paths; and (2) it is not possible to extend the model by adding more 'sets' without adding to the number of 'ranks' in the model. (A 'rank' consists of all the sets of a given—possibly transfinite—type. 'Ranks' are cumulative types; i.e., every set of a given rank is also a set of every higher rank. It is a theorem of set theory that every set belongs to some rank.) A statement that refers only to sets of less than some given rank—say, to sets of rank less than $\omega X2$—will be called a statement of 'bounded rank'. I ask the reader to accept it on faith that the statement that a certain graph G is a *standard* model for Zermelo set theory can be expressed using no 'non-nominalistic' notions except the '$\square$'.

If S is a statement of bounded rank and if we can characterize the 'given rank' in question in some invariant way (invariant with respect to standard models of Zermelo set theory), then the statement S can easily be translated into modal-logical language. The translation is just the statement that if G is any standard model for Zermelo set theory—that is, any standard concrete model—and G contains the invariantly characterized rank in question, then necessarily S holds in G. (It is trivial to express 'S holds in G' for any *particular* S without employing the set-theoretic notion of 'holding'.) Our problem, then, is how to translate statements of *un*bounded rank into modal-logical language.

The method is best indicated by means of an example. If the statement has the form $(x)(\exists y)(z)Mxyz$, where M is quantifier-free, then the translation is this:

[If G is any standard concrete model for Zermelo set theory and if P is any point in G, then it is possible that there is a graph G' that extends G (i.e., G is a subgraph of G') and a point y in G' such that G' is a standard concrete model for Zermelo set theory and such that
(if G'' is any graph that extends G' and such that G'' is a standard concrete model for Zermelo set theory and if z is any point in G'', then $Mxyz$ holds in G'')].

Obviously this method can be extended to an arbitrary set-theoretic statement.

So much for technical matters. I apologize for this brief lapse into technicality, but actually this was only the merest sketch of the technical development, and this much detail is necessary for my discussion. The real question is this: what, if any, is the philosophical significance of such translations?

If there be any philosophical significance to such translations—and I don't claim a great deal—it lies in this: I did not assume that any standard concrete model for Zermelo set theory is maximal. Indeed, I would be inclined to say that no concrete model could be maximal—nor any *non*-concrete model either, as far as that goes. Even God could not make a model for Zermelo set theory that it would be *mathematically* impossible to extend, and no matter what 'stuff' he might use. Yet I succeeded in giving a clear sense to statements about 'all sets' (clear relative to the notions I assumed to start with) *without* assuming a maximal model. In metaphysical language, it is not necessary to think of sets as one system of objects in some one possible world in order to follow assertions about all sets.

Furthermore, in construing statements about sets as statements about standard concrete models for set theory, I did not introduce possible concrete models (or even possible worlds) as objects. Introducing the modal connectives '□', '◊', '⊰' is not introducing new kinds of objects, but rather extending the kinds of things we can say about ordinary objects and sorts of objects. (Of course, one *can* construe the statement that it is possible that there is a graph G satisfying a condition C as meaning that *there exists a possible graph G satisfying the condition C*; that is one way of smoothing the transition from the modal logic picture to the mathematical objects picture.)

The importance of Zermelo set theory and of the other set theories based upon the notion of 'rank' lies in this: we have a strong intuitive conviction that whenever A's are possible, so is a structure that we might call 'the family of all sets of A's'. Zermelo set theory assumes only this intuition and the intuition that the process of unioning such structures can be extended into the transfinite. Of course, this intuitive conviction *may* be mistaken; it could turn out that Zermelo set theory has no standard models (even if Zermelo set theory is consistent—e.g., the discovery of an ω-inconsistency would show that there are no standard models). But so could the intuitive conviction upon which number theory is based be mistaken. If we wish to be cautious, we can assume only predicative set theory up to some 'low' transfinite type. (It is necessary to extend predicative type theory 'just past' the constructive ordinals if we wish to be able to define *validity* of schemata that contain the quantifiers 'There are infinitely many x such that' and 'There are at most a finite number of x such that', e.g.) Such a weak set theory may well give us all the sets we need for physics, and also the basic notions of validity and satisfiability that we need for logic, as well as arithmetic and a weak version of classical analysis. But the fact that we do have an intuitive conviction that standard models of Zermelo set theory, or of other set theories based upon the notion of 'rank', are *mathematically possible structures* is a perfectly good reason for asking what statements necessarily hold in such structures—for example, for asking whether the continuum hypothesis necessarily holds in such structures.

The real significance of the Russell paradox, from the standpoint of the modal logic picture, is this: it shows that *no* concrete structure can be a standard model for the naïve conception of the totality of all sets; for any concrete structure has a possible extension that contains more 'sets'. (If we identify sets with the points that represent them in the various possible concrete structures, we might say: it is not possible for all *possible* sets to exist in any one world!) Yet set theory does not become impossible. Rather, set theory becomes the study of what must hold in, for example, any standard model for Zermelo set theory.

IX

THE CONSISTENCY OF FREGE'S *FOUNDATIONS OF ARITHMETIC*

GEORGE BOOLOS

Is Frege's *Foundations of Arithmetic* inconsistent? The question may seem to be badly posed. The *Foundations*, which appeared in 1884, contains no formal system like those found in Frege's *Begriffsschrift* (1879) and *The Basic Laws of Arithmetic* (vol. 1, 1893; vol. 2, 1903). As is well known, Russell showed the inconsistency of the system of the *Basic Laws* by deriving therein what we now call 'Russell's paradox'. The system of the *Begriffsschrift*, on the other hand, can plausibly be reconstructed as an axiomatic presentation of second-order logic, which is therefore happily subject to the usual consistency proof, consisting in the observation that the universal closures of the axioms and anything derivable from them by the rules of inference are true in any one-element model.[1] Since the *Foundations* contains no formal system at all, our question may be thought to need rewording before an answer to it can be given.

One might nevertheless think that, however reworded and badly posed or not, it must be answered yes. The *Basic Laws*, that is, the system thereof, *is* inconsistent, and is widely held to be a formal elaboration of the mathematical programme outlined in the earlier *Foundations*, which contains a more thorough development of its programme than one is accustomed to find in programmatic works. Thus the inconsistency which Russell found in the later book must have been latent in the earlier one.

Moreover, the characteristic signs of inconsistency can be found in the use Frege makes in the *Foundations* of the central notions of 'object', 'concept', and 'extension'. Objects fall under concepts, but some extensions—numbers, in particular and crucially—contain concepts, and these extensions

---

First published in Judith Jarvis Thomson (ed.), *On Being and Saying: Essays for Richard Cartwright* (Cambridge, Mass.: MIT Press, 1987), 3–20. Reproduced by permission of MIT Press.

[1] I. S. Russinoff, 'Frege's Problem about Concepts', Ph. D. diss., MIT, Department of Linguistics and Philosophy, 1983.

themselves are objects, according to Frege. Thus, although a division into two types of entity, concepts and objects, can be found in the *Foundations*, it is plain that Frege uses not one but two instantiation relations, 'falling under' (relating some objects to some concepts) and 'being in' (relating some concepts to some objects), and that both relations sometimes obtain reciprocally: the number 1 is an object that falls under 'identical with 1', a concept that is in the number 1. Even more ominously (because of the single negation sign), the number 2 does not fall under 'identical with 0 or 1', which is in 2. Thus the division of the *Foundations'* entities into two types would appear to offer little protection against Russell's paradox.

It is not only Russell's paradox that threatens. Recall that Frege defines 0 as the number belonging to the concept 'not identical with itself'.[2] If there is such a number, would there not also have to be a number belonging to the concept 'identical with itself', a *greatest* number? Cantor's paradox also threatens.

It is therefore quite plausible to suppose that it is merely through its lack of formality that the *Foundations* escapes outright inconsistency and that, when suitably formalized, the principles employed by Frege in the *Foundations* must be inconsistent.

This plausible and, I suspect, quite common supposition is mistaken, as we shall see. Although Frege freely assumes the existence of needed concepts at every turn, he by no means avails himself of extensions with equal freedom. With one or two insignificant but possibly revealing exceptions, which I discuss later, the *only* extensions whose existence Frege claims in the central sections of the *Foundations* are the extensions of higher-level concepts of the form 'equinumerous with concept F'. (I use the term 'equinumerous' as the translation of Frege's *gleichzahlig*.) It turns out that the claim that such extensions exist can be consistently integrated with existence claims for a wide variety of first-level concepts in a way that makes possible the execution of the mathematical programme described in sections 68–83 of the *Foundations*. Indeed, I shall now present a formal theory, FA ('Frege Arithmetic') that captures the whole content of these central sections and for which a simple consistency proof can be given, one that shows *why* FA is consistent.

FA is a theory whose underlying logic is standard axiomatic second-order logic written in the usual Peano–Russell logical notation. FA could have been presented as an extension of the system of Frege's *Begriffsschrift*. Indeed, there is some evidence that Frege thought of himself as translating *Begriffsschrift* notation into the vernacular when writing the *Foundations*. Not only

---

[2] Plurals find happy employment here, as elsewhere in the discussion of concepts: e.g., instead of 'the number belonging to the concept "horse"', one can say 'the number of horses'. 0 is thus *defined* by Frege to be the number of things that are not self-identical. And Frege was right!

does the later work abound with allusions and references to the earlier, along with repetitions of claims and arguments for its significance; when Frege defines the ancestral in section 79, he uses the variables x, y, d, and F in exactly the same logical roles they had played in the *Begriffsschrift*.

FA is a system with three sorts of variable: first-order (or object) variables $a, b, c, d, m, n, x, y, z, \ldots$; unary second-order (or concept) variables $F, G, H, \ldots$; and binary second-order (or relation) variables $\phi, \psi, \ldots$. The sole non-logical symbol of the language of FA is η, a two-place predicate letter attaching to a concept variable and an object variable. (η is intended to be reminiscent of $\in$, and may be read 'is in the extension'. Frege's doctrine that extensions are objects receives expression in the fact that the second-argument place of η is to be filled by an object variable.) Thus the atomic formulas of FA are of the forms Fx (F a concept variable), $x\phi y$, and $F\eta x$. Formulas of FA are constructed from the atomic formulas by means of propositional connectives and quantifiers in the usual manner.

Identity can be taken to have its standard second-order definition: $x = y$ if and only if $\forall F(Fx \leftrightarrow Fy)$. Frege endorses Leibniz's definition ('... potest substitui ... salva veritate') in section 65 of the *Foundations*, but does not actually do what he might easily have done, namely, state that Leibniz's definition of the identity of x and y can be put: y falls under every concept under which x falls (and vice versa).

The logical axioms and rules of FA are the usual ones for such a second-order system. Among the axioms we may specially mention (i) the universal closures of all formulas of the form

$$\exists F \forall x(Fx \leftrightarrow A(x)),$$

where $A(x)$ is a formula of the language of FA not containing F free; and (ii) the universal closures of all formulas of the form

$$\exists \phi \forall x \forall y(x\phi y \leftrightarrow B(x, y)),$$

where $B(x, y)$ is a formula of the language not containing ϕ free. Throughout sections 68–83 of the *Foundations* Frege assumes, and needs to assume, the existence of various particular concepts and relations. The axioms (i) and (ii) are called comprehension axioms; these will do the work in FA of Frege's concept and relation existence assumptions.

The sole (non-logical) axiom of the system FA is the single sentence

Numbers: $\forall F \exists ! x \forall G(G\eta x \leftrightarrow F \text{ eq } G)$,

where F eq G is the obvious formula of the language of FA expressing the equinumerosity of the values of F and G, namely,

$$\exists \phi [\forall y(Fy \to \exists ! z(y\phi z \wedge Gz)) \wedge \forall z(Gz \to \exists ! y(y\phi z \wedge Fy))].$$

Here the sign η is used for the relation that holds between a concept G and the extension of a (higher-level) concept under which G falls; before we used the term 'is in' for this relation and 'contains' for its converse. In section 68 Frege first asserts that F is equinumerous with G if and only if the extension of 'equinumerous with F' is the same as that of 'equinumerous with G', and then defines the number belonging to the concept F as the extension of the concept 'equinumerous with the concept F'. Since Frege, like Russell, holds that existence and uniqueness are implicit in the use of the definite article, he supposes that for any concept F, there is a unique extension of the concept 'equinumerous with F'. Thus the sentence Numbers expresses this supposition in the language of FA; it is the sole non-logical assumption[3] utilized by Frege in the course of the mathematical work done in sections 68–83.

How confident may we be that FA is consistent? Recent observations by Harold Hodes and John Burgess bear directly on this question. To explain them, it will be helpful to consider a certain formal sentence, which we shall call Hume's principle:

$\forall F\, \forall G(NF = NG \leftrightarrow F\,\text{eq}\,G)$.

Hume's principle is so called because it can be thought of as explicating a remark that Hume makes in the *Treatise* (I. iii. 1, par. 5), which Frege quotes in the *Foundations*:

> We are possest of a precise standard by which we can judge of the equality and proportion of numbers. . . . When two numbers are so combined, as that the one has always an unite answering to every unite of the other, we pronounce them equal.

The symbol N in Hume's principle is a function sign which, when attached to a concept variable, makes a term of the same type as object variables; thus $NF = NG$ and $x = NF$ are well-formed. Taking $N \ldots$ as abbreviating 'the number of $\ldots$ s', we may read Hume's principle: the number of F's is the number of G's if and only if the F's can be put into one–one correspondence with the G's. (As Hume said, more or less.)

In his article 'Logicism and the Ontological Commitments of Arithmetic',[4] Hodes observes that a certain formula, which he calls '(D)' is satisfiable. He writes:

(D) $\begin{array}{c}\forall X \exists x \\ \forall Y \exists y\end{array} (x = y \leftrightarrow X\,\text{eq}\,Y)$

is satisfiable. In fact if we accept standard set theory, it's true.

(I have replaced Hodes's '$(Q_E z)(Xz, Yz)$' by 'X eq Y'. The label '(D)' is missing from the text of his article.) Branching quantifiers, which are notoriously

---

[3] It is non-logical by my lights, though not, of course, by Frege's.
[4] *Journal of Philosophy*, 81/3 (1984): 138.

hard to interpret, may always be eliminated in favour of ordinary function quantifiers. Eliminating them from (D) yields the formula $\exists N \exists M \forall X \forall Y$ ($NX = MY \leftrightarrow X$ eq Y). Now (D) is satisfiable if and only if Hume's principle is satisfiable. For if (the function quantifier equivalent of) (D) holds in a domain U, then for some functions N, M, $\forall X \forall Y(NX = MY \leftrightarrow X$ eq $Y)$ holds in U; $\forall Y(Y$ eq $Y)$ holds in U, so does $\forall Y(NY = MY)$, and therefore so does Hume's principle $\forall X \forall Y(NX = NY \leftrightarrow X$ eq $Y)$. Conversely, Hume's principle implies (D). Thus a bit of deciphering enables us to see that Hodes's claim is tantamount to the assertion that Hume's principle is satisfiable.

Hodes gives no proof that (D), or Hume's principle, is satisfiable. But Burgess, in a review of Crispin Wright's book *Frege's Conception of Numbers as Objects*,[5] shows that it is. He writes:

Wright shows why the derivation of Russell's paradox cannot be carried out in N^- [Wright's system, obtained by adjoining a version of Hume's principle to second-order logic], and ought to have pointed out that the system is pro[v]ably consistent. (It has a model whose domain of objects consists of just the cardinals zero, one, two, . . . and aleph-zero.)[6]

It will not be amiss to elaborate this remark. To produce a model M for Hume's principle that also verifies all principles of axiomatic second-order logic, take the domain U of M to be the set $\{0, 1, 2, \ldots, \aleph_0\}$. To ensure that M is a model of axiomatic second-order logic, take the domain of the concept variables to be the set of all subsets of U, and similarly take the domain of the binary (or, more generally, n-ary) relation variables to be the set of all binary (or n-ary) relations of U, that is, the set of sets of ordered pairs (or n-tuples) of members of U.

To complete the definition of M, we must define the function f by which the function sign N is to be interpreted in M. The *cardinality* of a set is the number of members it contains. U has the following important property: *the cardinality of every subset of* U *is a member of* U. (Notice that the set of natural numbers *lacks* this property.) Thus we may define f as the function whose value for every subset V of U is the cardinality of V. We must now see that Hume's principle is true in M.

Observe that an assignment s of appropriate items to variables satisfies $NF = NG$ in M if and only if the cardinality of $s(F)$ equals the cardinality of $s(G)$ and satisfies F eq G in M if any only if $s(F)$ can be put into one–one correspondence with $s(G)$. Since the cardinality of $s(F)$ is the same as that of $s(G)$ if and only if $s(F)$ can be put into one–one correspondence with $s(G)$, every assignment satisfies ($NF = NG \leftrightarrow F$ eq G) in M, and M is a model for Hume's principle.

---

[5] Aberdeen: Aberdeen University Press, 1983.
[6] *Philosophical Review*, 93/4 (1984): 638–40. The text of the review has 'probably consistent', which is an obvious misprint.

A similar argument shows the satisfiability of Numbers. Let the domain of M again be U, and let M specify that η is to apply to a subset V of U and a member u of U if and only if the cardinality of V is u. Then Numbers is true in M. (On receiving the letter from Russell, Frege should have immediately checked into Hilbert's Hotel.)

(It may be of interest to recall the usual proof that the comprehension axioms (i) are true in standard models (like M) for second-order logic. Let $A(x)$ be a formula not containing free F, and let s be an assignment. Let C be the set of objects of which $A(x)$ is true, and let s' be just like s except that $s'(F) = C$. Since $A(x)$ does not contain free F, s' satisfies $\forall x(Fx \leftrightarrow A(x))$) and s satisfies $\exists F \forall x(Fx \leftrightarrow A(x))$). Similarly for the comprehension axioms (ii).)

There is a cluster of worries or objections that might be thought to arise at this point. Does not the appeal to the natural numbers in the consistency proof vitiate Frege's programme? How can one invoke the existence of the numbers in order to justify FA? There is a quick answer to this objection: you mean we *shouldn't* give a consistency proof? More fully: we are simply trying to use what we know in order to allay all suspicion that a contradiction is formally derivable in FA, about whose consistency anyone knowing the history of logic might well be quite uncertain. We are not attempting to show that FA is true.

But there is perhaps a more serious worry. At a crucial step of the proof of the consistency (with second-order logic) of the formal sentence called Hume's principle, we made an appeal to an informal principle connecting cardinality and one–one correspondence which can be symbolized as— Hume's principle. (We made this appeal when we said that the cardinality of $s(F)$ is the same as that of $s(G)$ if and only if $s(F)$ can be put into one–one correspondence with $s(G)$.) Should this argument then count as a *proof* of the consistency of Hume's principle? What assurance can any argument give us that a certain sentence is consistent, if the argument appeals to a principle one of whose formalizations is the very sentence we are trying to prove consistent?

The worry is by no means idle. We have attempted to prove the consistency of Hume's principle by arguing that a certain structure M is a model for Hume's principle; in proving that M is a model for Hume's principle, we have appealed to an informal version of Hume's principle. A similar service, however, can be performed for the notoriously inconsistent naïve comprehension principle $\exists y \forall x(x \in y \leftrightarrow \ldots x \ldots)$ of set theory. By informally invoking the naïve comprehension principle, we can argue that all of its instances are true under the interpretation I under which the variables range over all sets that there are and $\in$ applies to a, b if and only if b is a set and a is a member of b. Let $\ldots x \ldots$ be an arbitrary formula not containing free y. (By the naïve comprehension principle) let b be the set of just those sets satisfying ... x ... under I. Then for every a, a and b satisfy $x \in y$ under I if and only if a satisfies ... x ... under I. Therefore b satisfies $\forall x(x \in y \leftrightarrow ... x ...)$ under I, and $\exists y \forall x(x \in y \leftrightarrow ... x ...)$ is true under I. Thus I is a model of all instances of the naïve comprehension principle. (Doubtless Frege convinced himself of the truth of the fatal rule (V) of *Basic Laws* by running through some such argument.) Of course we can *now* see that, *pace* the principle, there is not always a set of just those sets satisfying ... x ... But how certain can we be that the proof of the consistency of Hume's principle and FA does not contain some similar gross (or subtle) mistake, as does the 'proof' just given of the consistency of the naïve comprehension principle?

Let us first notice that the argument can be taken to show not merely that FA is consistent, but that *it is provable in standard set theory* that FA is consistent. (Standard set theory is of course ZF, Zermelo–Fraenkel set theory.) The argument can be 'carried out' or 'replicated' *in* ZF. Thus, if FA is inconsistent, ZF is in error. (Presumably the word 'provably' in Burgess's observation refers to an informal, model-theoretic proof, which could be formalized in ZF, or to a formal ZF proof.) Thus anyone who is convinced that nothing false is provable in ZF must regard this argument as a proof that FA is consistent. Moreover, if ZF makes a false claim to the effect that FA, or any other formal theory, is consistent, then ZF is not merely in error but is itself inconsistent, for ZF will then certainly also make the correct claim that there exists a derivation of $\bot$ in FA. (Indeed, systems much weaker than ZF, for example, Robinson's arithmetic Q, will then make that correct claim.)

Something even stronger may be said. We shall show that any derivation of an inconsistency in FA can immediately be turned into a derivation of an inconsistency in a well-known theory called 'second-order arithmetic' or 'analysis', about whose consistency there has never been the slightest doubt. In the language of analysis there are two sorts of variables, one sort ranging over (natural) numbers, the other over sets of and relations on numbers. The axioms of analysis are the usual axioms of arithmetic, a sentence expressing the principle of mathematical induction ('Every set containing 0 and the successor of every member contains every natural number'), and, for each formula of the language, a comprehension axiom expressing the existence of the set or relation defined by the formula.[7] If ZF is consistent, so is analysis; but ZF is stronger than analysis, and the consistency of analysis can be proved in ZF. It is (barely) conceivable that ZF is inconsistent; but unlike ZF, analysis did not arise as a direct response to the set-theoretic antinomies, and the discovery of the inconsistency of analysis would be the most surprising

---

[7] A standard reference concerning analysis is sect. 8.5 of J. R. Shoenfield's *Mathematical Logic* (Reading, Mass.: Addison-Wesley, 1967).

mathematical result ever obtained, precipitating a crisis in the foundations of mathematics compared with which previous 'crises' would seem utterly insignificant.

Let us sketch the construction by which proofs of $\bot$ in FA can be turned into proofs of $\bot$ in analysis. The trick is to 'code' $\aleph_0$ by 0 and each natural number z by $z + 1$ so that the argument given may be replicated in analysis. It is easy to construct a formula $A(z, F)$ of the language of analysis that expresses the relation 'exactly z natural numbers belong to the set F'. Simply write down the obvious symbolization of 'There exists a one–one correspondence between the natural numbers less than z and the members of F'. Let Eta(F, x) be the formula

$[\neg \exists z A(z, F) \,\&\, x = 0] \lor [\exists z(A(z, F) \,\&\, x = z + 1)]$.

Then, since $\exists!x\,\text{Eta}(F, x)$ and

$[\exists x(\text{Eta}(F, x) \land \text{Eta}(G, x))) \leftrightarrow F \,\text{eq}\, G]$

are theorems of analysis, so is the result

$\forall F\, \exists!x \forall G(\text{Eta}(G, x) \leftrightarrow F \,\text{eq}\, G)$

of substituting Eta(G, x) for $G\eta x$ in Numbers, as the following argument, which can be formalized in analysis, shows. Let F be any set of numbers. Let x be such that Eta(F, x) holds. Let G be any set. Then Eta(G, x) holds if and only if F eq G does. And since F eq F holds, x is unique. Of course each of the comprehension axioms of FA is provable in analysis under these substitutions, since they turn into comprehension axioms of analysis. Thus a proof of $\bot$ in FA immediately yields a proof of $\bot$ in analysis.

It is therefore as certain as anything in mathematics that, if analysis is consistent, so is FA. Later we shall see that the converse holds. (A sketch of a major part of the proof of the converse was given by Frege, in the *Foundations*. Of course.) The connection between FA and Russell's paradox is discussed later. Since the possibility that analysis might be inconsistent at present strikes us as utterly inconceivable, we may relax in the certainty that neither Russell's nor any other contradiction is derivable in FA.

We now want to show that the definitions and theorems of sections 68–83 of the *Foundations* can be stated and proved in FA, *in the manner indicated by Frege*. I am not sure that it is possible to appreciate the magnitude and character of Frege's accomplishment without going through at least some of the hard details of the derivation of arithmetic from Numbers, in particular those of the proof that every natural number has a successor, but readers who wish take it on faith that the derivation can be carried out in FA along a path *very* close to Frege's may skim over some of the next seventeen paragraphs. Do not forget that it is Frege himself who has made formalization of his work routine.

In the course of replicating in FA Frege's treatment of arithmetic, we shall of course make definitional extensions of FA. For example, as Frege defined the number belonging to the concept F as the extension of the concept 'equinumerous to F', so we introduce a function symbol N, taking a concept variable and making a term of the type of object variables, and then define $NF = x$ to mean $\forall G(G\eta x \leftrightarrow F \text{ eq } G)$; the introduction of the symbol N together with this definition is of course licensed by Numbers. It will also prove convenient to introduce terms $[x: A(x)]$ for concepts: $[x: A(x)]t$ is to mean $A(t)$; $F = [x: A(x)]$ is to mean $\forall x(Fx \leftrightarrow A(x))$; $[x: A(x)]\eta y$ is to mean $\exists F(F = [x: A(x)] \land F\eta y)$; $[x: A(x)] = [x: B(x)]$ is to mean $\forall x(A(x) \leftrightarrow B(x))$, etc. The introduction of such terms is of course licensed by the comprehension axioms (i).

Sections 70–3 provide the familiar definition of equinumerosity. In §73, Frege proves Hume's principle. Note that the comprehension axioms (ii) provide the facts concerning equinumerosity needed for this theorem to be provable. Once Hume's principle is proved, *Frege makes no further use of extensions.*[8, 9]

In §72 Frege defines 'number': 'n is a number' is to mean 'there exists a concept such that n is the number which belongs to it'. In parallel, we make the definition in FA: $Zx \leftrightarrow \exists F(NF = x)$. In §74 Frege defines 0 as the number belonging to the concept 'not identical with itself'; we define in FA: $0 = N[x: x \neq x]$. The content of §75 is given in the easy theorem of FA:

$\forall F \forall G([\forall x \neg Fx \rightarrow ((\forall x \neg Gx \leftrightarrow F \text{ eq } G) \land NF = 0)] \land [NF = 0 \rightarrow \forall x \neg Fx])$.

In §76 Frege defines 'the relation in which every two adjacent members of the series of natural numbers stand to each other'.[10] Correspondingly, we define nSm (read 'n succeeds m'):

$\exists F \exists x \exists G(Fx \land NF = n \land \forall y(Gy \leftrightarrow Fy \land y \neq x) \land NG = m)$.

---

[8] See Charles Parsons, 'Frege's Theory of Number', in his *Mathematics in Philosophy: Selected Essays* (Ithaca, NY: Cornell University Press, 1983), 164. Parsons writes: 'The proof of Peano's axioms can be carried out on the basis of this axiom not only in Frege's own formal system but also in Russell's theory of types and in the other systems of set theory constructed to remedy the paradoxes, which of course showed Frege's system inconsistent.' He does not consider the effect of adjoining the axiom to the system of the *Begriffsschrift*.

[9] In his estimable *Frege's Conception of Numbers as Objects* (Aberdeen: Aberdeen University Press, 1983). Wright sketches a derivation of the Peano axioms in a system of higher-order logic to which a version of Hume's principle is adjoined as an axiom. Wright discusses the question of whether such a system would be consistent, attempts to reproduce various well-known paradoxes in such a system, is unsuccessful, and concludes on p. 156 that 'there are grounds, if not for optimism, at least for a cautious confidence that a system of the requisite sort is capable of consistent formulation'. Wright's instincts are correct, as Hodes and Burgess have seen. It may be of interest to note that FA supplies the answer to a question raised by Wright on p. 156 of his book. It is a theorem of FA that the number of numbers that fall under none of the concepts of which they are the numbers is *one*. (Zero is the only such number.)

[10] Note that, although Frege here introduces the expression 'folgt in der natürlichen Zahlenreihe unmittelbar auf' for the *succeeds* relation, he will define 'finite' number only at the end of §83.

¬0Sa immediately follows in FA from this definition: zero succeeds nothing. In §77 Frege defines the number 1. We make the corresponding definition: $1 = N[x: x = 0]$. $1S0$ is easily derived in FA.

The theorems corresponding to those of §78 are proved without difficulty:

(1) $\quad aS0 \to a = 1$,

(2) $\quad NF = 1 \to \exists x Fx$,

(3) $\quad NF = 1 \to (Fx \wedge Fy \to x = y)$,

(4) $\quad \exists x Fx \wedge \forall x \forall y (Fx \wedge Fy \to x = y) \to NF = 1$,

(5) $\quad \forall a \forall b \forall c \forall d (aSc \wedge bSd \to (a = b \leftrightarrow c = d))$,

(6) $\quad \forall n(Zn \wedge n \neq 0 \to \exists m(Zm \wedge nSm))$.

Although Frege and we have now defined 'number', defined 0 and 1, proved that they are different numbers, proved that 'succeeds' is one–one, and proved that every non-zero number is a successor, 'finite number', that is, 'natural number', has not yet been defined; nor has it been shown that every natural number has a successor.

In §79 Frege defines the ancestral of ϕ, 'y follows x in the ϕ-series', as in the *Begriffsschrift*. Thus in FA we define $x\phi *y$:

$$\forall F(\forall a(x\phi a \to Fa) \wedge \forall d \forall a(Fd \wedge d\phi a \to Fa) \to Fy).$$

§80 is a commentary on §79. At the beginning of §81 Frege introduces the terminology 'y is a member of the ϕ-series beginning with x' and 'x is a member of the ϕ-series ending with y' to mean: either y follows x in the ϕ-series or y is identical with x. Frege uses the phrase 'in the series of natural numbers' instead of 'in the ϕ-series' when ϕ is the converse of the succeeds relation. In FA we define mPn to mean nSm, $m < n$ to mean $mP*n$, and $m \leq n$ to mean $m < n \vee m = n$. Frege defines 'n is a finite number' only at the end of §83. In FA we define Fin n to mean $0 \leq n$.

In §§82 and 83 Frege outlines a proof that every finite number has a successor. He adds that, in proving that a successor of n always exists (if n is finite), it will have been proved that 'there is no last member of this series'. (He obviously means the sequence of finite numbers.) This will certainly have been shown if it is also shown that no finite number follows itself in the series of natural numbers; in §83 Frege indicates that this proposition is necessary and how to prove it.

Frege's ingenious idea is that we can prove that every finite number has a successor by proving that if n is finite, the number of numbers less than or equal to n—in Frege's terminology 'the number which belongs to the concept "member of the series of natural numbers ending the n" '—succeeds n.

FREGE'S *FOUNDATIONS OF ARITHMETIC* 195

Frege's outline can be expanded into a proof in FA of: Fin $n \to N[x: x \leq n]Sn$. Since $ZN[x: x \leq n]$ is provable in FA, so is (Fin $n \to \exists x(Zx \wedge xSn)$).

In §82 Frege claims that certain propositions are provable; the translations of these into FA are aSd & $N[x: x \leq d]Sd \to N[x: x \leq a]Sa$ and $N[x: x \leq 0]S0$. Frege adds that the statement that for finite n the number of numbers less than or equal to n succeeds n then follows from these by applying the definition of 'follows in the series of natural numbers'.

$N[x: x \leq 0]S0$ is easily derived in FA: $xP *y \to \exists aaPy$ follows from the definition of the ancestral; consider $[z: \exists aaPz]$. Since $\neg 0Sa$ and $1S0$ are theorems, so are $\neg aP0$, $\neg aP*0$, $x \leq 0 \leftrightarrow x = 0$, and $N[x: x \leq 0] = N[x: x = 0]$, from which, together with the definition of 1, $N[x: x \leq 0]S0$ follows.

But the derivation of aSd & $N[x: x \leq d]Sd \to N[x: x \leq a]Sa$ is not so easy. Frege says that, to prove it, we must prove that $a = N[x: x \leq a$ & $x \neq a]$; for which we must prove that $x \leq a$ & $x \neq a$ if and only if $x \leq d$, for which in turn we need Fin $a \to \neg a < a$. This last proposition is again to be proved, says Frege, by appeal to the definition of the ancestral; it is the fact that we need the statement that no finite number follows itself, he writes, that obliges us to attach to $N[x: x \leq n]Sn$ the antecedent Fin n.

An interpretive difficulty now arises: it is uncertain whether or not Frege is assuming the finiteness of a and d in §82. Although he does not say so, it would appear that he must be assuming that d, at least, is finite, for he wants to show (aSd & $N[x: x \leq d]Sd \to N[x: x \leq a]Sa$) by showing $aSd \to (x \leq a$ & $x \neq a \leftrightarrow x \leq d)$. Without assuming the finiteness of a and d, he can certainly show $aSd \to \forall x(x < a \leftrightarrow x \leq d)$. However, $\neg a < a$, or something like it, is needed to pass from $x < a$ to $(x \leq a$ & $x \neq a)$, and Frege would therefore appear to need Fin a. But since Fin 0 is trivially provable and $\forall d \forall e(dPa$ & Fin $d \to$ Fin a) easily follows from propositions 91 and 98, $(xPy \to xP * y)$ and $(xP * y$ & $yP * z \to xP * z)$, of the *Begriffsschrift*, Frege's argument can be made to work in FA, provided that we take him as assuming that d (and therefore a) is finite. Let us see how.

From propositions 91 and 98, $dPa \to (xP * d \vee x = d \to xP * a)$ easily follows. We also want to prove $dPa \to (xP * a \to xP * d \vee x = d)$, for which it suffices to take $F = [z: \exists ddPz$ & $\forall d(dPz \to xP * d \vee x = d)]$, and show $(xP * a \to Fa)$ by showing, as usual, $(xPb \to Fb)$ and $(Fa$ & $aPb \to Fb)$.

$(xPb \to Fb)$: suppose xPb. Then the first half of Fb is trivial; and if dPb, then by §78(5) of the *Foundations*, $x = d$, whence $xP * d \vee x = d$. As for $(Fa$ & $aPb \to Fb)$, suppose Fa and aPb. The first half of Fb is again trivial; now suppose dPb. By §78(5), $d = a$. Since Fa, for some c, cPa, and then $xP * c \vee x = c$. Since cPa and $d = a$, cPd. But then by propositions 91 and 98, $xP * d$, whence $xP * d \vee x = d$. Thus $(xP * a \to Fa)$, whence $dPa \to (xP * a \to xP * d \vee x = d)$ and $dPa \to (xP * a \leftrightarrow xP * d \vee x = d)$ follow. Call the second (*).

We must now prove

(∗∗) Fin $a \to \neg aP * a$.

Since $\neg 0P * 0$, it suffices to show $0P * a \to \neg aP * a$. We readily prove ($0Pb \to \neg bP * b$) and ($\neg aP * a \,\&\, aPb \to \neg bP * b$). If $0Pb$ and $bP * b$, then by (∗), $bP * 0 \vee b = 0$, whence by propositions 91 and 98, $0P * 0$, impossible; if $\neg aP * a$, aPb, and $bP * b$, then by (∗), $bP * a \vee b = a$, whence by propositions 91 and 98, $aP * a$, contradiction.

Combining (∗) and (∗∗) yields

$dPa \,\&\, \text{Fin } a \to ((xP * a \vee x = a) \,\&\, x \neq a \leftrightarrow xP * d \vee x = d)$.

Abbreviating, we have

$dPa \,\&\, \text{Fin } a \to (x \leq a \,\&\, x \neq a \leftrightarrow x \leq d)$

and then by Hume's principle

$dPa \,\&\, \text{Fin } a \to N[x: x \leq a \,\&\, x \neq a] = N[x: x \leq d]$.

Thus, if Fin d, $N[x: x \leq d]Sd$, and dPa, then Fin a and aSd; since $a \leq a$,

$N[x: x \leq a]SN[x: x \leq a \,\&\, x \neq a] = N[x: x \leq d]$;

since aSd, by §78(5), $N[x: x \leq d] = a$, and therefore $N[x: x \leq a]Sa$. Since Fin 0 and $N[x: x \leq 0]S0$, we conclude

Fin $n \to (\text{Fin } n \,\&\, N[x: x \leq n]Sn)$,

whence Fin $n \to N[x: x \leq n]Sn$.

OK, stop skimming now. One noteworthy aspect of Frege's derivation of what are in effect the Peano postulates is that so much can be derived from what appears to be so little. Whether or not Numbers is a purely logical principle is a question that we shall consider at length in what follows. I now want to consider the status of the other principles employed by Frege, which, having argued the matter elsewhere, I shall assume are properly regarded as logical. Frege shows these principles capable of yielding conditionals whose antecedent is the apparently trivial and in any event trivially consistent Numbers and whose consequents are propositions like $\forall m(\text{Fin } m \to \exists n(Zn \wedge nSm))$. The consequents would 'not in any wise appear to have been thought in' Numbers; thus these conditionals at least look synthetic, and Frege himself would appear to have shown the principles and rules of logic that generate such weighty conditionals to be synthetic. But if the principles of Frege's logic count as synthetic, then a reduction of arithmetic to logic gives us no reason to think arithmetic analytic. There is a criticism of Kant to which Frege is nevertheless entitled: Kant had no conception of this sort of analysis and no idea that content could be thus created by deduction.

The hard deductions found in the *Begriffsschrift* and the *Foundations* would make evident, if it were not already so, the utter vagueness of the notions of *containment* and of *analyticity*. Even though *containment* appears to be closed under obvious consequence, it is certainly not closed under consequence; there is often no saying just when conclusions stop being contained in their premises.

In particular, the argument Frege uses to prove the existence of successors—show by induction on finite numbers n that the number belonging to the concept $[x: x \leq n]$ succeeds n—is a fine example of the way in which content is created. 'Through the present example', wrote Frege in the *Begriffsschrift*,

we see how pure thought . . . can, solely from the content that results from its own constitution, bring forth judgments that at first sight appear to be possible only on the basis of some intuition. This can be compared with condensation, through which it is possible to transform the air that to a child's consciousness appears as nothing into a visible fluid that forms drops.

That successors appear to have been condensed by Frege out of less than thin air may well have heightened some of its readers' suspicions that the principles employed in the *Foundations* are inconsistent.

On the other hand, Frege's construction of the natural numbers foreshadows von Neumann's well-known construction of them, the consistency of which was never in doubt. Frege defines 0 as the number of things that are non-self-identical; von Neumann defines 0 as the set of things that are non-self-identical. Frege shows that n is succeeded by the number of numbers less than or equal to n; von Neumann defines the successor of n as the set of numbers less than or equal to n. Peano arithmetic based on the von Neumann definition of the natural numbers can be carried out (interpreted) in a surprisingly weak theory of sets sometimes called 'general set theory', the axioms of which are:

Extensionality: $\forall x \forall y (\forall z (z \in x \leftrightarrow z \in y) \rightarrow x = y)$,
Adjunction: $\forall w \forall z \exists y \forall x (x \in y \leftrightarrow x \in z \vee x = w)$, and all
Separation axioms: $\forall z \exists y \forall x (x \in y \leftrightarrow x \in z \wedge A(x))$.

There is a familiar model for general set theory in the natural numbers: $x \in y$ if and only if starting at zero and counting from right to left, one finds a 1 at the xth place of the binary numeral for y. It is obvious that extensionality, adjunction, and separation hold in this model. Thus it has been clear all along that something *rather* like what Frege was doing in the *Foundations* could consistently be done.

The results of the *Foundations* that the series of finite numbers has no last member and that the 'less than' relation on the finite numbers is irreflexive complement those of the *Begriffsschrift*, whose main theorems, when applied to the finite numbers, are that 'less than' is transitive (§98) and connected

(§133). Much more of mathematics can be developed in FA than Frege carried out in his three logic books. (It would be interesting to know how much of the *Basic Laws* can be salvaged in FA.) Since addition and multiplication can be defined in any of several familiar ways and their basic properties proved from the definitions, the whole of analysis can be proved (more precisely, interpreted) in FA. (The equi-consistency of analysis and FA can be proved in primitive recursive arithmetic.) Thus it is a vast amount of mathematics that can be carried out in FA.

Instead of discussing this rather familiar material, I want instead to take a look at certain strange features of FA, one of which was alluded to earlier. Frege defined 0 as the number belonging to the concept 'not identical with itself'? What is the number belonging to the concept 'finite number'? Frege introduces the symbol ∞_1 to denote the latter number, shows that ∞_1 succeeds itself, and concludes that it is not finite. But, although Frege does not consider the former number and hence does not deal with the question of whether the two are identical, it is clear that he must admit the existence of such a number. The statement that there is a number that is the number of all the things there are (among them itself) is antithetical to Zermelo–Fraenkelian doctrine, but as a view of infinity it is not altogether uncommonsensical. The thought that there is only one infinite number, *infinity*, which is the number of all the things there are (and at the same time the number of *all* the finite numbers), is not much more unreasonable than the view that there is no such thing as infinity or infinite numbers. In any event, the view is certainly easier to believe than the claim that there are so many infinite numbers that there is no set or number, finite or infinite, of them all.

But can we decide the question of whether these numbers are the same? Not in FA. $N[x: x = x] = N[x: \text{Fin } x]$ is true in some models of FA, for example, the one given, and false in others, as we can readily see. Let U' be the set of all ordinals $\leq \aleph_1$, and let η be true of V, u ($V \subseteq U', u \in U'$) if and only if the (finite or infinite) cardinality of V is u. Numbers is then true in this structure, Fin x is satisfied by the natural numbers, $N[x: \text{Fin } x]$ denotes $\aleph_0$, but $N[x: x = x]$ denotes $\aleph_1$. $N[x: x = x] = N[x: \text{Fin } x]$ is thus an undecidable sentence of FA. Of course, so is $\exists x \neg Zx$, but $N[x: x = x] = N[x: \text{Fin } x]$ is an undecidable sentence about numbers. From Frege's somewhat sketchy remarks on Cantor, one can conjecture that Frege would have probably regarded $N[x: x = x] = N[x: \text{Fin } x]$ as false.

I now turn to the way Russell's paradox bears on the philosophical aims of the *Foundations*. My view is a more or less common one. As a result of the discovery of Russell's paradox, our idea of logical truth has changed drastically, and we now see arithmetic's commitment to the existence of infinitely many objects as a greater difficulty for logicism than Russell's paradox itself.

FREGE'S *FOUNDATIONS OF ARITHMETIC*

But is not Frege committed to views that generate Russell's paradox? Does he not suppose that every predicate determines a concept and every concept has a unique extension? In §83 he says: 'And for this, again, it is necessary to prove that this concept has an extension identical with that of the concept "member of the series of natural numbers ending with d".' In §68 he mentions the extension of the concept 'line parallel to line a'. And the number belonging to the concept F is defined as the extension of the concept 'equinumerous with the concept F'. How, in view of his avowed opinions on the existence of extensions, can he be thought to escape Russell's paradox?

The first quotation can be dealt with quickly, as a turn of phrase. Had Frege written 'to prove that an object falls under this concept if and only if it is a member of the series of...', it would have made no difference to the argument. The extension of the concept 'line parallel to the line a' is used merely to enable the reader to understand the point of the definition of number. (These are the insignificant but possibly revealing exceptions mentioned to the claim that the only extensions to whose existence Frege explicitly commits himself in §§68–83 are those of concepts of the form 'equinumerous to the concept F'.) Thus, if there is a serious objection to Frege's introduction of extensions of concepts, it must concern the definition of numbers as extensions of concepts of the form 'equinumerous with the concept F'.

And of course there is one. According to Frege, for every concept F there is a unique object x, an 'extension', such that for every concept G, G bears a certain relation, 'being in', designated by η, to x if and only if the objects that fall under F are correlated one–one with those that fall under G; that is, Numbers holds. And although the language of FA, in which Numbers is expressed, is not one in which the most familiar version of Russell's contradiction $\exists x \forall y(y\eta x \leftrightarrow \neg\, y\eta y)$ is a well-formed sentence, it is not true that Frege is now safe from all versions of Russell's paradox.

For consider rule (V) of Frege's *Basic Laws*:

$$\forall F\, \exists!x \forall G(G\eta x \leftrightarrow \forall y(Fy \leftrightarrow Gy)),$$

which yields an inconsistency in the familiar way.

Suppose rule (V) true. By comprehension, let $F = [y: \exists G(G\eta y \land \neg Gy)]$. Then for some x,

(∗) $\forall G(G\eta x \leftrightarrow \forall y(Fy \leftrightarrow Gy))$.

Since $\forall y(Fy \leftrightarrow Fy)$, by (∗), $F\eta x$. If $\neg Fx$, then $\forall G(G\eta x \rightarrow Gx)$, whence Fx; but if Fx, then for some G, $G\eta x$ and $\neg Gx$, whence by (∗), $\neg Fx$, contradiction.

Or consider the simpler

SuperRussell: $\exists x \forall G(G\eta x \leftrightarrow \exists y(Gy \land \neg G\eta y))$.

Suppose SuperRussell true. Let x be such that for every G, $G\eta x$ if and only if $\exists y(Gy \wedge \neg G\eta x)$. By comprehension, let $F = [y: y = x]$. Then, $F\eta x$ iff $\exists y(Fy \wedge \neg F\eta y)$, iff $\exists y(y = x \wedge \neg F\eta y)$, iff $\neg F\eta x$, contradiction.

SuperRussell and rule (V) are sentences of the language of FA about the existence of extensions every bit as much as Numbers is. Just as Numbers asserts the existence (and uniqueness) of an extension containing just those concepts that are equinumerous with any given concept, so SuperRussell asserts the existence of an extension containing just those concepts that fail to be in some object falling under them, and rule (V) asserts the existence (and uniqueness) of an extension containing just those concepts under which fall the same objects as fall under any given concept. Frege must deny that SuperRussell and rule (V) are principles of logic—if he maintains that the comprehension axioms are principles of logic. Principles of logic cannot imply falsity. But then Frege cannot maintain both that every predicate of concepts determines a higher-level concept and that every higher-level concept determines an extension and would thus appear to be deprived of any way at all to distinguish Numbers from SuperRussell and rule (V) as a principle of logic.

Too bad. The principles Frege *employs* in the *Foundations* are consistent. Arithmetic can be developed on their basis in the elegant manner sketched there. And although Frege couldn't and we can't supply a reason for regarding Numbers (but nothing bad) as a logical truth, Frege was better off than he has been thought to be. After all, the major part of what he was trying to do—develop arithmetic on the basis of consistent, fundamental, and simple principles concerning objects, concepts, and extensions—can be done, in the way he indicated. The threat to the *Foundations* posed by Russell's paradox is to the philosophical significance of the mathematics therein and not at all to the mathematics itself.

It is unsurprising that we cannot regard Numbers as a purely logical principle. Consistent though it is, FA implies the existence of infinitely many objects, in a strong sense. Not only does FA imply $\exists x \exists y(x \neq y)$, $\exists x \exists y \exists z(x \neq y \wedge x \neq z \wedge y \neq z)$, etc., it implies $\exists F(\text{DedInf } F)$, where DedInf F is a formula expressing that F is Dedekind-infinite, for example, $\exists x \exists G(\neg Gx \wedge \forall y(Fy \leftrightarrow Gy \vee y = x) \wedge F \text{ eq } G)$. In logic we ban the empty domain as a concession to technical convenience but draw the line there. We firmly believe that the existence of even two objects, let along infinitely many, cannot be guaranteed by logic alone. After all, logical truth is just *truth no matter what things we may be talking about and no matter what our (non-logical) words mean*. Since there might be fewer than two items that we happen to be talking about, we cannot take even $\exists x \exists y x \neq y$ to be valid.

How then, we might now think, *could* logicism ever have been thought to be a mildly plausible philosophy of mathematics? Is it not obviously demonstrably inadequate? How, for example, could the theorem

$$\forall x \neg x < x \land \forall x \forall y \forall z (x < y \land y < z \rightarrow x < z) \land \forall x \exists y x < y,$$

of (one standard formulation of) arithmetic, a statement that holds in no finite domain but which expresses a basic fact about the standard ordering of the natural numbers, be even a 'disguised' truth of logic?[11] The axiom of infinity was soon enough recognized by Russell as both indispensable to his programme and as damaging to the claims that could be made on behalf of the programme; and it is hard to imagine anyone now taking up even a small cudgel for $\exists x \exists y x \neq y$.

I have been arguing for these claims: (1) Numbers is no logical truth; and therefore (2) Frege did not demonstrate the truth of logicism in the *Foundations of Arithmetic*. (3) Logic is synthetic if mathematics is, because (4) there are many interesting, logically true conditionals with antecedent Numbers whose mathematical content is not appreciably less than that of their consequents. To these I want to add: (5) Since we have no understanding of the role of logic or mathematics in cognition, the failure of logicism is at present quite without significance for our understanding of mentality. Had Frege succeeded in eliminating the non-logical residue from his *Foundations*, the question would remain what the information that arithmetic is *logic* tells us about the cognitive status of arithmetic. But Frege's work is not to be disparaged as a (failed) attempt to inform us about the role of mathematics in thought. It is a powerful mathematical[12] analysis of the notion of natural number, by means of which we can see how a vast body of mathematics can be deduced from one simple and obviously consistent principle, an analysis no less philosophical for its rigour, profundity, and surprise.

A fantasy: after the *Begriffsschrift* Frege writes, not *The Foundations of Arithmetic*, but another book with the same title whose main claim is that, since arithmetic is deducible by logic alone from the triviality 'the number of *F*'s is the same as the number of *G*'s if and only if the *F*'s can be correlated one-one with the *G*'s', arithmetic is analytic, not synthetic, as Kant supposed. Frege then argues for the analyticity of $NF = NG \leftrightarrow F$ eq G on the ground that both halves of the biconditional have the same content, express the same thought. He considers an attempted defence of Kant: since the existence of an object can be inferred from $NF = NF$, $NF = NF$ must be regarded as synthetic, and therefore so must $NF = NG \leftrightarrow F$ eq G. Frege replies that $7 + 5 = 7 + 5$ is analytic.

[11] See Paul Benacerraf, 'Logicism: Some Considerations', Ph.D. diss., Princeton University, Department of Philosophy, 1960.
[12] See Paul Benacerraf, 'The Last Logicist', in P. French, T. Uehling, and H. Wettstein (eds.), *Midwest Studies in Philosophy*, vol. 6 (Minneapolis: University of Minnesota Press, 1981), 17–35.

If Frege had abandoned one of his major goals—the quest for an understanding of numbers not as objects but as 'logical' objects—taken as a starting-point the self-evident and consistent $\forall F \forall G(NF = NG \leftrightarrow F \text{ eq } G)$, and worked out the consequences of this one axiom in the *Begriffsschrift*, he would have been wholly justified in claiming to have discovered *a* foundation for arithmetic. To do so would have been to trade a vain philosophical hope for a thoroughgoing mathematical success. Not a bad deal. He could also have plausibly claimed to demonstrate the analyticity of arithmetic. (Of course his own work completely undermines the interest of such a claim.)

Perhaps the saddest effect of Russell's paradox was to obscure from Frege and us the value of Frege's most important work. Frege stands to us as Kant stood to Frege's contemporaries. *The Basic Laws of Arithmetic* was his *magnum opus*. Are you sure there's nothing of interest in those parts of the *Basic Laws* that aren't in prose?[13]

---

[13] The papers by Paul Benacerraf, Harold Hodes, and Charles Parsons cited have been major influences on this one. I would like to thank Paul Benacerraf, Sylvain Bromberger, John Burgess, W. D. Hart, James Higginbotham, Harold Hodes, Paul Horwich, Hilary Putnam, Elisha Sacks, Thomas Scanlon, and Judith Jarvis Thomson for helpful comments. Research for this paper was carried out under grant SES-8607415 from the National Science Foundation.

X

ARITHMETICAL TRUTH AND HIDDEN HIGHER-ORDER CONCEPTS

DANIEL ISAACSON

§0. INTRODUCTION

The incompleteness of formal systems for arithmetic has been a recognized fact of mathematical life for over half a century, and much has been said about it. Even so, I want still to raise some issues in this area, to advocate certain ways of looking at the phenomenon. In particular, I want to focus attention on the status of Peano arithmetic, and on the nature of those true statements in the language of arithmetic which are unprovable in it.

Gödel's construction of an undecidable sentence, given a formal system strong enough for numeral-wise expressibility of rather weak arithmetic, shows that each such formal system for arithmetical truth must admit proper extensions. This situation suggests that the choice of any particular formal system must be provisional, subject to an eventual mathematical need to go beyond it. At the same time, Peano arithmetic seems a natural and intrinsically important axiomatization. Does this impression arise through the historical accident of what people happened to arrive at first, or does it reflect rather some underlying conceptual fact?

This paper explores a viewpoint on which Peano arithmetic indeed occupies an intrinsic, conceptually well-defined region of arithmetical truth. The idea is that it consists of those truths which can be perceived as true directly from the purely arithmetical content of a categorical conceptual analysis of the notion of natural number. The truths expressible in the (first-order) language of arithmetic which lie beyond that region are such that there is no way by which their truth can be perceived in purely arithmetical terms. Via the phenomenon of coding, they contain essentially hidden higher-order, or

First published in The Paris Logic Group (eds.), *Logic Colloquium '85* (Amsterdam: North Holland, 1987), 147–69. Reproduced here by permission of the author.

The author has taken this opportunity to correct two misprints from the original publication, to make some further revisions to the text, and to amplify and update the references.

infinitary, concepts. On this perspective, Peano arithmetic may be seen as complete for finite mathematics.

§1. THE STARTING-POINT: PROPERTIES WHICH CHARACTERIZE THE CONCEPT OF NATURAL NUMBER

The nineteenth century saw tremendous development in the conceptual basis of mathematical knowledge. This was most notable, and pressing, in the case of real analysis and the understanding of the continuum. The development of correct definitions of limit, continuity, the derivative, the distinction between convergence and uniform convergence, isolation of the least upperbound and cut property of the reals allowed better understanding of results already obtained. It was also now possible to obtain new results not accessible earlier on the old conceptual basis (such as, most famously, the existence of a continuous but nowhere differentiable real-valued function of a real variable). In the later part of the nineteenth century this enterprise of rigorization came also to be applied to that most basic part of mathematics, the theory of whole numbers.

The fact that arithmetic was dealt with in this way only after the calculus had been reflects the differing nature of the demand for conceptual understanding of the reals and of the natural numbers. The drive to understand more deeply the basic, characteristic properties of the reals arose directly out of demands of mathematical practice. In the case of the natural numbers there was no doubting the truth of the results being obtained. The essential contrast is between finite and infinite. Each natural number is a finite entity, while a single real number is itself an infinite structure (whose infinity can be analysed in various ways, for example, by an infinitely proceeding process of increasingly sharp approximation). Success in analysing the mathematical structure of that infinity showed how the real numbers, and functions on them, could be thought of as built up by set-theoretic operations on the natural numbers. Hence the momentum of the foundational enterprise combined readily with philosophical curiosity to impel the attempt to understand the basic nature of the natural numbers.

In some ways nothing could be more transparent. The natural numbers are what arise in the process of counting. We mark the beginning of that process by counting 1, and the further natural numbers arise by iteration of the process of counting a next one (by a slight increase of sophistication, we may think of 0 as counting the empty collection, and in this way as constituting the smallest natural number). Yet, however much this simple account might seem plausibly to articulate our actual understanding, it shows a basic circularity, namely, that the notion of successive iterations is tantamount to the notion of natural number, which must mean that it cannot constitute the conceptual basis of the analysis of our grasp of this notion. Dedekind addressed himself specifically to this circularity (as he recounts in his letter to Keferstein (Dedekind 1890)), and arrived at the following non-circular characterization of the structure of the sequence of natural numbers: it consists of the smallest collection closed under a successor operation taking distinct elements to distinct elements and which contains an element not the successor of any element[1]. The notion of smallest such collection is expressed by the condition that the natural numbers consist of the intersection of all such collections (Dedekind 1890: 101; Dedekind 1888: ch. 6). Thus we arrive at the following explicit definition of the property of being a natural number:

$$N(x) \equiv_{df} \forall X(0 \in X \,\&\, \forall y(y \in X \to s(y) \in X) \to x \in X)$$

where (1) $\forall x(s(x) \neq 0)$, and (2) $\forall x \forall y \,(x \neq y \to s(x) \neq s(y))$). The correctness of this definition to characterize the intended structure is dependent on the existence of an inductive set, i.e. one containing 0 and closed with respect to the injective map s with 0 not in its range. Dedekind was well aware of this point, and in his theorem 66 attempted a proof, based on the notion 'possible object of thought'. How convincing such an existence proof could be is disputable, and one may prefer, rather, to take the required conclusion as given by an axiom.

From this definition the principle of induction for natural numbers follows:

(3) $\qquad \forall X(0 \in X \,\&\, \forall y(y \in X \to s(y) \in X) \to \forall x(x \in X))$,

where the elements of the domain satisfy the condition $N(x)$ given above. This analysis/definition of the natural numbers suggests and validates the axiomatization of arithmetic by the three axioms (1), (2), and (3).

Having characterized the structure of the succession of natural numbers, Dedekind then dealt with their arithmetic by establishing (in 1888, §126) the unique existence of functions defined by (primitive) recursion, which he applied to obtain the standard operations of addition, multiplication, and exponentiation. He also used this result to demonstrate the categoricity of this characterization of the natural numbers (§§132–4) (i.e. for any two structures

---

[1] The analyses of natural number by both Frege (1884) and Dedekind (1888) are, while independently arrived at, essentially identical as to the key idea (Dedekind's notion of chain, and Frege's notion of 'y follows in the ϕ-series after x'). My discussion here is in terms of Dedekind's analysis, both because the motivating line of development followed by Dedekind is the one I wish also to follow (as opposed to Frege's motivation in terms of ontology), and because Dedekind establishes the categoricity of his analysis.

satisfying these axioms, there is an isomorphism between them, which exists by a primitive recursion defining the transformation of one structure into the other). From categoricity it follows that these axioms have as logical consequences (what is true in every structure in which the axioms are true) exactly those sentences in the first-order language of arithmetic true in the structure of the natural numbers.

§2. TRANSITION TO A FIRST-ORDER AXIOMATIC SYSTEM: PEANO ARITHMETIC (PA)

The situation so far described sounds a thorough success. However, two features of it are problematic. The first is that this categorical axiomatization cannot be fully formalized. It relies upon the logic for quantifiers ranging over all subsets of the domain of individuals, and the very success of the axiomatization, its completeness as a consequence of categoricity, means that in light of the Gödel incompleteness theorem there can be no consistent formal system which recursively enumerates the logically valid formulae for these second-order quantifiers. Indeed, categoricity itself already establishes this result, since completeness with respect to logical validity implies compactness, and compactness implies non-categoricity of any axiomatization of arithmetic.[2]

The other difficulty has to do with impredicativity. We noted earlier that the characterization could only succeed in case there exists a set containing 0 and closed under succession. But in fact there must be a set consisting *precisely* of the natural numbers. If the smallest inductive set contained extraneous elements appropriately structured, then the axioms would still be true, but the structure in which they were true would not be isomorphic to the natural numbers. The given characterization of the natural numbers can only be known to succeed if the structure we are attempting to characterize is recognized to be an element in the second-order domain. A variety of responses to these difficulties are possible. I want here to follow one of these, namely, the transition to the corresponding first-order axiomatic system, Peano arithmetic.

To some extent a philosophical presumption has developed in favour of

[2] Note that there can be formal systems which are second-order, in the sense of having two sorts of variables: one of the sort to range over individuals, the other to range over classes or properties or relations of individuals. In so far as such a system is formal (that is to say, recursively enumerable), it is equivalent to a two-sorted first-order system in which the second-order variables may, *but need not*, be interpreted as ranging over *all* classes or properties or relations (as the case may be) of individuals. A formal second-order system for arithmetic is never categorical, just as no first-order system is, though it may be deductively stronger than a corresponding first-order system (e.g. may prove the Gödel sentence for the first-order formal system).

first-order axiomatizations (see e.g. Quine 1970: ch. 5). However, the expressive weakness of first-order languages strikes me as decisive against the possibility of working exclusively within first-order logic.[3] None the less, there is something of intrinsic importance in a first-order language, since it is one in which, for arithmetic, the objects we talk about by it are the natural numbers, and that is just what we do in doing arithmetic. It is natural then to look for a first-order formal system which expresses as much as possible of the insight into the structure of the natural numbers and their arithmetic given by Dedekind's analysis.

The first two axioms of Dedekind's analysis are, as they stand, first-order. What about the axiom of induction, with its quantification over sets of natural numbers? The way we make use of it in our study of the natural numbers is in establishing by mathematical induction that some particular numerical property applies to all the natural numbers. The purely arithmetical content of the induction principle (as distinct from its full conceptual content) is thus given by a *scheme* which tells us that induction holds for any particular numerical property. Numerical properties are expressed by free-variable formulae in a language for arithmetic. So we come to the question, what language should that be?

The formulation of Dedekind's analysis was in terms of successor and the distinguished element 1 (for Dedekind, 0 for us). As Dedekind's proof of the recursion theorem shows, all the usual arithmetic of the natural numbers can then be established on this basis. By contrast, if we restrict ourselves to those properties of natural numbers expressible only by quantification over the natural numbers themselves, then a system with only 0 and s as primitive is extremely weak, in particular, not strong enough to prove the existence of arithmetical functions defined by primitive recursion. However, if we take as primitive the two most familiar and basic arithmetical functions, namely, addition and multiplication, as given by their recursion equations, we then do arrive at a situation where all the further primitive recursive (and indeed general recursive) functions are expressible. This fact, which is a kind of first-order correlate to Dedekind's recursion theorem, was established by Gödel (1931: theorem VII), via arithmetical coding of finite sequences. In this way we arrive at a prima-facie justification of interest in the formal system of Peano arithmetic, PA, for which the first-order language has as its primitives

---

[3] This expressive weakness follows as a consequence of what is often cited as one of the main reasons why we should work within first-order logic, viz. the completeness of formal systems of first-order logic with respect to logical validity. The familiar point is that compactness follows from completeness, and from compactness we have it that the first-order truths of arithmetic will hold in structures not isomorphic to the natural numbers, showing thereby that we cannot express fully in a first-order language the understanding we have of the structure of the natural numbers.

0, s, +, ·, and has as axioms the first two Dedekind axioms, plus the recursion equations for + and ·, that is:

$$x + 0 = x \qquad x \cdot 0 = 0$$
$$x + s(y) = s(x + y) \qquad x \cdot s(y) = x \cdot y + x$$

and the scheme of induction with respect to properties of the natural numbers expressible in this language. In this way Peano arithmetic may be seen as an intrinsic system, arising in a natural way when investigating the general nature of natural numbers. Is there a sense in which it is canonical as a first-order system for arithmetic?

Other second-order analyses could have been used. For example, the natural numbers can be characterized by a second-order formulation of the least number principle, for which the corresponding first-order scheme is equivalent to the first-order scheme of induction. Could Peano arithmetic be canonical in the sense that it is equivalent to any first-order system which arises by forming a scheme of first-order substitution instances from a purely universal second-order sentence (sentences of this form are said to be Π_1^1) which serves as part of a categorical characterization of the natural numbers?

It seems that this condition is too strong. Consider, for example, a second-order sentence:

(∗) $\qquad \forall X(\forall y(\forall x(x \prec y \rightarrow x \in X) \rightarrow y \in X) \rightarrow \forall x(x \in X))$

expressing transfinite induction with respect to an arithmetically expressed primitive recursive well-ordering $\prec$ on the natural numbers of order-type $\geq \varepsilon_0$, where ε_0 is the smallest ordinal such that transfinite induction on an ordering of the natural numbers of that order type cannot be proved in Peano arithmetic (see refs. in n. 8 below). The second-order principle (∗) expresses the condition that we have exactly the natural numbers well-ordered by $\prec$, and so gives a categorical characterization of **N**, albeit now with a highly complex structure on it. Evidently the corresponding system generated by turning (∗) into a first-order scheme must be stronger than PA, in light of the unprovability in PA of transfinite induction of order-type ε_0.

In the present context, this case is not worrying, since principle (∗) is stronger than such apprehension as we have just of the natural numbers themselves, requiring as it does an irreducible notion of well-ordering. Alex Wilkie (1987) has obtained a result which can be interpreted as showing essentially that PA is the weakest first-order system arising from any categorical Π_1^1-characterization of the natural numbers. The question of the intrinsicness of Peano arithmetic might then be explored by assessing the conceptual content of the various categorical Π_1^1-characterizations of the natural numbers, looking to see whether those which yield Peano arithmetic are recognizable as conceptually equivalent to Dedekind's analysis of the notion of natural number, and whether those which yield stronger first-order systems require some conceptual element which goes beyond our grasp of the natural numbers. Such a project lies beyond the scope of this essay. Alex Wilkie's extremely interesting paper (1981) 'On Discretely Ordered Rings in which Every Definable Ideal is Principal', in which he shows that a resulting first-order system is (essentially) equivalent to Peano arithmetic, offers a sharply developed particular case apt for such consideration.

The attempt to recognize a conceptually significant boundary between first- and higher-order notions for arithmetic may be called into question by the predicativist viewpoint (I am grateful to Solomon Feferman for drawing my attention to this possibility). 'By *predicativity* is meant that part of (abstract) mathematical thought which is implicit in our conception of the natural numbers' (Feferman 1979: 68). That conception is the same as Dedekind's starting-point: 'Our conception of the natural number sequence N is as *generated* from an *initial number* 0 by repeated application of a successor operation $a \mapsto a'$ ' (ibid. 70). The project disallows use of any higher-order operations not themselves justified by that initial arithmetical conception, by which restriction the inherent impredicativity of Dedekind's analysis is to be avoided. However, higher-order operations are not excluded by it as such, and in particular it is argued that functionals determined by primitive recursion on numerical arguments can be predicatively justified. Nothing I have said counts against the possibility or the intrinsic interest of carrying through such an analysis. The question would be, rather, does that analysis count against the stability of arriving at a first-order axiomatic system?

In considering that question, I am *not* claiming that PA could itself constitute an adequate conceptual basis for our understanding of the concept of natural number. Far from it, I consider that we can only arrive at such a system on the basis of some higher-order understanding. The system PA arises as constituting the purely arithmetical content of our full understanding of the concept of natural number, where that understanding is implicitly and inherently higher-order.

If in pursuing the predicativist programme one were motivated by a firm conviction that impredicativity is *per se* illegitimate, then the Dedekind analysis would not be allowed as an acceptable conceptual basis for the account of arithmetical truths I am attempting to develop. But I believe that one can still arrive at PA as the first-order system given directly from the analysis of the concept of natural number, even where that analysis is carried through predicatively.

So the question must be not: is PA conceptually strong enough to analyse the concept of natural number? to which the answer must always be no, but rather: can we motivate focusing attention on an axiomatic system for arithmetic which is first-order? It seems to me that there is an intrinsic point of interest in working within a domain in which our only objects of quantification are the natural numbers themselves.

In the following section I turn to considerations which offer a characterization of the domain of mathematics captured in that first-order axiomatization.

§3. PEANO ARITHMETIC AS THE MATHEMATICS OF FINITE STRUCTURES

Peano arithmetic is equivalent to a natural theory of purely finite structures, namely ZF$^-$, the axiom system of Zermelo–Fraenkel set theory without the axiom of infinity and with the negation of the axiom of infinity. As is well known, the standard model for ZF$^-$ is the collection of hereditarily finite (pure) sets. The significance of this equivalence for the present discussion is twofold. First, it is natural to consider arithmetic as essentially the mathematics of that paradigm finite process, counting finite collections. To see that Peano arithmetic is equivalent to a natural axiomatization of finite structures encourages the belief that a right understanding of this domain has been reached. The second point concerns the account to be given of the incompleteness phenomenon in relation to the intrinsic position of PA. The suggestion is that those mathematical truths expressible in the language of arithmetic but not provable in PA contain 'hidden higher-order concepts', where what is hidden is revealed by the recognition of the phenomenon of coding. What I mean here by 'higher-order' includes the standard usage for quantification over sets of individuals in distinction to first-order quantification over the individuals themselves. But I also mean to include in this phrase something of the notion infinitary, in the sense of presupposing an infinite totality, and the applicability of such a notion in this context is suggested by the results being looked at in this section.

The possibility of formulating Peano arithmetic in a theory of the finite (for which I shall sometimes use the adjective 'finitary', hoping it will be clear and acceptable that I am thus using the word with a different sense from Hilbert), begins in a reworking of the Dedekind analysis of natural number, looked at in §2. This approach goes back to Zermelo, in his 1909 paper 'Sur les ensembles finis et le principe de l'induction complète'; see also Parsons 1987.

The basic idea of Dedekind's analysis, as we noted, is to stipulate that an object of our theory is a natural number just in case it belongs to every inductive set. Such sets are necessarily infinite. However, we can also correctly stipulate that x is a natural number by the condition that x belongs to the smallest set which contains 0 and is closed under successor *except* when applied to x, that is, a set of the form $\{0, s(0),..., s^n(0)\}$. Thus:

$N(x) \equiv_{df} \forall X(0 \in X \ \& \ \forall y(y \in X \ \& \ y \neq x \to s(y) \in X) \to x \in X)$.

This formulation is due to Michael Dummett, as reported by Hao Wang in his paper 'Eighty Years of Foundational Studies' (1958: 52–3). The existential condition required in order for this definition to succeed (corresponding to the existence requirement of an inductive set in the case of Dedekind's analysis), is evidently:

$\forall x \exists X(0 \in X \ \& \ \forall y(y \in X \ \& \ y \neq x \to s(y) \in X))$.

It is also clear that on this existential basis the definition works when the second-order quantifiers range only over all *finite* subsets of the domain (weak second-order logic).

I want to digress briefly to consider this definition in relation to the charge of impredicativity. That this definition picks out all the natural numbers, given the preceding existential condition, depends on a feature of the range of the second-order quantifier comparable to the impredicative requirement in the full second-order analysis that the second-order domain contain a set consisting exactly of 0 and the elements obtained from it by finite iteration of the successor function. This definition will pick out just the natural numbers in case for each natural number n the second-order domain contains a set consisting precisely of 0 and what is obtained from it by up to n-fold iteration of the successor function, and the domain of second-order quantification contains only *finite* sets. In the strictest sense, this condition falls short of impredicativity, in that the very set being defined is not required to lie within the range of the quantifiers of the definition. However, as we see, an exact representation of the natural number sequence must occur as elements of the domain, and I am inclined, therefore, to consider that the weak second-order definition does not fare significantly better on the score of avoiding impredicativity than the one based on full second-order logic.

This analysis provides a categorical characterization of the natural numbers, on the basis of which the full higher-order principle of induction is derivable. As before, this success shows that the system obtained cannot be given by a recursive set of axioms. Peano arithmetic as an axiom system is obtained from this analysis along lines similar to those described in the previous section.

Having arrived at PA by this route, there is then a very natural construction within ZF⁻, interpreting 0 as the empty set and the successor function in the von Neumann way as $a \mapsto a \cup \{a\}$, by which the axioms of PA are all provable. The construction is essentially similar to the usual interpretation of number theory in ZF, but renders explicit the finiteness of the set theory required to yield Peano arithmetic.

Quite strikingly, there is a converse to this result, that is to say, not only is PA interpretable in ZF⁻, but ZF⁻ is fully interpretable in PA. The interpretability of ZF⁻ in PA can be established on the basis of coding each hereditarily finite set by a natural number, as first shown by Wilhelm Ackermann (1937), by the following definition of an $\in$-relation among the natural numbers: $n \in m \equiv_{df}$ for some finite $a \subset \mathbf{N}$, $m = \sum_{k \in a} 2^k$ and $n \subset \in a$. For example, $\{0, 4, 7\}$ is coded by $2^7 + 2^4 + 2^0 = 145$. A delightful feature of this coding is that if the number is written in binary notation, the sequence of symbols '0' and '1' is then simply the characteristic function for the finite set coded by that number, beginning at the right with 0. Thus $145 = 10010001$. We take 0 to code the empty set, so $1 = 2^0$, which codes $\{0\}$, can be taken to code $\{\phi\}$. And so we have a bi-unique coding of hereditarily finite (pure) sets by natural numbers. For example, 145 codes $\{\phi, \{\{\{\phi\}\}\}, \{\phi, \{\phi\}, \{\{\phi\}\}\}\}$. On this coding, each of the axioms of ZF⁻ is translated into a true statement of arithmetic provable in PA. Thus Peano arithmetic can be established on the basis of finite set theory, and is itself, in a very natural way, a theory of the hereditarily finite sets.

§4. INCOMPLETENESS: THE FIRST EXAMPLE, VIA DIAGONALIZATION, AND THE PHENOMENON OF CODING

The incompleteness of all formal axiomatizations for arithmetical truth seems to have come as a surprise to mathematicians generally, including those mathematicians and logicians concerned particularly with the relationship between mathematics and formal systems, most notably David Hilbert. Gödel presented his proof of incompleteness specifically for the system of *Principia Mathematica*, and he notes its applicability to Zermelo–Fraenkel set theory. But it was also made clear in Gödel's account that his construction was applicable to any system strong enough to deal with the basic form of recursion.

It is in a way surprising, after the fact, that this result took Hilbert and his school so much by surprise. It was Hilbert, after all, who made the basic move required for it of realizing that the manipulation of symbols of a formal system should be seen as being of the same character as the computational manipulations of arithmetic. From this conceptual basis Hilbert gave an ingenious mathematical formulation to a programme for establishing that the full, infinitary range of mathematics was consonant with that part of mathematics which offers certain and absolute constraint, namely, finitary computations. This came down in effect to establishing the condition that if a purely universal first-order sentence is derivable in a given formal system for arithmetic, then that formula holds (which is to say that every numerical instance of the formula holds). Such a property for a sentence in the language of a given formal theory is called a reflection principle for that sentence with respect to the theory. Purely universal sentences in a first-order language for arithmetic are said to be Π_1^0, so the condition in question is called Π_1^0-reflection for the given system. Π_1^0-reflection for a given formal system is equivalent to the deductive consistency of the formal system (that Π_1^0-reflection implies consistency is immediate from, e.g., the instance of reflection for the formula $\forall x(x \neq x)$; an argument that consistency implies Π_1^0-reflection is sketched by Hilbert (1928; p. 474 in repr.). The programme of establishing these results was envisaged as proceeding within informal, intuitive, finitary mathematics.

Gödel saw that a further step was possible. If the syntactic manipulations of a formal system could be viewed as part of finitary mathematics, as akin to the elementary calculations of arithmetic, then one might be able actually to map the investigation of these syntactic manipulations into the arithmetic itself, and so establish it *within* formal systems of arithmetic. Gödel's primitive recursive arithmetization of syntax showed that indeed this could be done. On this basis one could then, as Gödel did, diagonalize on the provability predicate, proceeding analogously to the liar paradox by way of the heuristic formulation 'This sentence is unprovable', to obtain a true arithmetical statement unprovable in the given formal system.

Let us consider this result as applied to the formal system PA. Indisputably, it shows that PA is incomplete for mathematical truth expressible in the language of arithmetic. Does this situation count decisively against any claim for the intrinsic character of PA? One may try to say that it does not by observing that the Gödel sentence is from the point of view of usual mathematics rather peculiar. It arises not by working from arithmetical properties of the natural numbers, but by reflecting about an axiomatic system in which those properties are formalized. In a certain way, it might even, be said not to be arithmetical. It is not saying something about the natural numbers; rather, it is 'about' the statement itself. In that way the Gödel incompleteness phenomenon is not an arithmetical incompleteness. That is my viewpoint in this paper.

The contrasting viewpoint would hold that expressibility in the language of arithmetic renders the statement genuinely arithmetical. That the Gödel sentence says of itself 'This very sentence is underivable in PA' is, on this view, *merely* heuristic, and it is highly misleading to put it in such terms, as is sometimes done when the result is being expounded. It 'says' nothing about itself. What it asserts is that a certain universal relation holds of all natural numbers. Now it is crucially true, and obvious, that the Gödel sentence for PA in the formal language of arithmetic and the English sentence 'This very sentence is underivable in PA' are of fundamentally different character (in particular, of course, the question of whether or not that sentence in English is derivable in PA does not arise). None the less, it seems to me a significant fact, which cannot be brushed aside as *merely* heuristic, that the Gödel sentence *does* say something like what we also express with this sentence of English. The only way to *see* the arithmetical truth of the Gödel sentence is in terms of its connection with the situation we describe by that English sentence. And the situation described by that sentence goes essentially beyond arithmetic itself.

The key technique of Gödel's proof is the use of coding, the coding of syntactic relations and properties by properties and relations of natural numbers. At least in the case of Gödel sentences (I will consider some other undecidable sentences in the next section), the understanding of these sentences rests crucially on understanding this coding and our grasp of the situation being coded. The phenomenon of coding reveals fixed links between two situations or facts, one in the structure of arithmetic, the other in the realm of syntax of a formal system. These facts, and the link between them, are revealed by the description of the coding, but their existence is not dependent on being described.

We might consider whether, in view of its truth and independence from PA, we should adopt the Gödel sentence for PA, call it G, as a new axiom of arithmetic. Such a move would be unnatural. An axiom in this context should be an evident truth, in the terms in which it is expressed. But the truth of this statement, as a statement of arithmetic, is not directly perceivable. PA + G would not constitute, in this way, a purely arithmetical extension of PA. The Gödel sentence thus offers an instance of the general thesis of this paper that any axiomatic extension of Peano arithmetic must be motivated by considerations for establishing its truth which rely essentially on non-arithmetical notions.

Hilbert noted and made essential use, both technical and philosophical, of the similarity between formal manipulation of symbols and elementary computation on the natural numbers. He considered that this similarity offered a uniform account of the nature of these formal manipulations. Gödel's discovery of the phenomenon of coding shows that the account of formal syntax in these terms cannot be uniform, and that some truths of arithmetic must be seen in terms of their link to syntactic properties, rather than the other way round, as Hilbert had envisaged would always be possible. In these terms, it may be said that Gödel's discovery of incompleteness for arithmetical formal systems reveals not so much their deductive weakness, as rather the structural expressiveness of arithmetic. The arithmetic of the natural numbers can mimic quite other situations. If the truths in the language of arithmetic which express these mimic situations are to be seen as true, that will depend not on the principles which generate our understanding of the natural numbers, but on those which apply to the situation which is mimicked, and which reveal the coded connection between them.

§5. CONSIDERATION OF SOME FURTHER EXAMPLES

Attention has focused in recent years on some extremely interesting examples of arithmetical truths unprovable in Peano arithmetic which are thought of as much more genuinely and purely mathematical than the Gödel sentences. They include the study by Kirby and Paris (1982) of Goodstein's (1944) theorem, the Paris–Harrington (1977) variant of the finite Ramsey theorem, and Friedman's finite version of Kruskal's theorem (Smorynski 1982). The general thesis of this essay is tested and supported by these particular cases.

(a) *Goodstein's theorem* offers at first sight a particularly compelling example of genuinely arithmetical incompleteness in PA. It is easy to grasp what the theorem asserts about the natural numbers: namely, that the sequence of natural numbers generated by successive substitution in the exponential base representation of a number and subtraction is infinite (see Henle 1986: 45–8, 91–3, 137–9, for a readily accessible exposition of Goodstein's theorem). One can understand the situation it describes with just the sophistication which has become standard at the level of elementary school with the 'new maths' fascination with change of base for integer notation. It is possible to give a simple and perspicuous, purely mathematical demonstration of its truth for which the required mathematics is available in any undergraduate course on set theory which develops the notion of ordinal as far as the Cantor normal form theorem. It is in these terms evident that the result follows from ordinal induction of order type ε_0.

The converse also holds. Indeed, Goodstein's interest in studying these sequences of natural numbers was as a way of giving arithmetical expressions to ordinal inductions of order types $\leq \varepsilon_0$. Hence, by the adequacy of ε_0-ordinal induction for proving the consistency of PA, combined with Gödel's second incompleteness theorem, Goodstein's theorem must be unprovable in PA. Its unprovability can also be analysed more directly, as was done by Kirby and Paris (1982) using the model-theoretic technique of indicators and results of Ketonen and Solovay (1981) (see also Buchholz and Wainer 1987 and Takeuti 1987: 128–30). These two ways of seeing the independence of Goodstein's theorem reveal different features of it relevant to the general thesis of this essay.

The fact that it codes ε_0-induction tells us that there is no way to perceive the truth of Goodstein's theorem which does not also establish the correctness of ε_0-induction.[4] The question raised by Goodstein's theorem in this context comes then to the following: is it possible that we should manage to establish Goodstein's theorem using *only* purely arithmetical notions, that is, without the use of any 'higher-order' notions, such as 'arbitrary subset', or 'well-ordering', or 'sound axiomatization of arithmetical truth'? Could we have some basis just within our understanding of arithmetic on the natural numbers for taking Goodstein's theorem as an axiom of true arithmetic? What is at least clear is that the way in which we do know that Goodstein's theorem is true is not such a basis. Thus Goodstein's theorem as it stands does not refute the thesis of this essay. I do not see how to establish the stronger claim, that there is *no* such way. But still, I draw support for the viewpoint I am urging here because at first Goodstein's theorem seems to be a particularly likely counter-example to it, which then turns out, on closer consideration of what we do know about it, not to be one.

The second heuristic point for the general thesis under consideration which emerges from this example relates to the theme developed in §3, on Peano arithmetic as the mathematics of finite structures. Goodstein's theorem is of the form $\forall x \exists y \phi(x, y)$. The Kirby–Paris method of demonstrating its independence by the use of non-standard models shows the existence of a model of Peano arithmetic in which $\forall x \exists y \phi(x, y)$ is true, an initial segment of which is also a model of PA, and in which there are elements such that any witness to the true existential $\exists y \phi(a, y)$ lies beyond the initial segment (the element a must, of course, be non-standard, since for each standard n, PA $\vdash \exists y \phi(\bar{n}, y)$). The elements of an end extension of a given model of PA are infinite with respect to the initial segment model, in the sense that for that model they lie beyond all the natural numbers. One can then think of this demonstration of independence as modelling the fact that the process of generating Goodstein sequences goes essentially beyond finite arithmetic. Any particular calculation of a Goodstein sequence is finite. But the process as a whole is infinitary. In the initial segment which is a model of PA, everything

---

[4] See Isaacson 1992: 113–15 for arguments to show that ε_0-transfinite induction cannot be established arithmetically.

is true that is given by the theory of purely finite sets. Goodstein's theorem can be seen to fail in that situation, but then to hold in an infinitary extension.

(b) *The Paris–Harrington sentence.* Ramsey's theorem, in both its finite and infinite forms, concerns partition properties of sets of natural numbers (see Graham *et al.* 1980: ch. 1 for exposition). The finite version of Ramsey's theorem is provable in Peano arithmetic, coded suitably as to be expressed in the language of arithmetic (or it can be expressed more directly and proved in ZF-). Paris and Harrington (1977) found that a (seemingly) slight variation in the theorem made it, while still true, unprovable in PA (the variation consists in the requirement that the homogeneous set given by the original Ramsey theorem should satisfy the further condition that it be 'relatively large', in the technically stipulated sense that the cardinality of the set should be greater than the least number in the set). See also Takeuti 1987: 130–44 for a detailed exposition of the proof that the Paris–Harrington sentence is not derivable in Peano arithmetic. This result was hailed as what mathematicians had been looking for since the Gödel incompleteness theorems, namely, 'a strictly mathematical example of an incompleteness in first-order Peano arithmetic, one which is mathematically simple and interesting and does not require the numerical coding of notions from logic' (Jon Barwise, Editor's Note to Paris and Harrington 1977: 1133).

I do not dissent from the positive enthusiasm of this assessment. A question which it suggests for the present perspective is whether the adjective 'mathematical' could be replaced by 'arithmetical'? Is this a strictly *arithmetical* example of an incompleteness in first-order Peano arithmetic? I am using the word 'arithmetical' here as meaning both mathematical (in the sense that Barwise uses the term) and being *about* the arithmetic of the natural numbers. To mean by arithmetical 'expressible in the language of arithmetic', as one might, would beg, or simply obliterate the question I am trying to raise here. The issue of this essay can be seen as the question whether there *is* such a sense to 'arithmetical' *different* from expressibility in the language of arithmetic.

The criterion I have in mind is that a truth expressed in the (first-order) language of arithmetic is arithmetical just in case its truth is directly perceivable on the basis of our (higher-order) articulation of our grasp of the structure of the natural numbers *or* directly perceivable from truths in the language of arithmetic which are themselves arithmetical. The analysis of the number concept as discussed in §§2–4 seems to me to render the axioms of Peano arithmetic arithmetical, in the sense that their truth is directly perceivable so expressed, and on this basis the second clause renders the theorems of PA arithmetical (though not quite unproblematically; I shall say something in §6 about the possible non-arithmetical nature of some theorems of PA). It seems to me reasonably evident that the examples we have so far considered, the Gödel sentence, Goodstein's theorem, and the Paris–Harrington sentence, are not arithmetical in the first sense. The difficulty comes in having grounds for considering that the second condition does not apply, of being confident that no as yet unknown proof could be given in terms of recognizably arithmetical truths.

What can we say specifically about the Paris–Harrington sentence? Its proof is an easy consequence of the infinite Ramsey theorem: if the Paris–Harrington sentence is false, the set of counter-examples to it can be given the structure of a finitely branching infinite tree, and the infinite branch which is shown to exist by the König infinity lemma produces a counter-example to the infinite Ramsey theorem (see Paris and Harrington 1977: 1135). This argument, by its reliance on the infinite Ramsey theorem and the König infinity lemma, evidently goes beyond finite arithmetic. What about the possibility that some other proof might exist which constitutes a purely arithmetical basis for perceiving this truth, as happens in the case of the finite Ramsey theorem? (The finite Ramsey theorem, like the Paris–Harrington sentence, can be proved from the infinite Ramsey theorem by use of the compactness theorem, or König infinity lemma—see Graham *et al.* 1980: 16, but in this case there is also a direct proof of it by an argument using mathematical induction which can be formalized in Peano arithmetic; see ibid. 7–9.) The following theorem (3.1 in Paris and Harrington 1977) renders it unlikely that these could be any proof of the Paris–Harrington sentence which is arithmetical in the sense at issue here: the Paris–Harrington sentence is provably equivalent in PA to the soundness of Peano arithmetic with respect to purely existential first-order sentences. (Sentences of this form are designated as Σ_1^0 and soundness of PA with respect to them is the Σ_1^0-reflection principle for PA.) Σ_1^0-reflection is an expression of the soundness of PA as an axiomatization of arithmetical truth, and in these terms looks beyond the natural numbers themselves to our capacity to consider the correctness of our analysis of their fundamental properties. It is in fact a strong enough form of reflection not only to tell us that our formalization is consistent, but also to express that these are the intended notions, via the 1-consistency of PA (a special case of Gödel's notion of ω-consistency).[5] I construe this result as revealing something of the implicit (hidden) higher-order content of the (first-order) Paris–Harrington sentence. The Paris–Harrington sentence expresses a strong reflective property *about* the whole formalization of arithmetic, and

---

[5] For an account of the notions of ω-consistency and 1-consistency and a proof that 1-consistency is equivalent to Σ_1^0-reflection, see Smorynski 1977: 851–2.

is thereby implicitly higher-order, and so, in the sense I am trying to make clear, non-arithmetical.[6]

(c) *Friedman's finitization of Kruskal's theorem.* See Smorynski 1982 for a highly accessible account of this result, and Simpson 1985: §3 for a more general treatment; see also Takeuti 1987: 144–7. The finite version of Kruskal's theorem is expressible in the language of arithmetic and unprovable not only in Peano arithmetic, but also in predicative analysis, a strong extension of Peano arithmetic, whereas both Goodstein's theorem and the Paris–Harrington sentence are provable in predicative analysis, and already in weaker systems. The finite form of Kruskal's theorem constitutes, in this way, a particularly compelling case of the situation represented by the previous two examples, of a truth expressible in the language of arithmetic that is not arithmetically true.[7] (By its very strength, the finite form of Kruskal's theorem does not test so well the claim that PA itself is complete for arithmetical truth, which is done more sharply by considering true sentences unprovable in PA whose truth can be established by a relatively weak extension of PA.)

We noted above that the Paris–Harrington sentence can be proved from the infinite Ramsey theorem by a compactness argument. Similarly, the finite form of Kruskal's theorem (for finite sequences of trees) can be proved from Kruskal's theorem for infinite sequences of (finite) trees, via compactness. Such a proof cannot be arithmetical in the sense of this essay, as it must include a proof of the infinite Kruskal theorem, which cannot even be stated in a language interpreted over a domain of finite objects (it requires a quantifier ranging over infinite sequences of finite trees), so, *a fortiori*, cannot be proved in such a language. There is then the question as to whether some other proof of the finite form of Kruskal's theorem could be found which would establish its conclusion as arithmetical in the sense at issue here. It seems to me clear, from the strength of finite Kruskal theorem taken as an axiom, that there can

---

[6] One might ask, how can 1-consistency guarantee that the theory is of the standard model? After all, when formalized and expressed mathematically, the applicability of the Gödel phenomenon means that 1-consistency itself is subject to non-standard interpretation. Indeed, this is the case. There are ω-consistent (and so, *a fortiori*, 1-consistent) theories which have no standard model (Kreisel 1955; for a heuristic account of this result, see Isaacson 1992: n. 17). The point, for present purposes, is that the notion of 1-consistency has the required expressive power on our understanding of it as intuitive, non-formal mathematics.

[7] Cf. Smorynski 1982: 'The construction needed to establish [the Paris–Harrington theorem] is too complex; in fact, the function given by the construction grows so fast that FNT [= PA] cannot handle it. The independence of [Friedman's finite form of Kruskal's theorem] shares this feature with that of [the Paris–Harrington theorem]—with, of course, the difference that [Friedman's finite form of Kruskal's theorem] is independent of far stronger theories, has a far more complex construction underlying its validity, and is witnessed by a far more rapidly growing function' (pp. 186–7, or 391–2 in reprint).

be no such proof. In particular, the finite Kruskal theorem implies the well-ordering of a system of notation for the ordinals less than Γ_0, the smallest impredicative ordinal (see Smorynski 1982: 187–8). A characteristic of Γ_0 is that it cannot be obtained by a process of generation 'from below', and indeed its existence presupposes that of the first uncountable ordinal (ibid. 185). Accordingly, any attempt to establish its truth must proceed from essentially set-theoretic principles which are beyond anything required to establish the structure of the natural numbers, a well-ordering of order type ω.[8]

§6. HIGHER-ORDER CONCEPTS WITHIN PEANO ARITHMETIC

I have been attempting in this paper to assess the non-arithmetical character of statements in the language of arithmetic which are true but unprovable in PA in terms of their coding of mathematical situations whose description requires use of higher-order concepts. A serious challenge to stability of such an assessment arises from the observation that many statements in the language of arithmetic which *are* provable in PA code assertions of the same character as those I have been terming higher-order. Examples include transfinite induction of order type α for $\alpha < \varepsilon_0$ (see n. 8), consistency of subsystems of PA, such as PR (primitive recursive arithmetic) or PA_n (where the scheme of induction is restricted to formulae of logical complexity bounded by n), and so on. How can it be that the coded presence of such notions renders truths expressed in the first-order language of arithmetic non-arithmetical when they are unprovable in PA, but does not have this effect when the statement in question *is* provable in PA? But if it did have that effect, then the idea of provability in Peano arithmetic marking a natural boundary to arithmetical truth would be called into question.

The situation seems to me not so drastic as this. I am concerned with the way in which arithmetical truths can be established. The point about the examples of truths unprovable in PA considered in this paper is that they are, in each case, shown to be true by an argument in terms of truths concerning some higher-order notion, and in each case also a converse holds, so that the only way in which the arithmetical statement can be established is by an

---

[8] The point here is not simply that Γ_0 is ordinally greater than ω, since the first-order scheme of induction of order-type ω allows derivation of the schemes of induction for certain greater order types, in particular for any order-type of the form ω_n, for each natural number n, where $\omega_0 = \omega$, $\omega_{n+1} = \omega_n^\omega$. On the other hand, the scheme of induction with respect to the limit of the ordinals ω_n, ε_0, is not provable from ω-induction (these results were first obtained by Gentzen 1943; for a more recent exposition see Takeuti 1987; 125–6, 147–54). I have claimed that ε_0-transfinite induction cannot be established arithmetically, see n. 4 above.

argument which establishes the higher-order truth. The relationship of coding constitutes a rigid link between the arithmetical and the higher-order truths, which pulls the ostensibly arithmetical truth up into the higher-order. In cases such as, for example, $TI(\omega^\omega)$, an arithmetical sentence coding transfinite induction of order type ω^ω, there is similarly a rigid linkage between two kinds of truths. But in these cases, the linkage pulls the ostensibly higher-order truth down into the arithmetical. The statement in the language of arithmetic has a derivation in PA, and following through that derivation gives the basis for perception of that truth as true in arithmetic purely on the basis of directly perceivable arithmetical truths.

This answer may seem unrealistic, on the grounds that there can be cases where the higher-order perspective is essential for *actual* conviction as to truth of the arithmetically expressed sentence. One may know that a derivation in PA must exist but, if generated, would be so long as to be unsurveyable. This might be true, for example, in the case of Con(PR), the arithmetical statement coding the consistency of primitive recursive arithmetic, or $TI(\alpha)$ for $\alpha < \varepsilon_0$ but large in comparison with ω. There is a theorem of Gödel (1936), on the lengths of proofs, which points to this sort of situation as a systematic phenomenon: for each computable function f, there correspond infinitely many formulae ϕ in the language of arithmetic such that each ϕ is provable both in PA and PA^2 (axiomatic second-order arithmetic) but where if j is the length (measured in number of formulae of which it consists) of the shortest proof of ϕ in PA and k is the length of the shortest proof of ϕ in PA^2, then $j > f(k)$. The higher-order perspective can be essential, then, for shortening an otherwise unsurveyable proof.

This point seems to me a serious one, and in some ways I am inclined to accept rather than to resist the force of it. It comes down to the issue of the extent to which it seems relevant and legitimate to appeal to the notion of an operation being performable 'in principle', a notoriously difficult matter, the answer to which depends critically on the context and purpose of the thing to be done. If one is prepared to countenance a notion of being 'in principle' derivable in PA, then the present problem disappears. One might consider that this move is legitimate, as enabling one to define precisely a theoretical boundary, to which mathematical practice approximates. However, I have in my discussion been considering provability in terms of providing a basis for perceiving the truth of a given statement. In these terms, a proof in PA of a given proposition being infeasibly long has to be taken seriously. If one does so, then within the arithmetically expressible truths of mathematics, we must think of the boundary between those which are purely arithmetical and those which are essentially higher-order as running somewhat inside the collection of those for which derivations in PA exist. But in thus narrowing the boundary of the arithmetically provable to a proper subset of those sentences for which there exist derivations in PA, the considerations which favour doing so also dictate a correspondingly narrower domain of the arithmetically true. With these two notions shrinking thus together, the general thesis, that Peano arithmetic is complete with respect to purely arithmetical truth, stands.

§7. CONCLUSION

The point of this essay is essentially conceptual. It is concerned with what our attitude should be toward the phenomenon of incompleteness of formal systems for arithmetical truth. The term 'incompleteness' suggests that the formal system in question *fails* to offer a deduction which it *ought* to. The contrasting attitude, which I have been attempting to explore here, is to see these arithmetically expressible independent truths as exemplifying the expressive richness of finite structures, perceivable through the phenomenon of coding. Coding is not something which we do, but rather recognize as existing, thereby *discovering* rigid links between truths of ostensibly different character. These links belong to the mathematics of the structures being studied.

The formal system on which I have focused attention is Peano arithmetic. I have been attempting to explore its conceptual stability in light of the apparently destabilizing effect, through extensibility, of the phenomenon of incompleteness. I have offered heuristic and conceptual support for the viewpoint that PA is the strongest natural first-order system for arithmetic, and that there is a sense in which it is complete with respect to purely arithmetical truth. That sense can be expressed by the claim that any true extension of it must be based on considerations in terms of higher-order concepts beyond those needed to characterize our grasp of the structure of the natural numbers. Known cases of true sentences in the language of arithmetic unprovable by PA are such that we could not take them as axioms in an extension of PA, given only their arithmetical formulation. Rather, we must look to some more comprehensive mathematical theory, and to the link between the arithmetical statement and a corresponding statement in the more comprehensive theory, for our mathematical confidence in its truth. It is in this sense that truths in the language of arithmetic which lie beyond what is provable in Peano arithmetic must be perceived as true in terms of hidden higher-order concepts.[9]

---

[9] I am extremely grateful for opportunities I have had to present material which is developed here, and to members of my audiences on these occasions for helpful responses. I was stimulated to

REFERENCES

Ackermann, Wilhelm (1937), 'Die Widerspruchsfreiheit der allgemeinen Mengenlehre', *Mathematische Annalen*, 114: 305-15.

Buchholz, Wilfried, and Wainer, Stan (1987), 'Provably Computable Functions and the Fast Growing Hierarchy', in Stephen G. Simpson (ed.), *Logic and Combinatorics* (Providence, RI: American Mathematical Society), Contemporary Mathematics Series, 65: 179-98.

Dedekind, Richard (1888), *Was sind und was sollen die Zahlen?* (Brunswick: Vieweg); English trans. by W. W. Beman as 'The Nature and Meaning of Numbers', in Richard Dedekind, *Essays on the Theory of Numbers* (La Salle, Ill.: Open Court, 1901; repr. New York: Dover, 1963), 31-115.

—— (1890), letter to Keferstein, trans. in Jean van Heijenoort (ed.), *From Frege to Gödel: A Source Book in Mathematical Logic, 1879-1931* (Cambridge, Mass.: Harvard University Press, 1967), 98-103.

Feferman, Solomon (1979), 'A More Perspicuous Formal System for Predicativity', in Kuno Lorentz (ed.), *Konstructionen versus Positionen: Beiträge zur Diskussion um die Konstructive Wissenschaftstheorie* (Berlin: Walter de Gruyter), 68-93.

Frege, Gottlob (1884), *Die Grundlagen der Arithmetik: Eine logisch mathematische Untersuchung über den Begriff der Zahl* (Breslau: Wilhelm Koebner); English trans. by J. L. Austin (Oxford: Blackwell, 1950).

Gentzen, Gerhard (1943), 'Beweisbarkeit und Unbeweisbarkeit von Anfangsfällen der transfiniten Induktion in der reinen Zahlentheorie', *Mathematische Annalen*, 119: 140-61; English trans., 'Provability and Nonprovability of Restricted Transfinite Induction in Elementary Number Theory', in M. E. Szabo (ed.), *The Collected Papers of Gerhard Gentzen* (Amsterdam: North Holland, 1969), 287-308.

Gödel, Kurt (1931), 'Über formal unentscheidbare Sätze der Principia mathematica und verwandter Systeme I', *Monatshefte für Mathematik und Physik*, 38: 173-98; English trans. by Jean van Heijenoort, in Solomon Feferman *et al.* (eds.), *Kurt Gödel, Collected Works*, vol. 1 (Oxford: Oxford University Press, 1986), printed with original text facing, 144-95.

—— (1936), 'Über die Länge von Beweisen', *Ergebnisse eines mathematischen Kolloquiums*, 7: 23-4; English trans. by Stefan Bauer-Mengelberg and Jean van Heijenoort, in *Kurt Gödel, Collected Works*, vol. 1, with original text facing, 396-9.

---

pursue this theme by a lecture in Oxford by Laurie Kirby in January 1982, and first expounded some of these ideas, to an Oxford philosophy discussion group, the following month. Subsequent presentations have been to my philosophy of mathematics seminar, to the philosophy research seminar at St Andrew's, the Somerville Philosophical Society, the European summer meeting of the Association of Symbolic Logic, at Orsay in Paris, and the Center for the Study of Language and Information at Stanford. I am especially grateful for the invitation to speak at the ASL summer meeting (July 1985), the occasion for writing this paper, to appear in its Proceedings. I cannot name all those whose questions and comments and encouragement have benefited me, but I want especially to mention Jon Barwise, Oswaldo Chateaubriand, Burton Dreben, Solomon Feferman, Robin Gandy, Alexander George, Jocelyn Hawkins, Ruth Isaacson, Angus Macintyre, Dag Prawitz, and Philip Scowcroft. I owe an especially great debt in this paper to Alex Wilkie. His lectures in Oxford on models of arithmetic made work in this area accessible, and many of the thoughts I present here have developed from remarks of his (in particular, the emphasis on the link between Peano arithmetic and finite mathematics). His written comments on a draft of this paper have been illuminating and generous.

This paper is dedicated to the memory of my father, Robert Isaacson.

Goodstein, R. L. (1944), 'On the Restricted Ordinal Theorem', *Journal of Symbolic Logic*, 9: 33–41.

Graham, Ronald L., Rothschild, Bruce L., and Spencer, Joel H. (1980), *Ramsey Theory* (New York: John Wiley & Sons).

Henle, James M. (1986), *An Outline of Set Theory* (Heidelberg: Springer-Verlag).

Hilbert, David (1928), 'Die Grundlagen der Mathematik', *Abhandlungen aus dem mathematischen Seminar der Hamburgischen Universität*, 6: 65–85; English trans. by S. Bauer-Mengelberg and D. Follesdal, 'The Foundations of Mathematics', in Jean van Heijenoort (ed.), *From Frege to Gödel: A Source Book in Mathematical Logic 1879–1931* (Cambridge, Mass.: Harvard University Press, 1967), 464–79.

Isaacson, Daniel (1992), 'Some Considerations on Arithmetical Truth and the ω-Rule', in Michael Detlefsen (ed.), *Proof, Logic and Formalization* (London: Routledge), 94–138.

Ketonen, J., and Solovay, R. (1981), 'Rapidly Growing Ramsey Functions', *Annals of Mathematics*, 113: 267–314.

Kirby, Laurie, and Paris, Jeff (1982), 'Accessible Independence Results for Peano Arithmetic', *Bulletin of the London Mathematical Society*, 14: 285–93.

Kreisel, G. (1955), review of Leon Henkin, 'A Generalization of the Concept of ω-Consistency', *Mathematical Reviews*, 16: 103.

Paris, Jeff, and Harrington, Leo (1977), 'A Mathematical Incompleteness in Peano Arithmetic', in Jon Barwise (ed.), *Handbook of Mathematical Logic* (Amsterdam: North Holland), 1133–42.

Parsons, Charles (1987), 'Developing Arithmetic in Set Theory without Infinity: Some Historical Remarks', *History and Philosophy of Logic*, 8: 201–13.

Quine, W. V. (1970), *Philosophy of Logic* (Englewood Cliffs, NJ: Prentice-Hall).

Simpson, Stephen C. (1985), 'Nonprovability of Certain Combinatorial Properties of Finite Trees', in L. A. Harrington et al. (eds.), *Harvey Friedman's Research on the Foundations of Mathematics* (Amsterdam: North Holland), 87–117.

Smorynski, Craig (1977), 'The Incompleteness Theorems', in Jon Barwise (ed.), *Handbook of Mathematical Logic* (Amsterdam: North Holland), 821–65.

—— (1982), 'The Varieties of Arboreal Experience', *Mathematical Intelligencer*, 4: 182–9; repr. in Harrington et al. (eds.), *Harvey Friedman's Research on the Foundations of Mathematics*, 381–97.

Takeuti, Gaisi (1987), *Proof Theory*, 2nd edn. (Amsterdam: North Holland).

Wang, Hao (1958), 'Eighty years of Foundational Studies', *Dialectica*, 12: 465–97; repr. in Hao Wang, *A Survey of Mathematical Logic* (Peking: Science Press, 1962, and Amsterdam: North Holland, 1964), 34–56.

Wilkie, A. J. (1981), 'On Discretely Ordered Rings in which Every Definable Ideal is Principal', in C. Berline, K. McAloon, and J.-P. Ressayre (eds.), *Model Theory and Arithmetic (Proceedings, Paris, 1979/80)*, Springer Lecture Notes in Mathematics no. 890 (Heidelberg: Springer-Verlag), 297–303.

—— (1987), 'On Schemes Axiomatizing Arithmetic', in Andrew M. Gleason (ed.), *Proceedings of the International Congress of Mathematicians in Berkeley, California 1986* (Providence, RI: American Mathematical Society), 331–7.

Zermelo, Ernst (1909), 'Sur les ensembles finis et le principe de l'induction complète', *Acta Mathematica*, 32: 185–93.

XI

CONSERVATIVENESS AND INCOMPLETENESS

STEWART SHAPIRO

A common argument for mathematical platonism, due to Putnam[1] and Quine, is based on the role of mathematics in science. The main premiss is a statement to the effect that mathematics forms an essential part of virtually every scientific theory. That is, scientific theories of the material world are formulated in mathematical terms and have variables ranging over abstract mathematical entities, such as numbers. It follows from widely accepted principles of ontological commitment that such theories presuppose the existence of these abstract entities.

In a recent, provocative book entitled *Science without Numbers*, Hartry Field[2] attempts to refute this argument by rejecting and undermining its main premiss. Of course, Field does not deny that mathematics is *useful* in science, and, moreover, he admits that mathematics is a 'practical necessity' in the practice of science, but this is not to concede that mathematics is *essential* to science in the ontologically relevant way.

There are two aspects of Field's work. The first is to provide a 'nominalistic' formulation of each scientific theory, a formulation that does not refer to, or have variables ranging over, abstract entities. To this end, Field develops in detail a nominalistic formulation of Newtonian gravitational theory, a formulation which is to serve as a paradigm for other branches of science. The second aspect of Field's work is to account for how the addition of a mathematical theory to a nominalistic theory can be useful without presupposing that the relevant mathematical assertions are true or, indeed, that there is a subject-matter for the mathematics. Field argues that mathematical theories are *conservative* over nominalistic theories within science in that a nominalistic assertion of the science is a consequence of the combined theory only if it is a consequence of the nominalistic theory alone. Thus,

First published in *Journal of Philosophy*, 80/9 (1983): 521–31. Reproduced by permission of the *Journal of Philosophy* and Stewart Shapiro.

[1] Hilary Putnam, *Philosophy of Logic* (New York: Harper and Row, 1971).
[2] Princeton, NJ: Princeton University Press; Oxford: Blackwell, 1980.

mathematics can be useful in shortening derivations, but in principle it is dispensable.

The concern of the present essay is with the conservativeness notion. As Field notes, there is an ambiguity in its formulation which involves the distinction between semantic consequence and deductive consequence. I show that attention to this ambiguity, together with the differences between first-order and second-order languages, undermines Field's main argument. The conclusion is that for any reasonable physical and mathematical theories in Field's programme, either the mathematical theory is not conservative in the philosophically relevant way or the mathematics is not applicable to the physical theory in the usual way. As a preliminary, the two aspects of Field's programme are briefly presented.

Field's formulation of Newtonian gravitational theory postulates, and has variables ranging over, space-time points and regions, the latter regarded as classes (or 'Goodmanian sums') of space-time points. Such points and regions are taken to be theoretical *physical* entities, on a par with, say, electrons.[3] The nominalistic physics is formulated in a *second-order* language, whose first-order variables range over space-time points and whose second-order variables range over regions. In the closing chapter, a reformulation of the theory in a first-order language (with variables ranging over regions) is proposed. The physics is axiomatized in terms of certain primitive relations among space-time points. Examples of these include 'y Bet xz' interpreted as 'x, y, and z are co-linear and y is between x and z' and 'xy S-cong zw' interpreted as 'x and y are simultaneous, as are z and w, and the distance between x and y is equal to the distance between z and w'. The axioms entail that there are uncountably many space-time points and, in particular, that the system of space-time is isomorphic to R^4 (under the appropriate relations). This theory is thus incompatible with *finitism*, but Field maintains that it is compatible with nominalism in that it does not have variables ranging over abstract entities.

Conservativeness is a technical property between mathematical theories and scientific theories. Let N be a nominalistic theory within science and S a mathematical theory, such as set theory with ur-elements. (The ur-elements are to allow the combined theory to refer to sets of physical objects.) Ignoring

---

[3] Field regards space-time points as physical, non-abstract entities, because aspects of the collection of such points, such as its cardinality, depend on a physical (rather than mathematical) theory, and contingent properties of space-time points, such as having a (relatively) large gravitational force, are essential parts of causal explanations of observable phenomena. It might be noted that many reviewers of the book either take issue with the claim that space-time points are not abstract or question the very dichotomy between physical entities and abstract entities. (See, e.g., the reviews by D. Malament, *Journal of Philosophy*, 79 (1982): 523–34; M. Resnik, *Noûs*, 27 (1983): 514–19; and myself, *Philosophia*, 14 (1984): 437–44.) The issue is not important here.

a technical device Field employs (whose purpose is of no concern here[4]), the *conservativeness* of S over N is defined as follows (see p. 12):

Let A be any nominalistically statable assertion. Then A isn't a consequence of N + S unless A is a consequence of N.

Field informally argues that mathematics is conservative over physics in terms of the original motivation for doing mathematics and formulating mathematical theories. In addition, two formal arguments are presented in the appendix to chapter 1.

According to Field, the way mathematics is usually applied to science is that in the combined theory N + S each nominalistic statement A is proved to be equivalent to a statement A' formulated in the language of the mathematical theory S. The statement A' is called an *abstract counterpart* of A. The equivalences allow the scientist to appropriate the mathematical terminology and structures of S, such as the mathematical operations and inferences already developed. The conservativeness sanctions this process for a nominalistically-minded scientist. The use of mathematics is theoretically superfluous; hence it is ontologically harmless.

In the theory Field develops, the indicated equivalences are developed by proving (in the combined theory) that there is a structure-preserving function, a *representing homomorphism*, from the points of space-time to R^4. That is, one proves that there is a function f from the class of space-time points to the set of quadruples of real numbers such that, for example, if a, b, c, d are points and ab S-cong cd, then the temporal (or fourth) co-ordinate of $f(a)$ is equal to the temporal co-ordinate of $f(b)$, the temporal co-ordinate of $f(c)$ is equal to the temporal coordinate of $f(d)$, and the distance in R^4 between $f(a)$ and $f(b)$ is equal to the distance in R^4 between $f(c)$ and $f(d)$. On the basis of this representing homomorphism, one can 'translate' any statement about space-time points into an 'equivalent' statement about real numbers. It should be noted that the existence of this homomorphism is crucial for the application of mathematics to the nominalistic physics.

Field's programme for science is at least somewhat analogous to the Hilbert programme for mathematics. The 'contentual' part of, say, Newtonian gravitational theory is the nominalistic theory; the 'ideal' part is the mathematical theory which is adjoined. The role of conservativeness in the Field programme is analogous to the role of consistency in the Hilbert programme.[5] I show here that the Field programme has serious difficulties

---

[4] The purpose of the technical device is to 'separate' the ranges of the nominalistic variables and constants of N from the ranges of the mathematical variables and constants of S. In the present context, it may be assumed that this has been done.

[5] Actually, the Hilbert programme also requires conservativeness. It is easy to see, however, that in the case of the Hilbert programme, consistency and conservativeness are coextensive. The analogy between the Hilbert programme and the Field programme was suggested by Mark Steiner.

which are analogous to the difficulties with the Hilbert programme related to the Gödel incompleteness theorems.

There are two assumptions on the formal theories which are necessary for present purposes. The first is the common constraint that the physical theory N be recursively axiomatizable; the second is that it be provable in the mathematical theory S that N is consistent (through a suitable arithmetization). The latter is clearly reasonable since S is taken to include set theory, which, presumably, is stronger than any nominalistic theory of the physical world. Moreover, it is important for Field's purposes that it be provable in S that N has a model and, thus, that N is consistent.

I begin with the second-order theory that Field develops.[6] It is easily seen that the results apply to any similar second-order substitute. Following this, I turn to the prospects for a first-order version.

As formulated above (and by Field on p. 12), conservativeness is ambiguous as to whether it involves proof-theoretic derivability in N and N + S or semantic consequence in N and N + S. The ambiguous word is 'consequence'. To emphasize the distinction, I'll formulate each notion separately:

S is *semantically conservative* over N if and only if, for each nominalistically statable assertion A, if A is true in all models of N + S, then A is true in all models of N.

S is *deductively conservative* over N if and only if, for each nominalistically statable assertion A, if A is a theorem of N + S, then A is a theorem of N.

If the theories of N and S were both first-order, of course, there would be no ambiguity, since the Gödel completeness theorem entails that in such languages semantic consequence and deductive consequence are coextensive. The completeness theorem does not hold for second-order theories, however.[7]

As indicated by Field himself, only the semantic version of conservativeness has been established (see p. 40 and n. 30). Concerning the formal arguments (in the appendix to ch. 1), the first is explicitly semantic. The second invokes the completeness theorem, and thus applies only to first-order theories. Field notes, however, that a similar argument establishes the *semantic* conservativeness of the appropriate second-order theories. Although matters are not as straightforward, the informal argument for conservativeness

---

[6] It should be noted that the second-order language that Field develops has a *standard* semantics in which, for a given model, the second-order variables range over every subset of the domain. The present considerations depend on this. If one takes the same language with a Henkin semantics (in which, for a given model, the range of the second-order variables is a fixed subset of the power set of the domain), then the considerations brought below against the first-order case apply. In particular, the mathematics would be deductively conservative over the nominalistic physics (in light of the Henkin completeness theorem), but the mathematics cannot be applied to the physics in the way Field indicates. A second-order language with a Henkin semantics is, in effect, a two-sorted first-order language.

[7] See, e.g., G. Boolos and R. Jeffrey, *Computability and Logic*, 3rd edn. (Cambridge: Cambridge University Press, 1989), ch. 18.

(pp. 12–13) is also limited to the semantic version. In the course of this argument, Field asserts that if a body of nominalistic assertions is 'logically consistent', then 'it would seem that it must be possible and/or not *a priori* false, that such a . . . body of assertions about concrete objects . . . is true'. For second-order languages, this conditional is not correct unless 'logically consistent' is taken as the semantic 'satisfiable'. As is well known, in any (recursively axiomatized and sound) second-order deductive system for logic, there are sentences which are consistent in the sense that no contradiction can be derived from them, but which are not satisfiable. Such sentences *are* '*a priori* false' and *are not* 'possibly true' in the relevant senses of those terms.

According to Field, on the other hand, the role of mathematics in (nominalistically formulated) science is to shorten *deductions*. A few passages illustrate this:

. . . any inference from nominalistic premises to a nominalistic conclusion that can be made with the help of mathematics could be made (usually more long-windedly) without it. (p. x)

. . . the conclusions we arrive at [by adding the mathematical theory] are not genuinely new, they are already derivable in a more long-winded fashion from the [nominalistic theory] without recourse to the mathematical entities. (pp. 10–11)

We can use [the mathematical theory] as a device for drawing conclusions . . . much more easily than we could draw them by a direct proof . . . (p. 28)

For this purpose, *deductive conservativeness* of the mathematics over the physics is required. Let A be a nominalistic assertion proved in N + S. If S is not deductively conservative over N, then one cannot conclude that A can be proved in N alone, even more 'long-windedly'. The assertion A may not be derivable in N at all. Of course, if S is semantically conservative over N (and the deductive systems are logically sound), then one can conclude that A is a semantic consequence of N. In such a case, the situation would be that the addition of the mathematical theory allows the deduction of *new consequences* of N—consequences not deducible without the mathematics. The mathematics would not be theoretically superfluous or, at any rate, not be theoretically superfluous in the obvious or straightforward way.

I present a technical result: the presented (ur-elemented) set theory S is *not* deductively conservative over the nominalistic physics that Field develops. That is, there is a sentence θ formulated in the language of N such that S + N ⊢ θ but N ⊬ θ. Of course, given semantic conservativeness, θ is true in all models of N, but it is not deducible in N. The central idea of the result is that since the class of space-time points is isomorphic to R^4, it is possible to model the natural numbers in space-time and, in effect, to do arithmetic in N. To put it differently, the natural number structure is exemplified in the universe of space-time. The sentence θ is a Gödel sentence.

On page 65, Field indicates the definition in N of a *spatio-temporally equally spaced region*, a class of (discrete) points all of which lie on a single straight line and such that the distance between adjacent points is uniform.[8] Let $\psi(R,p)$ be a formula equivalent to 'R is an infinite, spatio-temporally equally spaced region containing p as end-point'.[9] We have $N \vdash \exists R \exists p \psi(R,p)$. Such pairs $\langle R,p \rangle$ are models of the natural numbers. Relative to R and p, one can construct a formula $\Sigma(x,y)$ equivalent to 'x and y are both in R, there is no point of R strictly between x and y, and if $x \neq p$ then x is strictly between p and y'. The formula Σ represents the 'successor relation' of $\langle R,p \rangle$. The analogues of addition and multiplication can also be defined.[10] It follows from N that if $\psi(R,p)$ holds, then $\langle R,p \rangle$ under Σ satisfies the axioms of Peano arithmetic.[11] Since (presumably) N is recursively axiomatized, the Gödel incompleteness theorems apply. Let $\mathrm{Con}_N(R,p)$ be the formula asserting the 'consistency' of N in terms of the points in R (analogous to the usual formulation in terms of the natural numbers). Finally, let θ be the sentence

$$\forall R \forall p [\psi(R,p) \to \mathrm{Con}_N(R,p)].$$

The relevant version of the incompleteness theorem entails that $N \nvdash \theta$. Notice that θ is a sentence in the language of the nominalistic physics. Its variables range over space-time points and regions.

A second presumption is that the consistency of N is provable in the set theory S. Let $\mathrm{Con}_N(\omega,0)$. Notice that it is provable in N + S that the formula $\psi(R,p)$ is 'categorical' and that any pair $\langle R,p \rangle$ satisfying ψ is isomorphic to the natural numbers $\langle \omega,0 \rangle$. The latter follows from the existence of a representing homomorphism from the points of space-time to R^4. Thus, we have

$$N + S \vdash \forall R \forall p [\psi(R,p) \to (\mathrm{Con}_N(\omega,0) \leftrightarrow \mathrm{Con}_N(R,p))].$$

Hence, $N + S \vdash \theta$. This refutes the deductive conservativeness of S over N.

---

[8] The definition of 'spatio-temporally equally spaced region' consists of a formula, with R free, equivalent to 'any three points in R are co-linear and for every point x of R which lies strictly between two points of R, there are points y and z in R such that (a) exactly one point of R lies strictly between y and z and that point is x, and (b) the distance between y and x is equal to the distance between x and z'.

[9] More formally, $\psi(R,p)$ is equivalent to 'R is a spatio-temporally equally spaced region. p is in R, p does not lie strictly between any two points of R, and, for every x in R, if $x \neq p$, then there is a point y in R such that x lies strictly between p and y'.

[10] Relative to $\langle R,p \rangle$, the statement that 'z is the "sum" of x and y' is equivalent to 'the distance between p and x is equal to the distance between y and z'. The statement that 'z is the "product" of x and y' is equivalent to 'either (i) $x = p$, and $z = p$; or (ii) $x \neq p$, and there is a region R such that (a) R̄ is spatio-temporally equally spaced, (b) R̄ is a subregion of R, (c) p, x, and z are all in R̄, (d) there is no point of R̄ that is strictly between p and x, and (e) the number of points in R between p and y is equal to the number of points in R̄ between p and z'. It might be noted that Field introduces cardinality comparisons for equally spaced regions in ch. 6.

[11] The elementary properties of the 'successor', 'addition', and 'multiplication' relations are easily verified. The induction property is a consequence of the completeness axiom for space-time.

A few brief remarks may be in order. The preceding result shows that for second-order theories, deductive conservativeness is not coextensive with semantic conservativeness. The gap can be substantial. Notice that if T is any categorical theory and S + T is satisfiable, then S is semantically conservative over T. Thus, for example, set theory is semantically conservative over second-order arithmetic. As is well known, however, in set theory one can deduce many arithmetic statements that are not theorems of arithmetic alone.

Of course, there is no indication that the deductive strength of Field's physics is insufficient. The sentence θ, taken as a statement about space-time points and regions, is rather obscure—it is not likely to form an essential part of the account of any phenomena to be explained by physics. It may be the case that every interesting or scientifically relevant theorem of N + S is a theorem of N. That is, attending to the set-theoretic hierarchy does allow the deduction of new consequences of N, but it may be that no new interesting or relevant consequences are obtained. The latter, however, remains to be shown (provided that a notion of 'scientific relevance' can be formulated). At any rate, the point of the result above is to refute Field's *argument* for the conservativeness of mathematics over physics in the philosophically relevant way.

A possible response to the result, perhaps, would be to maintain that semantic conservativeness is the more important notion. It may be suggested, for example, that one think of N as having a complete second-order logic. Considered this way, a given sentence is a 'theorem' of N if and only if it is a semantic consequence of the axioms. Of course, this would be to give up the presumption that N has a recursively axiomatized deductive system. Field at least suggests this view. He notes that the nominalistic physics 'does admittedly have a logic that one might find objectionable: it involves what may be called . . . *the complete logic of Goodmanian sums*, and this is not a recursively axiomatizable logic' (p. 38).

The onus of this response, however, is to show what sense a nominalist can make of the notion of 'second-order semantic consequence'. The usual formulation refers to models or interpretations of theories, items which are prima facie not available to a nominalist.[12] Moreover, it is still available to a platonist in the Quine/Putnam school to claim that in some cases a substantial amount of mathematics is necessary to obtain the second-order consequences of axioms (especially cases in which the axioms have only uncountable models). It would follow that the acceptance of a complete second-order logic has its own ontological commitments to mathematical entities.[13]

---

[12] This point is made in David Malament's review of Field's book.

[13] It may be that similar considerations have led to Quine's well-known view that second-order logic is 'set theory in disguise'. See his *Philosophy of Logic* (Englewood Cliffs, NJ: Prentice-Hall, 1970; repr. Cambridge, Mass.: Harvard University Press, 1986), ch. 5.

I turn to the first-order case. In light of the completeness theorem, Field's formal and informal arguments do establish the deductive conservativeness of mathematics over a first-order counterpart N' of the nominalistic physics. I show here, however, that the situation concerning the first-order version is actually worse. In particular, if N' is first-order, then one cannot prove (in N' + S)[14] the existence of a representing homomorphism from the points of space-time to R^4. (It does not matter whether S is first-order or second-order.) The importance of this follows from the fact that Field shows (correctly, I believe) that the representing homomorphisms constitute the crucial aspect of the application of mathematics to space-time. In sum, then, mathematics cannot be applied to first-order nominalistic physics in the usual way.

Field's first-order version of the nominalistic physics is developed in the final chapter. Its variables range over *regions* of space-time, and the language has a subregion relation. A *point* is defined to be a region that has no proper subregions.

It is thus straightforward to develop first-order counterparts to the formulae involved in the technical result above. In particular, let $\psi'(r,p)$ be equivalent to 'r is an infinite, spatio-temporally equally spaced region containing point p as end-point', and let $\text{Con}_{N'}(r,p)$ be a formula asserting the 'consistency' of N' in terms of the point subregions of r (under the appropriate counterparts to the successor, addition, and multiplication relations). As above, the analogue to the incompleteness theorem entails

$$N' \not\vdash \forall r \forall p [\psi'(r,p) \to \text{Con}_{N'}(r,p)].$$

However, the consistency of N' *is* provable in set theory. Let $\text{Con}_{N'}(\omega,0)$ be the usual formula asserting the 'consistency' of N' in terms of the natural numbers. Then $S \vdash \text{Con}_{N'}(\omega,0)$.

Suppose that one could prove in N' + S that there is a representing homomorphism from the 'points' of space-time to R^4. Then there would be a theorem of N' + S equivalent to 'If $\psi'(r,p)$, then $\langle r,p \rangle$ is isomorphic to $\langle \omega,0 \rangle$'. We would then have

$$N' + S \vdash \forall r \forall p [\psi'(r,p) \to \text{Con}_{N'}(r,p)].$$

This contradicts the (deductive) conservativeness of S over N'. Thus, it is not

---

[14] Since N' is formulated with axiom schemes, there is a potential ambiguity concerning the axioms of N' + S. At present, I take the axioms of N' + S to include the *axioms* of N', i.e., in N' + S, the instances of the axiom schemes of N' do *not* include terminology from the language of set theory. A similar restriction need not apply to any axiom schemes of S; i.e., in N' + S, instances of axiom schemes of S can include terminology from the language of physics. These policies coincide with Field's usage (see p. 17). Moreover, if (contrary to these policies) N' + S is taken to include instances of the axiom schemes of N' containing set-theoretic terminology, then, by an argument similar to that of the second-order case, S is neither deductively nor semantically conservative over N'.

a theorem of S + N' that there is a homomorphism from the points of space-time to R^4.

A semantic illustration of the result follows. The incompleteness theorem entails that there is a model M of N' containing elements a,b, such that M ⊨ $\psi'(a,b)$ but M ⊨ $\neg Con_{N'}(a,b)$. Clearly, M is a *non-standard* model of space-time. That is, the collection of 'space-time points' of M is not isomorphic to R^4. In fact, the system is non-Archimedean. Following a construction described by Field (in the appendix to ch. 1), one can show M to be a submodel of a model $\bar{M}$ of N' + S. (Recall that S is set theory with ur-elements.) Within $\bar{M}$, the set-theoretic hierarchy is standard. Thus, for example, in $\bar{M}$, R^4 is Archimedean. It follows that in $\bar{M}$, the space-time universe (the domain of M) is not isomorphic to R^4. Moreover, there can be no representing homomorphism from a non-Archimedean structure to an Archimedean structure. Thus, the existence of representing homomorphisms cannot be proven in N' + S.[15]

The result is somewhat general. For example, let T be any first-order axiomatization of arithmetic, and let S be a standard axiomatization of set theory with ur-elements. Then one cannot prove in T + S that the set of natural numbers is isomorphic to the set ω of finite ordinals.[16] As above, there are models of T + S in which the collection of finite ordinals is standard, but in which the model of T is non-standard.

In conclusion, if one wishes to use set theory to study or shed light on a particular theory (whose consistency can be proved in set theory) or structure

[15] It should be pointed out that in the closing pages (104–6), Field raises a technical result similar to the above, but he does not seem to realize its full applicability. The context at hand is the comparison of a first-order nominalistic physics N' with a first-order 'platonistic' physics P'. Field notes that one can formulate an analogue to arithmetic in the nominalistic language (using an arbitrary, infinite, equally spaced region with one end-point), and thus construct a Gödel sentence θ' for N'. He then states that θ' is (or should be) provable in P', and thus that there are nominalistically statable consequences of P' which are not provable in N'. He replies, first, that the sentence θ' is somewhat 'recherché', and thus it may still be the case that all interesting and scientifically relevant consequences of P' are consequences of N'. He then considers the possible rejoinder that a nominalistic physics ought to have, as consequences, all of the nominalistically statable consequences of P', even the recherché ones. In reply, it is pointed out that even the platonistic P' has a nominalistically statable Gödel sentence (which is presumably provable with the help of set theory), and therefore that a similar rejoinder applies to P'. He concludes that the only way to avoid the principle behind the rejoinder is to employ the original second-order theories. I have shown here, however, that this doesn't help. The consideration of 'nominalistic' Gödel sentences has consequences beyond the comparison of two first-order theories: in the second-order case, it undermines conservativeness; in the first-order case, it undermines applicability.

[16] The policies concerning the axioms of T + S are analogous to those stated in n. 14 above. In T + S, the instances of the axiom schemes of T do not include set-theoretic terminology, but instances of any axiom schemes in S may contain arithmetic terminology. Of course, as in n. 14, if one did allow instances of the axiom schemes of T to contain set-theoretic terminology, then one could prove that ω is isomorphic to the set of natural numbers in the usual way. In this case, S is not conservative over T.

(which can be modelled in set theory), then there is a trade-off between deductive conservativeness and applicability of the set theory in the usual or straightforward way. One cannot have both. The trade-off depends on the decision as to whether a first-order or a second-order axiomatization is employed for the original structure or theory. The theorems for the application of set theory—the existence of representing homomorphisms—are not possible unless all models of the original theory are 'standards'. This, in turn, requires a second-order axiomatization, and, in this case, deductive conservativeness does not hold.[17]

[17] I would like to thank Mark Steiner and George Schumm for several useful suggestions concerning this essay.

XII

IS MATHEMATICAL KNOWLEDGE JUST LOGICAL KNOWLEDGE?

HARTRY FIELD

Logicism is the claim that mathematics is part of logic. This claim flies in the face of Kant's denial that mathematics is analytic, that is, true by logic and definitions alone; and it seems to me that if mathematics is taken at face value, Kant is surely right.

The reason for this assessment is that mathematics, if taken at face value, makes existential assertions: it asserts, for instance, that there exist prime numbers greater than a million, and therefore that there exist numbers. Indeed, even Kant's example '5 + 7 = 12' makes an existential assertion, if understood in the usual way: it asserts not only that *if* there are x, y, and z such that $x = 5$ and $y = 7$ and $z = 12$, *then* $x + y = z$, but the further existential claim that there *are* such x, y, and z. So to argue against the idea that mathematics, if taken at face value, is true by logic and definitions alone, we only need argue that you can't get existential assertions out of logic and definitions alone.

And Kant did provide such an argument (though not in his discussion of mathematics). Anselm, Descartes, and others had argued that the existence of God is a matter of logic, of conceptual necessity: that it follows from the very concept of God that God exists. Kant argued that this can't possibly be correct, for logic (and logic together with definitions) can never categorically assert the existence of anything. Kant's argument for this principle is that contradictions usually stem from postulating one or more objects and making various assumptions about the postulated object or objects that are mutually inconsistent: for example, postulating a triangle, and then saying something else that implies that it has more than three sides. But there is never a contradiction if we reject the triangle—there is nothing there about which we have made contradictory assumptions. And, to quote Kant, 'the same holds true of the concept of an absolutely necessary being. If its existence

First published in *Philosophical Review*, 93 (Oct. 1984): 509–52. Reproduced by permission of the publisher and the author. © 1984 Cornell University Press.

is rejected, we reject the thing itself with all its predicates; and no question of a contradiction can then arise.'[1] He sums it up by saying 'I cannot form the least concept of a thing which, should it be rejected with all its predicates, leaves behind a contradiction'.[2] I think that this argument is rather persuasive. If it is correct, it cannot be contradictory to deny the existence of God; and it cannot be contradictory to deny the existence of numbers either, for they don't have the mysterious power of leaving behind a contradiction when their existence is rejected any more than God does.

One can quibble with this argument of Kant's for the principle that logic and definitions alone imply no existence assertions; nevertheless, the principle itself is a very compelling one. Perhaps when a person denies the existence of God or of numbers, what the person is saying is false or even 'metaphysically impossible' (whatever that means); but it is not itself a logical contradiction in any normal sense of 'logical contradiction'. Moreover, there is good reason not to depart from the normal sense of 'logic' by counting existence assertions as part of logic: doing so would tend to mask the fact that there is a substantive epistemological question as to how it is possible to have knowledge of the entities in question (God, numbers, etc.).[3]

So mathematics, taken at face value, cannot be reduced to anything reasonably called logic. In a sense of course the classical logicists did take mathematics at other than face value: they held that though it at first blush appears to concern specifically mathematical entities like natural numbers, real numbers, and tensors, it can really all be shown to be part of the theory of properties (or the theory of propositional functions or the theory of extensions of concepts or whatever). But the theory of properties or propositional functions or whatever to which the classical logicists hoped to reduce mathematics asserted the existence of a vast array of properties (or propositional functions), so the problem of how this can reasonably be regarded as logic recurs. (It is, indeed, a problem which eventually led one famous logicist,

---

[1] Kant, *Critique of Pure Reason*, trans. Norman Kemp Smith (New York: St Martin's, 1965), B622–3.

[2] Ibid. B623–4.

[3] Admittedly, there are also questions about how it is possible to know logical truths like 'If snow exists, then snow is white or snow is not white': but such questions seem much less gripping than questions about how we can know the existence of specific kinds of entities, and it seems very unlikely that any reasonable answer to the question of how logic is known would bring with it an answer to the question of how the existence of God or of numbers or of any other specific sorts of entities is known. See, e.g., Ch. I for a discussion of epistemological difficulties that our apparent knowledge of the existence of mathematical entities raises and which don't seem to be raised by straightforwardly logical knowledge. (In my paper 'Realism and Anti-Realism about Mathematics'. *Philosophical Topics*, 13 (1982): 45–69; repr. in Hartry Field, *Realism, Mathematics and Modality* (Oxford: Blackwell, 1989). I discuss this further and argue that contrary to what is sometimes claimed, the epistemological difficulties that Benacerraf discusses do not depend on assuming a causal theory of knowledge.) It would hardly be a solution to the problem that Benacerraf raises to say that we know that numbers exist because *logic guarantees* that they exist.

Frege, to abandon his logicism: 'it seems that [logic] alone cannot yield us any objects.... [So] probably on its own the logical source of knowledge cannot yield numbers.'[4]) If one is going to retain logicism (in conjunction with the Kantian requirement on logic that I have advocated), one is going to have to provide an interpretation of mathematics according to which mathematics does not really make existential assertions despite all appearances to the contrary. I do not regard the prospects of doing this in a plausible way as at all promising, so I will not be defending logicism.

Still, I think that the idea that mathematical knowledge is just logical knowledge is largely correct, for I want to maintain what might be called a *deflationist* position about mathematical knowledge. That is, I want to say that what separates a person who knows a lot of mathematics from a person who knows only a little mathematics is *not* that the former knows many and the latter knows few of such claims as those that mathematicians commonly provide proofs of (i.e., of those claims, such as that there are prime numbers greater than a million, which I have argued to be non-logical). Rather, in so far as what separates them is knowledge at all,[5] it is knowledge of various different sorts. Some of the knowledge that separates them is empirical knowledge (e.g., about what other mathematicians accept and what they use as axioms). Putting empirical knowledge aside, my claim is that the rest of the knowledge that separates those who know lots of mathematics from those who know only a little is knowledge of a purely logical sort—even on the Kantian criterion of logic according to which logic can make no existential commitments.

The epistemological advantages of such a view are obvious: it obviates the need of postulating mathematical knowledge that is not logical and hence that is presumably synthetic a priori; and (putting questions of a prioricity or a posterioricity aside) it obviates the need for postulating epistemological access to a special realm of mathematical entities. None the less, it is not at all obvious how any such deflationist view is to be worked out in detail, or how plausible it can be made. This paper is an attempt to survey some of the main problems that must be overcome in defending a deflationist view and to suggest ways of dealing with them.

---

[4] Frege, 'A New Attempt at a Foundation of Arithmetic', in Frege, *Posthumous Writings* (Chicago: University of Chicago Press, 1979), 278–9.

[5] Another thing that separates those who know lots of mathematics from those who know only a little isn't knowledge at all strictly speaking; it is *ability* ('knowledge how' as opposed to knowledge that): the ability to prove mathematical theorems, the ability to see the relevance of mathematical theorems to practical matters, and so forth. But only to the extent that the possession of such abilities depends on the possession of knowledge *that* does the possession of such abilities raise epistemological problems; that is why in the text I have focused only on the *knowledge* (knowledge *that*) which separates those who know lots of mathematics from those who know little.

I

The crudest attempt to state a deflationist position would be to say that all mathematical knowledge is really just knowledge that certain mathematical claims follow from certain other mathematical claims and bodies of claims: we can know, for instance, that the claim that there are primes greater than a million follows from the usual axioms of number theory. (This form of deflationism is reminiscent of, but importantly different from, what is sometimes called 'deductivism' or 'if-thenism'. Some of the differences will be discussed near the end of the essay.) This crude form of deflationism is difficult to believe: for in addition to knowing that certain claims follow from certain bodies of other mathematical claims, don't we also know the consistency of some of those bodies of mathematical claims? For instance, don't we know the consistency of various mathematical axiom systems? If one were to take the crude form of deflationism seriously, one would have to say that we can't really know that an axiom system is consistent: we can know only that the consistency of one axiom system follows from the axioms of another system which itself can't be known to be consistent (though of course its consistency can be known to follow from still a third axiom system). This strikes me as extremely implausible.

Fortunately, there is no need for a deflationist to be a crude deflationist, and so no need for him to take this line on consistency. For it would seem that knowledge that a certain axiom system is consistent (i.e., knowledge that some claim of the form 'p & -p' *doesn't follow* from the system) is every bit as much *logical* knowledge as is knowledge that a certain claim *does follow* from the system. The implausibility of the crude form of deflationism lies in its being forced to try to explain apparent knowledge of *what doesn't follow* in terms of knowledge of *what does follow*. But there is no point in trying to do this if both have equal claims to count as logical knowledge.[6]

A less crude form of deflationism, then, is the view that the only knowledge that differentiates a person who knows lots of mathematics from a person who knows only a little (aside from empirical knowledge of various sorts, such as the sort mentioned earlier) is

(i) knowledge that certain mathematical claims follow from certain other mathematical claims or bodies of claims,
(ii) knowledge of the consistency of certain mathematical claims or bodies of claims,

and other knowledge of a basically similar sort; and that all this knowledge is logical.

---

[6] As we will see in the next section, there are grounds which could lead some people (though they won't lead me) to deny that both have equal claims to count as logical knowledge.

Unfortunately, however, there is a powerful objection to this less crude form of deflationism (and to the cruder form as well); and that is that the knowledge cited in (i) and (ii) of the previous paragraph is *not* logical knowledge. For instance, the knowledge in (i)—that certain mathematical claims follow from certain others—isn't *logical* knowledge (i.e., knowledge of logical truths); it is *metalogical* knowledge, for it is knowledge *about the relation of logical consequence*. Now, the relation of logical consequence can be understood in either of two ways. First, it can be understood semantically; in that case, to say that A is a logical consequence of T is to say that in all models in which T is true, A is true as well. But, so understood, the knowledge that A is a consequence of T is knowledge about *all models*. Models are mathematical entities, so knowledge about logical consequence understood semantically is a special sort of mathematical knowledge. It is not knowledge of a logical truth, such as 'If snow exists, then snow is white or snow is not white'. So much for the semantic construal of logical consequence; but how about the syntactic construal, on which to say that A is a logical consequence of T is to say that there is a formal derivation of A from T (according to the rules of some specific formal system)? Clearly this is no better, for then knowledge that A is a consequence of T in the syntactic sense is knowledge *of the existence of formal derivations*. This cannot be logical knowledge: logic can't assert the existence of formal derivations any more than it can assert the existence of God. Indeed, it is mathematical knowledge, for formal derivations in the intended sense are the abstract objects dealt with in the mathematical theory of proof. (They aren't simply strings of symbols on paper, for A can be a consequence of T without there being a piece of paper on which someone has written a formal derivation of A from T.)

I have stated the objection as an objection to the claim that the kind of knowledge mentioned in (i) is logical knowledge: but clearly the objection holds equally well for the kind of knowledge mentioned in (ii). One might be tempted to conclude that the idea that there is no mathematical knowledge over and above logical knowledge is simply a mistake.

I want to resist this conclusion, and to see how to do this, it is worth nothing that even independently of the fact that you seem to need mathematical entities to define logical consequence and consistency, there is something else unintuitive about the idea that the only mathematical knowledge there is, is strictly speaking of form (i) or (ii) or something similar. For the knowledge cited in (i) and (ii) is metalinguistic; are we really to hold that all mathematical knowledge is metalinguistic? Indeed, doesn't the metalinguistic fact that one sentence follows from another depend on the fact that certain words appearing in these sentences are used as logical words by speakers of the language in question? If so, knowledge of what follows from what has a contingent element that the mathematical knowledge we were trying to convey presumably lacks. This then is another (admittedly less compelling) objection to the version of deflationism under discussion. And it seems clear that we can get around both objections simultaneously if we can find a way to 'semantically descend', that is, to state the sort of mathematical knowledge that the deflationist wants to focus on without going metalinguistic.

How this is to be done is clearest in the case of finite bodies of mathematical claims, and for the moment I will confine my attention to them. If A is the conjunction of all members of a body T of mathematical claims (e.g., the conjunction of all the axioms of a finitely axiomatized theory), then instead of saying that we have mathematical knowledge that this theory is consistent, why not simply say

(ii*) we know that $\Diamond A$

(where the modal operator '$\Diamond$' is to be read 'it is logically possible that' or 'it is logically consistent that'). Here the claims in T are used, not mentioned, so the contingency objection doesn't apply. And because they are used, not mentioned, the symbol '$\Diamond$' cannot be understood as a predicate that needs explanation in set-theoretic or proof-theoretic terms; it must be understood as an operator, and indeed an operator that is widely regarded as an operator of logic; consequently the earlier objection doesn't apply either.

The point I have made for (ii) applies of course to (i) as well. That is, instead of saying that the claim B follows from the body of claims whose conjunction is A, why not say

(i*) we know that $\Box(A \supset B)$,

where '$\Box$' ('it is logically true that') is of course defined as '$\neg \Diamond \neg$'.

So our third version of deflationism (the final version, apart from a slight alteration designed to handle theories that are not finitely axiomatized) is that what differentiates a person with lots of mathematical knowledge from a person with only a little (apart from differences in abilities (cf. note 5) and in empirical knowledge) is that the former but not the latter has lots of knowledge of the type (i*) and (ii*) and other knowledge of a basically similar sort. (One of the things that this last 'hedge' clause covers is modal knowledge not either of the form '$\Diamond A$' or '$\Box A$'; that is, the conditional knowledge that *if* it is consistent that A, *then* it is consistent that B seems like knowledge that a deflationist could perfectly well allow.[7])

---

[7] It should be noted that the modal knowledge which deflationism allows is knowledge of purely logical possibility—deflationism does not allow knowledge of mathematical possibility in any interesting sense. This makes deflationism very different from the viewpoint that Hilary Putnam calls 'mathematics as modal logic' in his paper 'What Is Mathematical Truth?', in *Philosophical Papers*, vol. 1 (Cambridge: Cambridge University Press, 1975), 60–78. According to 'mathematics as modal

In moving from (i) and (ii) to (i*) and (ii*), I of course have to accept the idea that the notion of logical possibility is an acceptable notion, and also I must accept that it really is a part of logic and not something that must be explained in terms of *entities* (e.g., models, formal derivations, or possible worlds). In addition, in accepting that (ii*) counts as logical knowledge (for suitable assertions A), I have to interpret the idea of logical knowledge (and logical truth) in a way broader than current orthodoxy permits. This I will now explain.

II

Consider the claim

(1) $\Diamond \exists x \exists y \, (x \neq y)$,

which says that it is logically possible that there be at least two objects in the universe. Is this a truth of logic? I think the natural answer is 'yes', and most of the people I have asked agree. None the less, on the usual approach to defining logical truth for modal logic (Kripke's approach[8]), (1) does not come out logically true. Indeed, it is a curious feature of Kripke's approach to defining logical truth for modal logic that *no* sentence of the form $\ulcorner \Diamond A \urcorner$ is logically true, except in the trivial case where A itself (and hence $\ulcorner \Box A \urcorner$) is logically true. To me, this seems quite unmotivated. It may be countered that while it is perhaps initially natural to regard (1) as a logical truth, Kripke's model-theoretic definition of logical truth for modal logic is also quite intuitive, and it is a consequence of this model theory that (1) not be logically true. My reply is that there is an alternative model-theoretic definition of logical

---

logic', 'the mathematician . . . makes no existential assertions at all. What he asserts is that certain things are possible and certain things are impossible—*in a strong and uniquely mathematical sense of "possible" and "impossible"* ' (p. 70; italics mine). Putnam claims that despite its making no existential assertions, mathematics as modal logic really states the same facts as does mathematics on its platonistic interpretation. I think that there is considerable plausibility in Putnam's claim. The reason is that in the 'strong and uniquely mathematical sense of "possible" ', 'it is possible that A' is an object-level analogue of 'A is consistent with any true mathematical theory'; something very akin to mathematical truth (and therefore to mathematical existence) is being sneaked into Putnam's possibility operator. Putnam's position apparently is that if you take ordinary set theory S and some non-standard alternative S' that is inconsistent with S (but internally consistent), then S is mathematically possible, but S' isn't. The deflationist viewpoint is different: S and S' are on a par in that $\Diamond A x_S$ and $\Diamond A x_{S'}$. Of course, S may be more useful than S' for various purposes, but if so, this requires an explanation. (See Sects. III and IV of this essay.) A deflationist cannot regard it as an acceptable explanation to say that S describes a mathematical possibility and S' does not. I believe that Putnam's view is that such an explanation *would* be acceptable and, indeed, would be the only explanation possible.

[8] Saul Kripke, 'Semantical Considerations on Modal Logic', *Acta Philosophica Fennica*, 16 (1963): 83–94.

truth for modal logic that is simpler and I think more natural than Kripke's, and which is much closer than Kripke's to the model-theoretic definition of logical truth for first-order logic (for instance, in involving no reference to 'possible worlds' or to any other entities not used in the model theory for first-order logic); and according to this alternative model-theoretic definition of logical truth, (1) comes out logically true. (The basic idea of this alternative method of defining logical truth has occurred to quite a few people, starting with Carnap. I describe the version of it that I favour in an Appendix).

Indeed, it is a consequence of the non-Kripkean approach to defining logical truth for modal logic that any assertion of the form $\ulcorner \Diamond A \urcorner$ that is true is logically true, and that any assertion of the form $\ulcorner \Diamond A \urcorner$ that is false is logically false. It is essential to the plausibility of this that '$\Diamond$' be read 'it is *logically* possible that'—it is *logical* possibility (not mathematical possibility or 'metaphysical possibility') that we are concerned to give the logic of. (See n. 7.) Exactly what the force of 'logically possible' is depends on further stipulation: it depends on what one takes non-modal logic to include. Some philosophers (e.g., Carnap) regard the non-modal logic to which we are adding '$\Diamond$' as including not only first-order logic properly so-called, but also 'meaning postulates' specifying 'logical' relations among predicates. Consequently, a non-modal sentence such as

(2) $\exists x$ (x is a bachelor & x is married)

would count as *logically* false for Carnap, and as a result,

(3) $\Diamond \exists x$ (x is a bachelor & x is married)

also comes out logically false. But I prefer not to follow Carnap in taking meaning relations among predicates to be part of logic. My preference is not based on any firm doctrines about analyticity; indeed, my preference is partly based on a desire to remain neutral about such issues. Mostly, however, my preference is based on simplicity: it is simpler to develop a basic modal logic that takes no account of meaning relations among predicates; once one has such a logic, it is easy to obtain from it a derivative logic that takes into account any relations among predicates that one cares to regard as meaning-postulates, if one so desires. If one adopts this strategy—and I shall—then (2) is not *logically* false; it is *logically* consistent that there be married bachelors (even though it may not be consistent with meaning-postulates that there be married bachelors). Now, if it is *logically* consistent that there be married bachelors, and '$\Diamond$' is read as an operator meaning 'it is *logically* consistent that', then (3) comes out true. Indeed, any sentence of the form $\ulcorner \Diamond \exists x (Px \& Qx) \urcorner$ where P and Q are atomic, comes out true; what else could you expect if meaning-relations among predicates are not taken into account? Moreover, it seems as if it ought to be part of the *logic* of the logical consistency operator '◊' that sentences of the form ⌜◊∃x (Px & Qx)⌝ are true. That is,

(4) ◊∃x (x is red & x is round)

should be not only true but logically true; and similarly for (3). The view that (3) shouldn't be logically true, indeed shouldn't even be true, results from giving to '◊' a sense not intended.[9]

I hope this gives some idea of (and some motivation for) the view of modality and of the logic of modality that I will be presupposing. For a bit more detail, see the Appendix.

Let us return to the issue of mathematical knowledge, and in particular to the version of deflationism arrived at in Section I. Part of the position arrived at there was (a) that mathematicians sometimes know things of form ◊Ax, where Ax is a conjunction of axioms of a theory; and (b) that this knowledge is logical knowledge. Now a minimum condition for (b) to hold is that ◊Ax must count as a logical truth. Ax itself won't be logically true, if it is a conjunction of axioms of a typical mathematical theory; so in order to adhere to the version of deflationism put forth in Section I, we clearly have to adhere to a non-Kripkean conception of logical truth according to which some non-trivial assertions of possibility are part of logic. The non-Kripkean conception of logical truth sketched above (or if you prefer, the more fully Carnapian variant sketched in n. 9) will do. Indeed, they have the feature that the modal assertion ◊Ax will be logically true if it is true at all. So there is no danger on these conceptions that we might know that ◊Ax and what we know not be logically true: if we know it, it's true, and so it's logically true.

Does this show that our knowledge that ◊Ax is logical knowledge? Not by itself—for it might be claimed that though ◊Ax is logically true, it is known by non-logical means.

There is a more interesting and a less interesting version of this claim. The less interesting version points out that much of our knowledge of possibility

[9] As remarked above, there need be no doctrinal difference between the view I have advanced, according to which (3) counts as logically true, and Carnap's view, according to which a sentence typographically like (3) counts as logically false. For we can represent the Carnapian view within the view I have advanced, by introducing Carnapian notions by definition into both the object language and the metalanguage. Thus 'is C-logically true' and 'is C-logically false' are to mean 'follows from ——' and 'is inconsistent with ——', where the blanks are to be filled by the 'meaning-postulates' for English, and '◊$_C$A' is to mean '◊ (A & ——)'. Then though (3) is still logically true (and hence C-logically true), still

(3$_C$) ◊$_C$∃x (x is a bachelor & x is married)

is C-logically false (and, indeed, logically false). Moreover, we may if we like agree with Carnap that it is C-logical truth and '◊$_C$' that are the philosophically more important notions. Whether or not one agrees with that philosophical claim, the procedure of focusing first on logical truth and on '◊' is of quite considerable technical convenience. This will be evident, for instance, when we come to formulate 'the Conditional Possibility Principle' later in this section.

is to some extent inductive. For instance, our knowledge that $\Diamond Ax_{NBG}$ (where NBG is von Neumann–Bernays–Gödel set theory and Ax_{NBG} is the conjunction of all its axioms) seems to be based in part on the fact that we have been unable to find any inconsistency in NBG. And, it can be claimed, this inductive element in our knowledge precludes that knowledge from being logical. Now, even this less interesting version of the claim that our knowledge of the form $\Diamond Ax$ is non-logical raises some interesting issues about the nature of logic, issues that may be relevant to the precise wording of the deflationist's claim. But there is no need to go into such matters, for it seems quite clear that the basic idea of deflationism cannot be undercut by pointing out that much of our knowledge of possibility has a partly inductive character. The basic idea of deflationism is that one is to avoid postulating knowledge of a realm of mathematical entities, and that one is to do this by saying that ordinary mathematical claims are not known to be true. The deflationist holds that what separates those who know a lot of mathematics from those who know only a little is various sorts of knowledge and abilities, none of which give rise to the philosophical problems that knowledge of a realm of mathematical entities gives rise to (or is commonly thought to give rise to). A good deal of the knowledge that separates those with lots of mathematical knowledge from those with only a little is empirical. (I mentioned this earlier, and will discuss it more fully near the end.) Other of this knowledge is, let us suppose, *straightforwardly* logical in that it involves no inductive elements. And other of this knowledge is knowledge of logical truths by partly inductive means. Perhaps the fact that this latter knowledge is partly inductive keeps it from being logical, and perhaps not. Perhaps it makes the knowledge empirical, perhaps not. I would incline toward answering both of these questions in the negative; but however they are answered, the fact that some of our knowledge of logical truths is partly inductive does not in any way support the claim that it is based on knowledge of a special realm of abstract entities or on knowledge of the truth of ordinary mathematics. Because of this, the fact that some of our knowledge of logical truths is in part inductive can't be used to argue against the essentials of the deflationist position.

As I've remarked, there is also a more interesting version of the claim that though $\Diamond Ax$ is logically true, it is known by non-logical means—a version which, if true, would genuinely count against deflationism. Consider what Frege said about knowledge of the consistency of mathematical theories in §95 of *The Foundations of Arithmetic*: 'Strictly, of course, we can only establish that a concept is free from contradiction by first producing something that falls under it.'[10] Obviously this is not literally correct—we can establish

---

[10] Trans. J. L. Austin (Oxford: Blackwell, 1959), 106e.

that the concept 'horse with wings' is free from contradiction without producing a horse with wings—but the position can be weakened without totally altering its spirit. The weakened version of Frege's claim grants that there is knowledge of possibility that does not arise from knowledge of actuality, but which arises instead from reflection on the logical form of concepts. But, it maintains, all such knowledge of possibility is conditional: one cannot attain categorical knowledge of possibility by this means alone. Rather, categorical knowledge of possibility can only be obtained either directly from knowledge of actuality, or indirectly, that is, from direct categorical knowledge of possibility in conjunction with conditional knowledge of possibility and other logical knowledge. So, for instance, reflection on the common logical form of 'horses with wings' and 'animals with tails' yields the conditional knowledge that *if* it is logically possible that there are animals with tails, *then* it is logically possible that there are horses with wings.[11] This knowledge together with the knowledge that there actually are animals with tails then yields knowledge that it is logically possible that there be horses with wings. The Fregean position is that all knowledge of possibility arises by some such means. (Of course, the knowledge of actuality on which knowledge of possibility is ultimately based may, on Frege's view, be a priori.)

If this Fregean position about knowledge of possibility were correct, then deflationism would be in deep trouble. For presumably we know (or at least have good reason to believe) the claim $\Diamond Ax_{NBG}$ and the claim $\Diamond Ax_R$ where R is the theory of real numbers; but, according to deflationism, we do not know (or have good reason to believe) the claims Ax_{NBG} and $\Diamond Ax_R$ would have to be based on conditional knowledge of possibility that arises by reflection on logical form, together with other logical knowledge plus knowledge of actuality. Now, one principle that I think a Fregean would grant is that if ϕ is nonmodal and ψ is a generalized substitution instance of it (i.e., is obtained from ϕ by substituting formulae for non-logical predicates or by uniformly restricting the ranges of all quantifiers and free variables or both[12]), then we can

---

[11] Here and throughout the rest of this section, the fact that we have not included meaning-relations among predicates as part of logic pays off.

[12] The restriction of quantifiers and free variables is to be by a formula $D(x_1)$ which may contain other free variables besides x_1; the formula $A_p(x_1, \ldots x_k)$ to be substituted for the k-place predicate p may likewise contain other variables free. The restriction on quantifiers and free variables is to be made only on the quantifiers and free variables of the original formula, not on any new ones introduced in an A_p or in D. [To be more formal: before performing the general substitution in a formula B, replace all bound variables in B or in D or in an A_p that occur (free or bound) in any other of the formulas by new variables. Then for any subformula X of B, associate an X^* as follows: if X is $p(v_1, \ldots, v_k)$, let X^* be $A_p(v_1, \ldots v_k)$; if X is $\neg Y$, or $Y \supset Z$, let X^* be $\neg Y^*$, or $Y^* \supset Z^*$; if X is $\forall v Y(v)$, let X^* be $\forall v(D(v) \supset Y(v))$. Finally, the generalized instance B^s of B is $D(y_1) \& \ldots \& D(y_n) \& B^*$, where $y_1, \ldots y_n$ are the variables free in B.] The possibility of B^s is in effect the possibility of its existential quantification, both with respect to the variables free in B (now restricted by D) and with respect to any other free variables introduced in the generalized substitution (which are unrestricted). It is

know that *if* ◊ψ, *then* ◊φ, by reflection on logical form alone.[13] If this Conditional Possibility Principle is granted to the Fregean, then by embedding real number theory R in set theory NBG, the Fregean can admit that we can know that

(i) If ◊Ax_{NBG}, then ◊Ax_R.

[For if R' is the set-theoretic assertion to which the conjunction of the axioms of real numbers 'reduce', then the knowledge in (i) arises from the knowledge that

(ii) □(Ax_{NBG} → R')

together with the knowledge that

(iii) if ◊R', then ◊Ax_R.

(ii) is knowledge of necessity rather than of possibility,[14] and on the Fregean view, this is unproblematic; and (iii) is knowledge that results by the Conditional Possibility Principle just given.] But it is essential to this example that NBG be at least as rich as R. From the Fregean standpoint, any reason to believe that ◊Ax_{NBG} has to rest either on a reason to believe ◊T for some *richer* theory T or else on a reason to believe Ax_{NBG}. The deflationist cannot allow that there is any reason to believe either Ax_{NBG} or any other mathematical theory. It is *compatible* with deflationism that there is an empirical theory T richer than NBG which can reasonably be believed: but (a) it is hard to believe that there is any plausible empirical theory in which NBG can be embedded, and (b) it is totally implausible that our reasons for believing that ◊Ax_{NBG} should rest entirely on reasons to believe any specific empirical theory. So from a Fregean standpoint, deflationism is simply not a viable position.

I don't find this fact terribly upsetting, however, because I don't think that the Fregean viewpoint has a great deal of plausibility. In the first place,

---

easy to see that if a generalized substitution instance of B has a model, so does B itself, so the Conditional Possibility Principle is validated by the semantics of the Appendix.

[13] This principle is valid as it stands on a free logic like that of Dana Scott's 'Existence and Description in Formal Logic', in R. Schoenman (ed.), *Bertrand Russell: Philosopher of the Century* (Boston: Little, Brown, 1967), 181–200. If one prefers a free logic like that of Tyler Burge's 'Truth and Singular Terms', *Noûs*, 8 (1974): 309–25, where each assertion of the form ⌜if $p(t_1, \ldots, t_n)$ then $\exists x (x = t_1) \& \ldots \& \exists x (x = t_n)$⌝ for atomic p is regarded as a truth of logic, then the principle must be modified slightly (say by redefining 'substitution instance' to mean a substitution instance in the normal sense conjoined with a clause of the form ⌜$\exists x (x = t)$⌝ for each term t in the sentence in which the substitutions are being made).

[14] Of course, knowledge of possibility is derivable from it: e.g., that ◊ (Ax_{NBG} → R') or that ◊Ax_{NBG} → ◊R'; presumably, however, the Fregean view is that knowledge of possibility is never problematic if it is derivable from knowledge of necessity. (I count a knowledge claim as involving knowledge of possibility if its formulation contains at least one positive occurrence of '◊' or at least one negative occurrence of '□'.)

consider a point I made earlier, that part of our reason for believing that $\Diamond Ax_{NBG}$ is the fact that we have been unable to derive any contradictions from Ax_{NBG}. I argued then that this was a point that a deflationist could consistently recognize; I now want to observe that there is a serious question as to whether an advocate of *the Fregean position* could recognize this point. For our inability to derive a contradiction from Ax_{NBG} certainly doesn't give us reason to think that Ax_{NBG}: if our reasons for believing that $\Diamond Ax_{NBG}$ had to be based on reasons for believing Ax_{NBG}, our inability to derive a contradiction from Ax_{NBG} would be irrelevant to our knowledge that $\Diamond Ax_{NBG}$.[15]

In the second place, much of the motivation for the Fregean position is lost when we move from the crude formulation that Frege actually gives for his position (in the quotation above from §95 of *The Foundations of Arithmetic*) to the more defensible formulation that I have given. The motivation for the crude Fregean position is that it provides a simple solution to an epistemological problem, the problem of explaining the source of our knowledge of possibility. The crude Fregean position is that there really is no problem here: the source of our knowledge of possibility is just knowledge of actuality. The

[15] It may seem that I am slurring over a complication. For it may seem that the fact that after persistent efforts we have not succeeded in deriving a contradiction from Ax_{NBG} doesn't in itself provide evidence for $\Diamond Ax_{NBG}$; rather, it provides evidence for a claim of *impossibility*, namely the impossibility of there being a derivation of a contradiction from $\Diamond Ax_{NBG}$. If this is right then we need to establish a connection between this impossibility claim and the possibility claim $\Diamond Ax_{NBG}$. Such a connection is established by the modal completeness theorem for first-order logic, discussed in the next section.

How does this affect the argument in the text against the Fregean viewpoint? It might initially be thought to undercut the argument: for there is nothing in the Fregean viewpoint to rule out acceptance of the impossibility of deriving a contradiction from NBG on the basis of failures to find such derivations; and once that impossibility claim is accepted, it would appear that an application of the modal completeness theorem would yield knowledge that $\Diamond Ax_{NBG}$. But of course the question is, how is the completeness theorem known? Knowledge of the modal completeness theorem is knowledge of possibility, and the proof of it sketched in the next section assumes that $\Diamond Ax_{NBG}$ (though doubtless we could make do with $\Diamond Ax_M$ for a mathematical theory M somewhat weaker than NBG). Consequently, from a Fregean viewpoint it is hard to see how one could ever apply the theorem unless one already knew of an actual structure in which NBG (or the hypothetical weaker theory M) could be embedded. But that would mean that there would be no chance of adding to the credibility of the claim $\Diamond Ax_{NBG}$ (or the claim $\Diamond Ax_M$, if an appropriate weaker M is found) by persistently trying to derive a contradiction from NBG (or M) and consistently failing.

We see, then, that describing the epistemological situation as in the opening paragraph of this note would not help the Fregean. It would, though, create a problem for the non-Fregean (whether the non-Fregean be a platonist or a deflationist): for if our failures to derive a contradiction from Ax_{NBG} only give reason to believe $\Diamond Ax_{NBG}$ if we presuppose modal completeness, and that requires $\Diamond Ax_{NBG}$, then such appeal to our failure to derive a contradiction is problematic from the non-Fregean position as well. Fortunately however, the more complicated description of the epistemological situation seems wrong. It depends on thinking that the record of failures to find a contradiction could only enter into the epistemological picture as a premise to an enumerative induction. The right way to look at the matter, though, is as an inference to the best explanation: the assumption that $\Diamond Ax_{NBG}$ is the most plausible explanation of the failure to find a contradiction in NBG. Modal completeness is irrelevant. (What is relevant is what I call in Section IV the 'weak modal soundness' of first order logic, but this is not something that one proves by $\Diamond Ax_{NBG}$.)

more defensible alteration of the Fregean position gives up this advantage: there is knowledge of possibility (not based solely on knowledge of necessity) that is not based on knowledge of actuality, but on 'reflection on relations of logical form'. The ability to 'reflect on relations of logical form' is supposed to allow us to know each instance of the schema '$\Diamond \psi \rightarrow \Diamond \phi$', where ϕ is non-modal and ψ is a generalized substitution instance of it; but is it so clear that any motivated account of how we 'reflect on logical form' so as to come to this knowledge wouldn't also provide an account of how we know categorically some claims of form $\Diamond \phi$? After all, any claim of the same logical form as

(S) $\Diamond$ (there are at least $10^{10^{10}}$ apples)

is also a logical truth. So why can't 'reflection on logical form' *show* that it is a logical truth? Why do we need to rest all our confidence in (S) on the claim that there actually are at least $10^{10^{10}}$ *somethings*? If we do need to rest knowledge of (S) on knowledge of actuality, that is rather surprising. What motivation is there for granting that we can have knowledge of possibility through 'reflection on logical form', but at the same time denying that knowledge of a simple possibility claim like (S) can be known by the same process?

Indeed, it seems to me that the Fregean position leads to quite counterintuitive consequences, for it seems clear to me that we have much more solid reason to believe (S) than to believe in the existence of $10^{10^{10}}$ entities of any kind. Certainly the claim that there are at least $10^{10^{10}}$ *physical* entities is not obvious. (I am inclined to believe it, for I am inclined to believe that regions of space are physical and that there are infinitely many of them; but I am much less confident of this than I am of (S).) And in my view, the claim that there are at least $10^{10^{10}}$ mathematical entities is far *less* obvious—in fact, I don't think there are *any* such entities—so I certainly wouldn't want to rest my belief in (S) on *that*. The idea that I should be as uncertain of (S) as I am of the claim that there are infinitely many physical or mathematical entities in the universe seems preposterous, and since the Fregean view has this consequence, I would need a much better motivation for that view before I could take it seriously.

The points I have made here for (S) arise for the claim $\ulcorner \Diamond Ax_{NBG} \urcorner$ as well. That is, if this claim is true, so is every other claim of the same logical form; so why can't 'reflection on logical form' (whatever exactly that is) or whatever other process or combination of processes one needs to account for modal reasoning give us reason to believe it? Indeed, the example of $\ulcorner \Diamond Ax_{NBG} \urcorner$ makes clear that the claim of even the crude Fregean view to be epistemologically pure was a hoax. The crude Fregean view was presented as having the epistemological advantage that it makes knowledge of possibility depend entirely on knowledge of actuality—but *mathematical* actuality is the only actuality that could work in the case of the claim $\ulcorner \Diamond Ax_{NBG} \urcorner$, and once this is seen, it is hard to see how there is an epistemological advantage. The problem of how I know that it is *logically possible* that Ax$_{NBG}$ is 'solved' on the Fregean view by saying that I know that there *actually are* the entities that Ax$_{NBG}$ says there are and that they are interrelated as Ax$_{NBG}$ says. I find it hard to grasp how anyone could know (or have reason to believe) that such platonic entities actually exist (as opposed to being merely logically possible): or how anyone could know (or have reason to believe) that if such entities do actually exist, then they are related in one way rather than in some other. Consequently, the idea that I could explain my knowledge (or my reason to believe) that ◊Ax$_{NBG}$ by reference to such knowledge of the platonic realm seems to me a total obfuscation of the real epistemological issues about knowledge of logical modalities.

III

Deflationism is, of course, a non-realist philosophy of mathematics: it holds that we cannot know (or have any reason to believe in) the existence of mathematical entities or the truth of ordinary mathematical claims; indeed, it would be natural to couple deflationism with the further claim that we have good reason to believe that there are no mathematical entities and hence that most ordinary mathematical claims are false.[16]

It has been widely held[17] that the most serious difficulty facing any non-realist philosophy of mathematics is *the problem of application*: how can one account for the utility that reasoning about mathematical entities has for disciplines other than mathematics, if mathematics isn't construed in a realistic fashion? Applied to deflationism, the problem is: how can one explain the applicability of mathematics to disciplines other than mathematics, without assuming that ordinary mathematical claims (including those claims that assert the existence of mathematical entities) are true?

In a book I wrote several years ago[18] I focused on one aspect of the problem of application: the problem of explaining the applicability of mathematics *to physical science* (and to everyday empirical reasoning), without assuming the truth of the mathematics that was being applied. But there are

---

[16] Strictly speaking, the existentially quantified assertions will all be false and the universally quantified ones all vacuously true.

[17] e.g., Frege says that 'it is applicability alone which elevates arithmetic from a game to the rank of a science' (*Grundgesetze der Arithmetik*, vol. 2, sect. 92, trans. Max Black in P. Geach and M. Black (eds.), *Philosophical Writings of Gottlob Frege* (Oxford: Blackwell, 1966), 187). The point has been most thoroughly developed by Hilary Putnam in *The Philosophy of Logic* (New York: Harper, 1971).

[18] Field, *Science without Numbers: A Defense of Nominalism* (Princeton: Princeton University Press; Oxford: Blackwell, 1980).

other aspects to the problem of application, and a main task of the rest of the essay will be to say something about one of the more pressing such aspects: the problem will be to explain the applicability of mathematics (in this case, proof theory and model theory) *to the study of logical reasoning*, without assuming the truth of the mathematics that is being applied.

Before turning to this main topic, I want to say something about the applicability of mathematics to physical science and to ordinary empirical reasoning. Any account of the usefulness of a mathematical theory in dealing with the physical world will say that this usefulness depends on two things:

(a) the fact that the mathematical theory is 'mathematically good';
(b) the fact that the physical world is such as to make the mathematical theory particularly useful in describing it.

Different accounts of the usefulness of mathematics in application to the physical world will differ as to how (a) and (b) are to be elaborated.

A deflationist account of the application of mathematics must involve two claims. As regards (a), it must say that 'mathematical goodness' does not involve truth, but only something less demanding, such as consistency.[19] (This is strictly inaccurate, involving an inappropriate semantic ascent. The more accurate formulation is that the deflationist must claim that in explaining the application of a mathematical theory, we do not need to assume the conjunction of its axioms, since that conjunction isn't logically true; he must claim that instead we need to assume only something weaker which *is* logically true, such as the result of prefixing the conjunction of the mathematical axioms with the modal operator '$\Diamond$'. For the moment I will forego such accuracy for the sake of naturalness.) As regards (b), the deflationist must be able to formulate the facts about the physical world that make the mathematical theory 'fit it', without assuming in the formulation any standard mathematics (either the mathematical theory whose usefulness is being explained or some other mathematical theory). For if we have to assume the truth of standard mathematics anywhere in our account—in (b) or in (a)—then a deflationist would have to hold that such an account was unknowable.

[19] Of course, a deflationist can and will recognize that not all consistent mathematical theories (and not all mathematical theories that are *strongly* consistent in the sense shortly to be defined) are of equal mathematical interest—just as the platonist can and will say that not all *true* mathematical theories are of equal mathematical interest. What makes a mathematical theory interesting is a complicated matter—richness in consequences is one factor, relevance to prior work in mathematics and in science is another, elegance is a third, and doubtless there are further factors still. There is no need to discuss such factors here, for they are ones that the platonist and the deflationist can agree on. What is important in the present context is the features of mathematical goodness that go beyond interestingness. For the platonist, a mathematical theory should not only be interesting, it should be *true*; and my question is, what serves the role for the deflationist that truth serves for the platonist?

So a deflationist has two tasks corresponding to (a) and (b) above. The task corresponding to (b) is by far the more difficult, but since I have treated it at length in my book and since it involves issues rather far removed from those of the rest of the essay, I will say no more about it here. The task corresponding to (a) is much easier. I have discussed it too in my book and in a more elementary paper;[20] but here I will need to summarize quickly what I said about it.

My conclusion was that a mathematical theory needn't be true to be good; and, indeed, if it were true, this wouldn't be enough for it to be good, for a good mathematical theory must have a property that might be called *strong consistency* or *conservativeness* and that doesn't follow from truth alone. To say that a mathematical theory M is strongly consistent is to say roughly that if you take any theory T that says nothing about mathematical entities, and add T to M, then if T is consistent, so is T + M.[21] Although strong consistency doesn't follow from truth, it does follow from *necessary* truth; I believe that the widespread view of mathematics as necessarily true shows an implicit recognition of the importance of strong consistency. Strong consistency, however, is weaker than necessary truth, for strongly consistent theories needn't be true at all. As for ordinary consistency, this is not in general a sufficient requirement on a mathematical theory. There is an important class of mathematical theories ('pure' mathematical theories—those dealing with mathematical entities alone) for which ordinary consistency entails strong consistency; so for such theories the only requirement is that they be consistent in the ordinary sense. But there is another important class of mathematical theories (e.g., certain versions of set theory with ur-elements which play an important role in the application of pure mathematical theories) for which ordinary consistency does not entail strong consistency; and for these theories strong consistency is the important notion. I have argued in the works mentioned that in explaining the application of mathematics to the physical

---

[20] 'Realism and Anti-Realism about Mathematics'. For a discussion of some more technical aspects of the issues surrounding (a) (including replies to some technical objections that have been raised), see my paper, 'On Conservativeness and Incompleteness', *Journal of Philosophy*, 82 (1985): 239–60. Both papers repr. in Field, *Realism, Mathematics and Modality*.

[21] More formally, M is strongly consistent if for any T, if T is consistent, then so is T* + M, where T* is the result of restricting all quantifiers in T to non-mathematical entities. If M is an impure mathematical theory, the domains of T* and M will overlap (e.g., if M is set theory with ur-elements, both domains will contain the non-mathematical ur-elements); that is part of the explanation why strong consistency doesn't reduce to ordinary consistency for typical impure mathematical theories.

The above definition of strong consistency is essentially the one given in my book. (In the book I followed the standard artifice of regarding logic as ruling out the empty domain; because of this, some extra complexity was needed in the definition of strong consistency, but it is not needed here, since the artifice has been dropped.) In the book I did not contemplate adding mathematics to theories that contain modal operators, for I was concerned only with adding them to physical theories, and I did not (and do not) want to allow modal operators into nominalized physics. If we do consider the addition of mathematics to a theory T with modal operators, we might want to complicate the definition of T*; but there is no need to go into that here.

world we never need assume that the mathematics is true; we need only assume that it is strongly consistent (i.e., conservative).

This sounds like an account of application that is congenial to the deflationist: the truth of standard mathematics needn't be assumed in the account. so there seems to be no problem in reconciling the application of mathematics with the deflationist's claim that the truth of standard mathematics can't be known.

I think that this point is correct in spirit, but there is a difficulty that must be faced. For I have defined strong consistency in terms of ordinary consistency, and ordinary consistency is usually defined in terms of the existence of models. So won't the assertion that a theory is strongly consistent be an assertion about the existence of models? If so, then even though strong consistency doesn't entail truth, it is still hard to see how a deflationist could ever claim to know any theory to be strongly consistent. And if he can't claim to know that, it is certainly awkward for him to maintain that the strong consistency of a theory is essential to its application.

Many people[22] have objected along these lines to the account of mathematics put forward in my book, and with considerable justification. But from what I have said so far in this essay, it should be clear in outline how I now want to handle the objection. I want to say that in explaining the application of a mathematical theory M to the physical world, it is not strictly accurate to say that we need to assume the strong consistency of the mathematical theory. Rather, what we must assume is a certain modal claim, one which bears the same relation to the claim that M is strongly consistent that $\Diamond Ax_M$ bears to the claim that M is consistent in the ordinary sense. The points I made several paragraphs back about the relation between strong consistency, ordinary consistency, truth, and necessary truth should really have been made at the object level: instead of saying that strong consistency is entailed by necessary truth and entails consistency, but neither entails nor is entailed by truth, I should have said that the claim Q that is the modal analogue of the strong consistency of M is entailed by $\Box Ax_M$, entails $\Diamond Ax_M$, and neither entails nor is entailed by Ax_M. Similarly, for the point about explaining the utility of M: instead of saying that we don't need to assume the truth of M, but only its strong consistency, I should have said that we don't need to assume Ax_M but only Q.

That, it should be clear, is how I want to handle the objection. But *can* I handle it in this way? There is a technical difficulty to doing so, for there is a

---

[22] e.g., Michael Resnik in a review of my book in *Noûs*, 27 (1983): 514–19; Charles Chihara, 'A Simple Type Theory without Platonic Domains', *Journal of Philosophical Logic*, 13 (1984): 249–83; and Michael Detlefsen, *Hilbert's Programme: An Essay on Mathematical Instrumentalism* (Dordrecht: Reidel, 1986).

technical difficulty in figuring out exactly how Q (the modal analogue of strong consistency) is to be formulated.

Before turning to this technical difficulty, I want to discuss an earlier technical difficulty that I mentioned in Section I: the difficulty about theories that are not finitely axiomatized. In Section I, I suggested that 'knowledge of the consistency of the theory of linear order' is really just modal knowledge: it is knowledge of the form ◊B, where B abbreviates the conjunction of the axioms of the theory of linear order. But suppose we consider a theory that is not finitely axiomatizable. Can't we know that such a theory is consistent? And how do we represent such knowledge as modal knowledge, given our inability to conjoin all of the infinitely many axioms?

There is more than one way to respond to this objection; my current inclination is to respond by introducing a further logical device that will allow us finitely to axiomatize theories which, without the device, can't be so axiomatized. The device I have in mind is what is called a 'substitutional quantifier'. The name seems to me misleading: it is not a quantifier at all, as quantifiers are normally understood; rather, it is simply a device for representing sufficiently regular infinite conjunctions in a finite notation.[23]

The non-finitely axiomatized theories we ordinarily use are theories with quite regular infinite collections of axioms. For the theories consist of finitely many axioms plus finitely many axiom *schemas*: schemas like 'For every formula F, $\ulcorner \exists z \forall y (y \in z \equiv F) \urcorner$ is to be an axiom'. To represent this finitely, we merely need conjoin all the infinitely many sentences in the language of the form $\ulcorner \exists z \forall y (y \in z \equiv F) \urcorner$; and we can do that if we have a substitutional quantifier 'Π' with formulas as the substitution class, for we simply say '$\Pi F \exists z \forall y (y \in z \equiv F)$'. (This is not a metalinguistic assertion but an infinitary conjunction of non-metalinguistic claims: it is no more metalinguistic than is a finite conjunction like '$\exists z \forall y (y \in z \equiv y$ is a cat$) \& \exists z \forall y (y \in z \equiv y$ is a dog$)$'.) Anything that we normally regard as a mathematical theory can be finitely axiomatized using such a device, so any knowledge that we possess of the consistency of mathematical theories can be represented in the form ◊B if we're allowed to use a substitutional quantifier in formulating B.

One advantage of this way of solving the technical difficulty about non-finitely axiomatized theories is that it solves the technical difficulty about the modal analogue of strong consistency as well. Strong consistency is defined in terms of ordinary consistency as follows:

(iv) for any theory T, if T is consistent, then so is T* + M;

---

[23] The view of substitutional quantification implicit in these remarks is set out in more detail in my review of Dale Gottlieb's book *Ontological Economy: Substitutional Quantification and Mathematics* (New York: Clarendon Press, 1980), in *Noûs*, 28 (1984): 160–5.

here T* is the result of restricting all the quantifiers of T to non-mathematical entities. What is the modal analogue of this? It will differ from (iii) in not being metalinguistic: instead of saying that if T is consistent, then so is T* + M, we will say that if $\Diamond Ax_I$, then $\Diamond((Ax_I)^* \& Ax_M)$. But then, how do we *generalize* such an object-level statement to all theories T? Again, we must invoke a substitutional quantifier (or some other form of infinite conjunction); we must say

(iii*) Π B (if $\Diamond B$, then $\Diamond(B^* \& Ax_M)$).

My solution to the technical difficulty raised several pages back, then, is that 'knowledge of conservativeness' is really just modal knowledge of the form (iii*). Such modal knowledge does not require knowledge that Ax_M (much less than $\Box Ax_M$); and this is all to the good, since Ax_M (and $\Box Ax_M$) entails the existence of mathematical entities, and consequently is not logically true. For certain mathematical theories, (iii*) will be stronger than the claim $\Diamond Ax_M$, but even so, it is a logical truth. And the application of the theory M to the physical world requires only the logical truth (iii*); it does not require a claim like Ax_M which asserts the existence of mathematical entities and hence is not logically true.

IV

So far I've argued that for lots of purposes where we might seem to need notions like consequence and consistency and strong consistency or conservativeness, we really need only modal analogues of these notions, and that this is good, because you can explain how facts involving the modal analogues are known without postulating knowledge of mathematical entities. In other words, in many contexts metalogical notions (at least notions of *semantic* metalogic) are dispensable in favour of corresponding object-level notions.

But an important problem remains for a deflationist: the problem of accounting for the utility of reasoning at the metalogical level rather than at the object level. The problem is especially striking for proof-theoretic reasoning, since here there seems to be no object-level analogue. An object-level assertion is one that makes no reference to sentences or formulae (or abstract analogues of them such as propositions); consequently, it can make no reference to axioms or rules of inference or formal derivations. It is hard to see how any such assertion could in any interesting sense be an analogue of an assertion that one sentence is (or is not) formally derivable from another, using a given system of logical axioms and logical inference rules.

How then are we to account for the utility of proof-theoretic reasoning? Traditionally, proof-theoretic concepts are defined in terms of mathematical

JUST LOGICAL KNOWLEDGE? 255

entities, with the result that proof-theoretic reasoning becomes reasoning about mathematical entities. If we accept the usual definitions of proof-theoretic concepts, then a deflationist cannot regard proof theory as a subject of which we can have any knowledge. So how can a deflationist account for its utility?

There seem to be two ways for the deflationist to try to solve this problem. The first is to reject the usual definition of proof-theoretic concepts, and provide alternative definitions which make no reference to mathematical entities. The idea would be to show that if proof-theoretic notions such as formal derivability are understood in terms of these alternative definitions, then claims about formal derivability (etc.) can be known consistently with deflationism, that is, consistently with there being no knowledge of mathematical entities.

One way to try to work out this first approach would be to take derivability to be some sort of modal notion. First we could try to get a sufficiently powerful theory of actual inscriptions, without introducing modality: in terms of such a theory, we could explain notions like 'e is a well-formed inscription', 'e and f are type-identical inscriptions', 'd is (an inscription that constitutes) a derivation (according to system F)', and various predicates of inscriptions that describe them structurally ('being an A-inscription', where A is an expression type). I believe that this part of the project could be worked out in first order logic (though some care is needed because there are no means here to make recursive definitions explicit). The second part of the project would be to make some modal extension: in it, we might hope, we could understand 'A is derivable' to mean 'it is possible that there is a derivation whose last line is an A-inscription', instead of (as the platonist would have it) as meaning that there actually exists a certain type of abstract sequence of abstract analogues of the symbols. This has considerable initial attraction as an account of the ordinary meaning of 'derivable'. I am, though, reluctant to introduce a new type of possibility beyond strictly logical possibility here, unless we can define it from strictly logical possibility plus other acceptable notions; and there are substantial difficulties that must be overcome for the project of doing this.[24] We could avoid these difficulties with additional logical devices

[24] For instance, there are at least two obstacles to taking the relevant sort of possibility to be consistency with the first order theory that one obtained in the first part of the project. The first obstacle arises from the fact that logical possibility is thoroughly anti-essentialist: this poses a problem for translating sentences in which 'derivable' occurs in the scope of a quantifier. ('He uttered an underivable inscription' would always be false on the naive translation.) The natural way to solve this problem would be to take all 'quantification in' to be substitutional. The second obstacle is that the consistency with axiomatic proof theory (even axiomatic platonistic proof theory) of the existence of a proof is not sufficient for provability in the normal sense: incompleteness theorems give cases of unprovable formulae where the assertion that there is a proof is consistent with proof theory. The natural way to solve this would be to say that for A to be provable, the existence of a proof of A must

like a substitutional quantifier, but these might make the appeal to modality unnecessary anyway. I will not pursue these matters here.

In any case, carrying out this programme wouldn't really solve the more general problem raised two paragraphs back. The problem was for the deflationist to account for the utility of proof theory without assuming the truth of mathematics. And in presenting the problem it was assumed that proof theory meant *platonistic* proof theory. The first deflationist approach to this problem says that there is a nominalistic proof theory that is just as good as platonistic proof theory, and that the nominalist has no difficulty in accounting for the utility of *that*. But unless more is said, it looks as if this is merely changing the subject from the original question, which was how the utility of the platonistic theory is to be explained.

My primary interest, then, will be with the second deflationist approach to the problem of explaining the utility of proof theory: the approach to the problem of explaining the utility of platonistic proof theory without assuming it true. I will explain shortly how this second approach can be carried out.

But first I want to shift attention from proof theory back to semantics. The problem with which I began this section—the problem of accounting for the utility of mathematics in metalogical reasoning—is a problem that arises for semantics as well as for proof theory (though it may initially seem less striking a problem in the case of semantics, since there we have object-level analogues of our metalogical notions). And again, there are two different approaches that one might be inclined to take.

The first approach would be to reject the usual model-theoretic definitions of semantic consequence and similar notions, and propose alternative definitions instead. Can this be done? In a sense it can. We can say

(5) B is a semantic consequence of Γ (where Γ is a finite list of sentences together with a finite list of schemas) if and only if $\Box$ (if all members of Γ are true, then B is true).

This definition would be objectionable if the word 'true' here were used in a 'transcendent' sense, that is, in a sense in which 'Snow is white' wouldn't have been true if 'white' had meant 'green'; for in that transcendent sense the modal claim on the right of (5) is going to be false even if B is a consequence of Γ in the usual sense. But let us understand 'true' instead in the 'immanent' sense,[25] that is, as applicable most directly to one's own language and as obeying there the principle that $\Box$ ('Snow is white' is true if and only if snow is white). We can define such an immanent sense of 'true' using substitutional

---

be compatible with a (nominalistic or platonistic) proof theory stated in a powerful logic that can rule out derivations that are not genuinely finite: say a logic with the quantifier 'there are only finitely many', or a logic with a substitutional quantifier or other device of infinite conjunction.

[25] This use of the terms 'transcendent' and 'immanent' was suggested by (but isn't quite the same as) Quine's use of these terms in *Philosophy of Logic*.

quantifiers: S is true if and only if $\Pi p\,(S = \text{`p'} \supset p)$. Or alternatively, one can follow Grover, Camp, and Belnap[26] and regard 'true' (or, at least, 'true' in the immanent sense) not as a predicate at all but (roughly speaking) as simply the means by which substitutional quantifiers with sentences as substituends are represented in English. In either case, it turns out that (5) is equivalent within standard mathematics to $\Box$ (if Ax_Γ, then B), where Ax_Γ is a conjunction of all the axioms in Γ (using substitutional quantifiers to conjoin the instances of the schemas). On this alternative to the usual way of defining consequence, what I earlier called the modal analogue of a claim that one thing was a consequence of another wouldn't really be an *analogue* at all; it would simply be what the consequence claim *means*.

I don't attach a great deal of philosophical significance to the possibility of defining semantic consequence in this non-standard way. My reason for mentioning it is only to point out that even if it is adopted, it leaves an important problem unsolved: namely, how is a deflationist going to explain the utility of *model-theoretic* definitions of semantic consequence? Even assuming that it is somehow better to define consequence modally in the way just indicated (a claim on which I take no stand), still model-theoretic semantics has proved enormously useful; and it is not immediately evident *why* it should be useful if consequence is really to be defined modally via an immanent notion of truth, or if consequence claims are to be rejected as unknowable and only modal analogues of them are to be claimed knowable. So a deflationist must give an account of why standard uses of model theory are legitimate even if model theory isn't true, just as he must give an account of why standard uses of platonistic proof theory are legitimate even if platonistic proof theory isn't true.

In order to provide such accounts, we must first ask to what uses modeltheoretic semantics and proof theory are standardly put; only then can we ask whether these uses are explainable from a deflationist viewpoint. I will not attempt an exhaustive account of the uses to which model-theoretic semantics and proof theory are put, but it seems to me that the central uses are as devices for finding out about logical possibility. Model theory is used for this purpose via the instances of the following two schemas: the *model-theoretic possibility schema*

(MTP)   If there is a model for 'A', then $\Diamond A$;

and the *model existence schema*

(ME)   If there is no model for 'A', then $\neg \Diamond A$.[27]

[26] D. Grover, J. Camp, and N. Belnap, 'A Prosentential Theory of Truth', *Philosophical Studies*, 27 (1975): 73–125.

[27] If A contains free variables, interpret 'model' in (MTP) and (ME) to mean a model together with an assignment function for the free variables.

And proof theory is used for this purpose via the *modal soundness schema*

(MS) If there is a proof of '-A' in F, then $-\Diamond A$,

which holds for any reasonable formal system F for any fragment of logic, and via the *modal completeness schema*

(MC) If there is no proof of '-A' in F, then $\Diamond A$,

which holds for certain areas of logic (i.e., certain types of sentence A) and certain sufficiently strong formal systems for those areas of logic.

From the platonist standpoint, all the instances of these four schemas are true (for the appropriate formal systems F, in the case of the last two schemas).[28] What *justification* might a platonist offer for these schemata? I'll focus on MTP and MS, since these seem the evidentially primary ones. (The other two follow from these, of course, by the classical completeness theorem.) In the case of MTP, each first-order instance (instance where the instantiating formula is non-modal) is almost immediate from an instance of the Conditional Possibility Principle of section 2 (together with NBG and an instance of $\ulcorner A \supset \Diamond A \urcorner$).[29] (I'll discuss the case where the instantiating formula is modal later.) In the case of MS, though, the situation is more delicate. To simplify the discussion, we'll pick an F in which in any application of a rule of inference, the formula being inferred is a logical consequence of the formula

---

[28] Actually there is some question as to whether we should expect (ME) to hold for arbitrary logics. Even for first-order logic, it seems somewhat surprising that (ME) holds (in the same way that the 'Skolem paradox' seems somewhat surprising): just as the universe of classes is too big to form a countable model, it is too big to form a class, and hence too big to form any model at all: so just as it seems somewhat surprising that the sentence Ax_{NBG} that formulates the theory of this universe of classes should have a countable model, it seems a bit surprising that it should have a model at all. But the classical completeness theorem shows that it does have a model if it is not formally inconsistent; this makes (ME) a set-theoretic consequence of (MS) in the case of first-order logic, which surely makes it platonistically acceptable in the case of first-order logic. Still, it may not be acceptable for arbitrary logics. It seems clear that if it fails for some logic—e.g., if there is a sentence A formulable in that logic which expresses enough about the intended model of NBG to preclude there being any class that can be a model of A—then the usual model-theoretic definition of consistency should be regarded as extensionally incorrect for that logic: the imagined sentence A is intuitively consistent, even if it has no model.

[29] For any formula B containing predicates $p_1, \ldots, p_k$, let $M_1, E_1, \ldots, E_k$ be variables not in B, and let B$ be the generalized substitution instance that results from B by restricting quantifiers and free variables by the predicate 'x is in M' and by replacing (for each i and each $v_1, \ldots, v_{n_i}$) '$p_i(v_1, \ldots, v_{n_i})$' by '$\langle v_1, \ldots v_{n_i}\rangle$ is in E_1'. Also, let B* be the existential closure of B$. (Example: if B is '$\forall y\, (p_1(y,z))$', B$ is '$z \in M\ \&\ (\forall y \in M)(\langle y,z\rangle \in E_1)$', and B* is '$\exists M \exists E_1 \exists z[z \in M\ \&\ (\forall y \in M)(\langle y,z\rangle \in E_1)]$'.) $\Diamond$B$ is equivalent to $\Diamond$B*; this plus the Conditional Possibility Principle yields $\Diamond B^* \supset \Diamond B$. So to get the instance of MTP we need only argue in NBG for 'If there is a model M and an assignment function s in which B comes out true, then $\Diamond B^*$.' But in fact we can inductively argue in NBG that if there is such an M and s then *actually* B*; so since possibility follows from actuality, we have the instance of MTP.

from which it is inferred.[30] A platonist might be tempted to argue for the validity of MS (as applied to F) by induction. More accurately, we could argue informally that for all sentences B, if there is a proof of B in F then ⌜□B⌝ is true, by induction on the length of the proof: certainly ⌜□B⌝ is true when B is a logical axiom of any reasonable F, and if B is directly inferable from $B_1, \ldots, B_n$ then ⌜□($B_1 \ldots B_n \supset B$)⌝ is true, and so using the induction hypothesis ⌜□B⌝ is true. In particular then, for all sentences A, if there is a proof of ⌜-A⌝ in F then ⌜◊A⌝is true; by disquotational properties of truth the instances of (MS) follow.

But this informal induction goes beyond the modal consequences of standard mathematics, if 'standard mathematics' is taken in the normal way: for in the induction I have utilized a notion of truth (for sentences in the modal language) that has not been defined. Indeed, besides the usual problem with defining truth, and hence carrying out the induction, that the Tarski indefinability theorem poses, there is a further problem: what would the recursion clause for formulae that begin with a logical possibility operator be? In the Appendix, in defining truth *in a model*, I use the clause

⌜◊A⌝ is true *in model M* ≡ there is a model in which A is true.

(For perspicuity I consider here only the case where A is a sentence.) It might be natural to say by analogy

(∗) ⌜◊A⌝ is true ≡ there is a model in which A is true.

But if 'standard mathematics' is taken as a first-order axiomatic theory, this won't do very well as part of a recursive definition of truth, for then ⌜◊A⌝ together with standard mathematics won't in general entail ⌜◊A⌝ is true⌝.[31] Reason: by Gödel's second incompleteness theorem, there are models of axiomatic set theory NGB in which the sentence 'NGB has a model' will be false. In such a model, '◊NGB' will be true (it's true in all models), but on the proposed definition of truth ' "◊NGB" is true' will be false in the model.

I am not denying, of course, that a platonist should accept the material equivalence (∗) (for sentences A not containing terms like 'true'—see note 31). Indeed, assuming that ' "A" is true' is equivalent to 'A' (for sentences not

---

[30] In the case of non-modal quantification theory, the system of Hunter's *Metalogic* (Berkeley: University of California Press, 1973) has this property: instead of continuing a rule of universal generalization it contains the principle that the universal generalizations of quantificational axioms are quantificational axioms. The analogous strategy can be used in modal systems: avoid a necessitation rule by taking necessitations of all axioms to be axioms.

[31] The sentences A involved here don't contain 'true' or related terms, so the lessons of the semantic paradoxes do not do anything to make this conclusion palatable.

containing 'true'), (*) is simply the conjunction of MTP and ME, which I have said a platonist ought to accept. The argument is, though, that there are models of set theory in which the left to right direction of (*) fails, so that *barring independent support of MS and hence of ME*, (*) does not have the kind of necessity we would like in a recursion clause of a definition of truth.

We could avoid this difficulty about the '◊A' clause by weakening MS to a non-modal soundness schema, i.e. by dropping the operator '◊' in it, and at the same time restricting its instances to non-modal formulae. But even this wouldn't help with the basic problem: arguing for even this restricted non-modal soundness inductively requires a truth predicate, and we don't have one by the Tarski indefinability theorem.

Indeed, the Gödelian example shows not only that there are problems in formalizing the intuitive proof of MS modally within standard mathematics: it shows that MS is not a consequence of standard mathematics alone, even for non-modal A and even in the strong modal logic given in the Appendix. Let CON_{NBG} be the statement that there is no F-proof of not-NBG (where F is a proof procedure for first order logic). Then an instance of MS—indeed, an instance with a purely first order sentence as the substituend for the schematic letter 'A'—is

If -CON_{NBG} then -◊NBG;

equivalently,

(**) If ◊NBG then CON_{NBG}.

By Gödel's second incompleteness theorem, CON_{NBG} is not derivable from NBG. So there is a first order model in which NBG holds and CON_{NBG} doesn't. But the model theory in the Appendix takes first order models to be modal models as well, so NBG doesn't imply CON_{NBG} modally either. But NBG does imply ◊NBG modally, so (**) cannot be a modal consequence of NBG.[32]

At this point in the original version of this essay I introduced a nonstandard sense of 'standard mathematics', employing substitutional quantifiers and an ω-rule governing them, from which MS does follow modally by an inductive proof like the one recently sketched (and in which (*) becomes a reasonable recursion clause for truth). It now seems to me, though, that this made the presentation of several points confusing, and the weight put on substitutional quantification may seem suspicious. It seems simpler just to say that a platonist just accepts MS, even though it is not provable in standard

---

[32] In fact, of course, even the weakening of MS that drops the possibility operator is not a consequence of standard mathematics.

JUST LOGICAL KNOWLEDGE? 261

mathematics from more elementary modal principles. It is worth remarking (though how important this is I'm not sure) that in the modal logic of the Appendix, the non-modal instances of MS are enough to generate (in standard mathematics in the usual sense) all instances of MTP and even of ME, including those where modal sentences are the substituends.[33]

I have been discussing the epistemological status, from a platonistic perspective, of MTP and MS and the other lettered schemata. It is clear that once these schemata are available, the platonist can use ordinary platonistic model theory and proof theory for finding out about possibility and impossibility. But how is this of any help to a deflationist, who denies that the existence of mathematical entities and the truth of mathematical theories can be known? Assume that we construe 'proof' and 'model' in the usual platonistic way. Then if there are no mathematical entities, MTP and MS are only *vacuously* true: they are useless as an aid to finding out about possibility and impossibility, because their antecedents can never be fulfilled. Admittedly, one could do a bit better by considering *concrete* proofs (made by actual physical inscriptions) and *concrete* models. By construing 'model' concretely, we'd get a 'weak MTP' that is non-vacuous (it would be little more than a restatement of the Conditional Possibility Principle, of course); but it would generate the logical possibility of rich structures only from controversial empirical premises. The 'weak MS' obtained by construing 'proof' concretely would be even more severely limited, since rigorous proofs in formal systems are rarely given for anything that is in the least complicated. Apparently, then, the deflationist has a problem.

In fact, though, the problem is easily solved. Instead of MTP and MS (or perhaps, in addition to their weak versions), the deflationist employs the following modal surrogates:

---

[33] For both MTP and ME, one uses an induction on the modal degree of the formula A that is the substituend (generalizing MTP and ME to formulae as in note 27). If A is of modal degree n + 1, let A* be the result of taking each sub-formula of form ◊B for which B is degree 0 (i.e., non-modal) and replacing it by 'B∨-B' if B has a model and by 'B & -B' otherwise. Now we've proved the degree 0 instances of MTP, and the degree 0 instances of ME follow by completeness from the degree 0 instances of MS, which we're assuming. Using these, we get (in S5) that if B non-modal then $\Box((\Diamond B) \equiv (\Diamond B)^*)$, and by a subinduction on the depth of the embedding of ◊B in A, $\Box(A \equiv A^*)$. Consequently, (1) $\Diamond A \equiv \Diamond(A^*)$. Also, another trivial subinduction shows that any model is a model of A if and only if it is a model of A*, and consequently (2) A has a model if and only if A* has a model. But (1) and (2) and MTP for A* (which is of modal degree n) yield MTP for A; and analogously for ME.

Incidentally, the reliance on the degree 0 instances of MS is essential, even for MTP: for instance, if ◊NBG but NBG has no model, then -◊NBG has a model (indeed, it is true in any model), but -◊-◊NBG by the S5 axiom, so we have a degree 1 violation of MTP. Still, a given instance of MTP, with A the substituend formula, can be proved from NBG plus n degree 0 instances of MS, where n is the number of occurrences of logical operators in A. It is easy to calculate what the required instances of MS are (inspection of the proof above shows how to do it), and in a typical case most of them will be provable in NBG anyway so they won't really need to be added.

(MTP#) If $\Box$(NBG ⊃ there is a model for 'A') then $\Diamond$A

and

(MS,#) If $\Box$(NBG ⊃ there is a proof of '-A' in F) then -$\Diamond$A.

From these and the classical completeness theorem, one can derive

(ME#) If $\Box$(NBG ⊃ there is no model for 'A') then -$\Diamond$A

and

(MC#) If $\Box$(NBG ⊃ there is no proof of '-A' in F) then $\Diamond$A,

in the case where first order sentences are the only substituends; indeed, ME# can be argued to hold also where modal sentences are substituends.[34] The deflationist can use the hatched schemata in pretty much the same way the platonist used the unhatched ones: to find out that A is, or is not, logically consistent, it suffices to derive a model-theoretic or proof-theoretic statement from standard mathematics.

Could it be claimed that the deflationist has less reason to believe MTP# and MS# than the platonist has to believe MTP and MS? I do not see how this could be made plausible. First let's consider the instances of MTP# in which the instantiating formula is non-modal. Earlier (note 29) I presented a platonistic argument for the corresponding instances of MTP, within standard mathematics; in effect, then, I showed that for non-modal A

$\Box$(If NBG & there is a model for 'A', then $\Diamond$A.)

From this it follows (in S4) that

(MTP*) If $\Diamond$(NBG & there is a model for 'A'), then $\Diamond$A.

From this and the assumption $\Diamond$NBG, one gets the instance of MTP#. So the platonist's argument from NBG to MTP yields an argument from $\Diamond$NBG to MTP#. The deflationist does of course need to assume $\Diamond$NBG in order to accept MTP#, but if my earlier arguments in Section II are correct, such consistency claims are ones that a deflationist can have perfectly good reason to believe. In any case, the deflationist's epistemological burden is strictly weaker than that of the platonist, who must believe not only that $\Diamond$NBG but that *actually* NBG.

The above reasoning can be extended to instances of MTP# where the instantiating formula is modal. In note 33 I presented a platonistic argument for an arbitrary instance of MTP: the argument was from NBG plus a few instances of MS (which ones being easily calculable from the substituend sentence in MTP). Letting NBG$_A$ be NBG with these added instances, the

---

[34] This will be clear from the remarks on MTP# below, in conjunction with note 33.

reasoning of the previous paragraph shows that the platonist's argument from NBG$_A$ to MTP yields a deflationist argument from $\Diamond$NBG$_A$ to MTP#. (Actually to a slight strengthening of MTP#, one obtained by replacing 'NBG' in it by 'NBG$_A$'.) Again, the claim that the deflationist needs, $\Diamond$NBG$_A$, is a logical consequence of the one that the platonist needs, and I don't see how it can be plausibly argued that the deflationist is in worse epistemological shape than the platonist here.

What about the epistemological status of MS#? In the case of MS, the reader will recall, the platonist had no hope of rigorously deriving it from NBG (even in the strong modal logic of the Appendix); nonetheless, MS is a claim that a platonist ought to accept as a primitive modal assumption. The deflationist, similarly, can accept MS# as a primitive modal assumption. Alternatively, the deflationist can derive each instance of MS# from *the possibility of* what the platonist assumes: i.e., from $\Diamond$(NBG + the corresponding instance of MS). (This is a trivial derivation in S5.) Either way, the deflationist seems to be in pretty good epistemological shape.

A platonist might respond to this by saying that in the case of MS we have an informal (and unformalizable) inductive argument in its favour; but that the deflationist has no such inductive argument in favour of sentences of form $\Diamond$(NBG + B) where B is an instance of MS, and so must ride piggyback on a platonist argument that he or she does not accept. I think, though, that this is wrong. To say that one accepts an informal inductive argument that cannot be formalized in one's theory is to say in effect that one accepts a stronger theory. In the case of the argument for MS, it might plausibly be argued that we are implicitly employing ordinary set theory in some modal framework, to which a truth predicate has been added. (Without the modal framework, the addition of a truth predicate suffices to prove non-modal soundness.)[35] If the modal framework is chosen appropriately—and the truth predicate is a truth predicate for modal sentences as well as non-modal ones—modal soundness should be derivable analogously. But now, if the platonist can appeal to such a powerful theory S, the deflationist can appeal to $\Diamond$AX$_S$.[36] And since the instances of MS follow from S, the instances of MS# follow from $\Diamond$AX$_S$, by the same argument as for MTP#.

Another alternative for the platonist who wants a proof of MS is to employ some sort of ω-rule. That is, even without a notion of truth we can give a general method for proving each instance of

---

[35] Here I assume that it is a truth predicate for sentences of set theory that don't contain it, that it is allowed to occur in the separation and replacement schemas, and that 'proof' in MS is understood as 'proof not employing the new truth predicate'.

[36] S won't be finely axiomatized; but we can avoid the use of substitutional quantifiers here, by using the theory consisting of the possibilization of each finite conjunction, as discussed in the postscript [not reprinted here].

(MS^k)   If there is a proof of '-A' of length k, then -◊A

(where k is any numeral and A any formula): exhaustively search the derivations of length k, and the result will either falsify the antecedent or yield a proof of the consequent. But without a truth predicate, we cannot in ordinary set theory get from these instances to the generalization that for all k MS^k holds. But if an ω-rule were adopted, a platonist could get this generalization from the various MS^k, or from the argument for their provability. My response of course is that if a platonist is allowed to employ an ω-rule, so is the deflationist. But if so, then the same sort of argument as above gets us from the platonist's 'derivation' of MS in the expanded logic to a deflationist's 'derivation' of MS$^\#$ in the expanded logic. However the platonist twists and turns in an effort to avoid taking MS as simply a primitive assumption, the deflationist can twist and turn too.

I conclude that the deflationist has no more difficulty in using platonistic model theory and proof theory in finding out about possibility and impossibility than does the platonist.

In Section III, I distinguished two 'problems of application': the problem of application of mathematics to the physical world and the problem of application of mathematics to the study of logical reasoning. I have just outlined a solution to the latter problem; but readers may be puzzled by a disanalogy between this solution and the solution offered in my book to the problem of application to the physical world. In solving both problems of application, I tried to legitimize a certain kind of instrumentalism about mathematics: I tried to argue that platonistic physics and platonistic metalogic were usable even if not true. But in *explaining* why the usability of these theories didn't depend on their truth, I had to do more work in the case of platonistic physics than in the case of platonistic metalogic. That is, the explanation given in my book for the legitimate usability of platonistic physics turns on the existence of a nominalistic physics. But the explanation I have just given for the legitimate usability of platonistic proof theory doesn't require the existence of a nominalistic proof theory. (I have stated that such a nominalistic proof theory may be possible, but my explanation of the usability of platonistic proof theory in finding out about possibility and impossibility did not turn on this.) What accounts for the difference?

The answer is that physics has an explanatory function: you need physical theories to explain physical phenomena. According to the form of nominalism I accept, one should not junk a platonistic explanation of a phenomenon unless there is a satisfactory nominalistic explanation to take its place. It is because of this principle that a satisfactory nominalistic formulation of physical theories is required. Now, the main role of platonistic proof theory is not explanatory. If I use proof theory as an aid to discovering whether B follows from A, it is not because the proof-theoretic principles are in any way explanatory of the fact that $\Box$ (A ⊃ B) or -$\Box$ (A ⊃ B); the proof theory is solely an instrument of discovery, and needn't be replaced by some other theory about which we must take a non-instrumentalist attitude.

I have not said that proof theory has *no* explanatory function, but only that its *central* function is not explanatory. There is a sense in which proof theory can be used to explain. Suppose I want to explain the historical fact that no one has ever produced a physical inscription which constitutes a formal derivation, in Kleene's system of logic, of an explicit contradiction. Intuitively, the reason that no one has ever produced such an inscription is that it is impossible that there be one; and here I think 'impossible' can be taken to mean 'logically inconsistent with certain assumptions we make about physical inscriptions'. Any codification of those assumptions is a theory of physical inscriptions, and from it we can explain the historical fact in question. One way to codify those assumptions about physical inscriptions is to formulate platonistic proof theory and then add a principle saying that there is a certain kind of homomorphism mapping physical inscriptions into expressions in the platonistic sense. This codification produces a platonistic explanation of the historical fact in question. I also think that assumptions about physical inscriptions which are adequate to the purpose at hand can be stated nominalistically (without using devices going beyond first order logic, in this case); if so, then we will have a nominalistic explanation of the historical fact. I take it, though, that the use of proof theory to explain such historical facts is of less importance than its use in finding out about possibility and impossibility; and in the latter uses, the platonistic proof theory does not serve an explanatory function, and so no nominalistic proof theory is required.

I have been discussing a disanalogy between my treatments of physics and of proof theory; how does model theory enter into the picture? That is, in explaining the legitimate usability of platonistic model theory, did we (as with platonistic physics) need to develop a nominalistic analogue of the platonistic theory? Or (as with the central applications of platonistic proof theory) did we not? This question is largely verbal: it depends on whether we regard modal logic itself as an analogue of platonistic model theory. If we do so regard it, then model theory is like physics, and the earlier sections of this essay were devoted largely to developing the nominalistic analogue as a necessary[37] prelude to explaining the legitimate usability of platonistic model theory. If we do not so regard it, then model theory is like proof theory in its most central

[37] The prelude is necessary, for as I argued in 'Realism and Anti-Realism about Mathematics', there is no way to explain the legitimate usability of metalogic by conservativeness alone if the underlying logic is taken to be non-modal.

applications: we did not need a nominalistic analogue of model theory because model theory doesn't serve to explain anything, but simply serves as a tool for enabling us to find out more easily about possibility and impossibility. The difference between these two viewpoints is merely a difference between ways of looking at what has been done: whatever the viewpoint, the argument earlier in this section shows how one can explain the legitimate usability of platonistic model theory without assuming its truth.

V

In this essay I have been advocating the view that all mathematical knowledge that isn't straightforwardly empirical is knowledge of a purely logical sort. By 'mathematical knowledge' here I do not mean knowledge of the claims of mathematics. According to the view I have advocated (deflationism), there *is* no mathematical knowledge in that sense. Rather, by 'mathematical knowledge' I mean the sort of knowledge that those who know a lot of mathematics have a lot of and those who know little mathematics have little of. If deflationism is false, this will include knowledge of mathematical claims; but whether deflationism is true or false, it will include a lot of knowledge that isn't knowledge of mathematical claims.

As hinted several times, some of the knowledge that separates those who know lots of mathematics from those who know only a little is straightforwardly empirical. A set theorist typically knows which axioms of set theory are generally accepted within the mathematical community, which theorems have been proved, which unsolved problems are generally regarded as important; and any algebraist knows that mathematicians have thoroughly developed a generalization of vector space theory in which the role that fields play in vector space theory is played by unitary rings. All these sorts of knowledge are empirical knowledge about the mathematical community. Besides such knowledge about the mathematical community, mathematicians typically have other empirical knowledge that non-mathematicians tend to lack; for example, typically there will be various complicated empirical claims about physical space which they know because they know them to follow from the empirical fact that physical space can be locally approximated as a Cartesian power of the real numbers. If one were to attempt a realistic account of all of the knowledge differences that separate a typical mathematician from a typical non-mathematician, I think that such differences in empirical knowledge would play a large role. So (contrary to what the title of this essay might suggest) a great deal of mathematical knowledge is the sort of straightforward empirical knowledge that no one could possibly regard as logical. The interesting question, however, concerns the mathematical knowledge that remains when this straightforward empirical knowledge is ignored. The deflationist claim that I have defended is that the only such knowledge there is, is purely logical—even on a conception of logic according to which logic can make no existence claims. (It would not seem to me to be terribly interesting to say only that such knowledge was logical on a broader conception of logic like that of the logicists—a conception on which logic guarantees the existence of a realm of platonic entities. See note 3.)

The deflationist view is reminiscent of, but importantly different from, a position that has been called 'deductivism' or 'if-thenism'. Deductivism is usually characterized[38] as the view that when someone asserts a typical mathematical statement (e.g., that there are infinitely many primes), what he or she really means is that this statement follows from a certain body of other mathematical statements. (*Which* body of other mathematical statements? The standard axioms of the field of mathematics in question, if the field has an accepted axiomatization; otherwise, some other body of claims implicit from the context. Deductivists tend to be a little vague about this, often entirely ignoring the situation where there is no generally accepted axiomatization.)

One of the major differences between this and deflationism is that deflationism, unlike deductivism, does not claim that mathematical claims mean anything other than what they appear to mean. Instead of saying that mathematical assertions don't mean what they appear to mean, the deflationist says that what they literally mean can't be known: the knowledge that underlies a mathematician's assertions is not what those assertions literally say. I'm afraid that many readers will still find this implausible, but it certainly seems to me less implausible than the claims about meaning made by the deductivist. (The implausibility of the deductivist position is especially evident in the case of a mathematical assertion A made in the absence of a generally accepted axiomatization. The deductivist must select some one body of other mathematical statements, and claim that what is meant in saying A is really that A follows from this other body of statements. But the deflationist, since he makes no claim about meaning, need not single out any one body of other mathematical statements as relevant: *no* bodies of other statements are relevant to what the assertion of A *means*, and *lots* of bodies of assertions are relevant to what the mathematician who asserted A *knows*, since a great many distinct pieces of knowledge of the interrelation of A with other mathematical claims may have been part of the motivation for asserting A.)

Indeed, the deflationist can easily handle a problem that is often thought to sink deductivism. Consider a mathematician asserting a claim which he

[38] See, e.g., Michael Resnik, *Frege and the Philosophy of Mathematics* (Ithaca, NY: Cornell University Press, 1980), ch. 3.

knows not to follow from previously accepted axioms: he intends it as a new axiom. What does he mean when he asserts this mathematical claim? If one takes deductivism entirely literally, then, according to the deductivist, the mathematician must mean either

(a) that the new axiom follows from the old axioms

or

(b) that the new axiom follows from the system that consists of the old axioms plus the new axiom.

But the mathematician doesn't believe (a), and (b) is totally trivial. Since the whole point of the deductivist's claims about what the mathematician means is to make what he means directly reflect part of the knowledge that led to the assertion, both of these alternatives are intolerable.

For a deflationist, on the other hand, the situation where the mathematician introduces a new axiom poses no special problem. The reason is that the deflationist does not accept the programme of trying to represent the knowledge that leads the mathematician to make the assertion in the meaning of the assertion. The kind of knowledge that typically leads a mathematician to assert a new axiom is clear enough: it is knowledge that the axiom (in conjunction with previously accepted axioms) has certain desirable consequences and doesn't seem to have undesirable ones. In other words, it is knowledge of the logical interrelations of the proposed axiom with other mathematical claims, which is just the sort of knowledge that the deflationist wants to allow anyway. So the situation where someone asserts a mathematical claim because of the consequences he knows it to have is no more of a problem for a deflationist than the situation where he asserts it because he knows it to be a consequence of previously accepted claims. In general, I think it would be extremely surprising if careful attention to the sociology of mathematical practice turned up features of that practice that couldn't plausibly be handled along deflationist lines.

VI

It is often supposed that one of the things that differentiates those who know a lot of mathematics from those who know only a little is that the former but not the latter have considerable knowledge of a realm of platonic entities such as sets, numbers, and tensors—entities that bear neither causal nor spatio-temporal relations to us or to anything we can observe. If this were correct, there would be a considerable problem in explaining how knowledge of such a realm could be attained. The strategy of this essay has been to point out various facts that aren't about such a platonic realm, facts which mathematicians typically know and non-mathematicians typically don't. These facts include empirical facts, like the facts mentioned early in Section V; and they include logical facts, like the facts about logical possibility that I have stressed in the bulk of the essay. I see no reason to believe that there is a further kind of fact—non-empirical and non-logical—that the mathematician also knows. Therefore, I see no reason to suppose that the mathematician has knowledge of the existence of mathematical entities or the truth of ordinary mathematical claims.

APPENDIX

In this Appendix I will sketch an alternative to Kripke's model theory for modal logic, one which will give the intuitively correct results about which sentences involving the 'logically possible' operator are logically true. The model theory will be, like Kripke's, platonistic, for it will presuppose a large body of pure set theory. What a deflationist should say about the status of a platonistic model theory such as this was discussed in Section IV. (The purpose of the model theory is not to confer intelligibility on the modal operator. In my view, logic stands on its own; it doesn't need model theory for its intelligibility. Indeed, it is hard to see how a logic could get its intelligibility from the model theory for it, since one would need the logic in understanding (e.g., in being able to reason from) the model-theoretic assertions.)

In the model theory for first-order logic, we say that a sentence is *logically true* if and only if it is *true in all models*. Here, a *model* consists of a set of objects (the entities that *exist in the model*) plus a stipulation as to which things if any the predicates are *true of* in the model, which things if any the names *denote* in the model, and so forth. We also need the notion of an *assignment function for a model*: it is a partial function that assigns entities that exist in the model to all, some, or none of the variables of the language.[39] Given a model M and an assignment function s for M, it is possible recursively to define what it is for a formula of the language to be *true in M relative to* s. We

[39] It is standard in first-order logic to restrict consideration to models in which at least one object exists and in which all names denote. When this restriction is made, assignment functions are taken to be total functions; i.e., they assign something to every variable. I have tacitly lifted this restriction in the body of the Appendix, since it would be more glaringly anomalous in modal logic than it is in first-order logic. (There are two ways of lifting the restriction that seem about equally reasonable, those of Scott and of Burge—see n. 13. Strictly speaking, the definition of 'model' given in the Appendix is applicable only to Burge's system, but nothing of substance would be altered if we complicated the definition slightly so as to apply to Scott's. In particular, the definition of 'model' for a modal logic based on Scott's free logic would be the same as for the non-modal Scott free logic.)

then define a formula to be *true in M* if and only if it is true in M relative to s for *every* assignment function s for M; and we define a formula to be *logically true* if and only if it is true in M for every model M.

How should we generalize this to modal logic? Kripke's approach is to keep the idea that logical truth is truth in all models, but to redefine the notion of model: in the case of the system S5 (which is the one of interest for present purposes), a model is to be a non-empty set of possible worlds, one of which is designated as actual; each possible world is determined by a set of objects that exist in that possible world, plus a stipulation as to what things in that world the predicates are true of in that world in that model, plus similar stipulations for names and other primitive vocabulary of the language. A sentence of the form $\ulcorner \Diamond A \urcorner$ will be true in a model just in case it is true in at least one possible world in the model. For $\ulcorner \Diamond A \urcorner$ to be logically true, it must be true in all models, *and hence in particular it must be true in all models in which there is only one possible world* (i.e., in which there are no possible worlds other than the actual world of the model). It is clear that there is no way that this can happen unless A itself is true in all models; that is, unless A itself is logically true. That is the curious feature of Kripke's definition of logical truth for modal logic that I noted at the beginning of Section II.

I propose an alternative way of generalizing the definition of logical truth for sentences of first-order logic to a definition appropriate to modal sentences. As on Kripke's approach, we are to retain the idea that logical truth is truth in all models. In addition, *we retain the definition of model used in first-order logic.* (We do not introduce possible worlds; rather, a model will be in effect just the 'actual world portion' of a Kripke model.) The *only* difference between the proposed definition of logical truth for modal logic and the usual definition of logical truth for first-order logic is that in recursively defining truth in a model, we need an extra clause that will handle formulae beginning with '$\Diamond$'.

Moreover, the rule for '$\Diamond$' will be a lot like the rule for '$\exists$' used in first-order logic. In first-order logic the rule for '$\exists$' is as follows:

$\ulcorner \exists x \, B \urcorner$ is true in M relative to s if and only if B is true in M relative to s^*, for some s^* that assigns something to the variable x and that is just like s except in what it assigns to the variable x.

Note that in this rule we quantify over assignment functions, leaving the model fixed. I propose that in our rule for '$\Diamond$' we quantify over models and assignment functions together:

$\ulcorner \Diamond B \urcorner$ is true in M relative to s if and only if B is true in M* relative to s^*, for some model M* and some assignment function s^* for M*.

If B is a sentence (i.e., has no free variables), all reference to assignment functions can be proved irrelevant. That is, in that case the rule reduces to

⌜◇B⌝ is true in M if and only if B is true in M* for some model M*.

It is clear that on this approach, unlike Kripke's, such sentences as

◇∃x ∃y (x ≠ y)

and

◇∃x (x is an electron)

will come out true in all models, and hence logically true.

The approach that I have sketched for defining logical truth for modal sentences has its roots in chapter 5 of Carnap's *Meaning and Necessity*;[40] though much of what Quine found abhorrent about Carnap's ideas has been dropped. In the first place, the approach I have sketched does not rely in any way on the idea of meaning. As I remarked in the text, the basic modal logic is formulated in such a way that it does not reflect 'meaning-relations among predicates' if such a notion be recognized. (Though if one does recognize such a notion, a derivative modal logic can be obtained which does reflect such relations.) In the second place, the treatment of free variables that I have given does not require the introduction of 'individual concepts', and it is thoroughly anti-essentialist in that no formula of the form ⌜◇B⌝ is true in a model with respect to one assignment function unless it is true in that model with respect to every other assignment function. (Again, however, the notion of logical possibility can be used to introduce derivative notions of possibility which are essentialist in various ways.) A third respect in which my views modify Carnap's (though in this case, not a modification that Quine would favour) is that Carnap's idea was to regard modal concepts as derivative from semantic concepts. On my view it is, if anything, the other way around.[41]

---

[40] Chicago: University of Chicago Press, 1956. An account even more similar than Carnap's to that given here is that of Nino Cocchiarella, 'On the Primary and Secondary Semantics of Logical Necessity', *Journal of Philosophical Logic*, 4 (1975): 13–27. But Cocchiarella's method of dealing with variables and their interaction with modal operators seems to me unacceptable: e.g., it leads to Ramsey's bizarre conclusion that 'It is possible that there are at least $10^{10^{10}}$ objects' is logically false if the world happens to contain less than $10^{10^{10}}$ objects, but is logically true if the world happens to contain at least $10^{10^{10}}$ objects. (Cf. the last section of Ramsey's paper 'The Foundations of Mathematics', in *The Foundations of Mathematics* (New York: Littlefield, Adams, 1960), 1–61.) Despite this difference between Cocchiarella's account and the one I have offered, most of Cocchiarella's philosophical remarks about logical truth in modal logic apply to my account as well as to his own. (Some of the others who have advocated something in this general ballpark are Richard Montague (*Formal Philosophy* (New Haven: Yale University Press, 1974), ch. 1); Jaako Hintikka ('Standard vs. Nonstandard Logic', in Evandro Agazzi (ed.), *Modern Logic* (Dordrecht: Reidel, 1981)); and Dana Scott ('On Engendering an Illusion of Understanding', *Journal of Philosophy*, 68 (1971): 787–807.)

[41] I would like to thank Tony Anderson, Paul Benacerraf, Charles Chihara, Geoff Joseph, Janet-Levin, David Lewis, Penelope Maddy, Colin McGinn, Bill Tait, Richard Warner, and an anonymous referee for helpful comments that have affected the final version. Much of the research was carried out under a summer research grant from the National Science Foundation (SES-8205264).

XIII

THE STRUCTURALIST VIEW OF MATHEMATICAL OBJECTS

CHARLES PARSONS

INTRODUCTION

By the 'structuralist view' of mathematical objects, I mean the view that reference to mathematical objects is always in the context of some background structure, and that the objects involved have no more to them than can be expressed in terms of the basic relations of the structure. The idea is succinctly expressed by Michael Resnik:

In mathematics, I claim, we do not have objects with an 'internal' composition arranged in structures, we have only structures. The objects of mathematics, that is, the entities which our mathematical constants and quantifiers denote, are structureless points or positions in structures. As positions in structures, they have no identity or features outside of a structure.[1]

Views of this kind can be traced back to the end of the nineteenth century, although it has often not been clear whether they were being offered as general views about mathematical objects, or as accounts of one or another particular kind of such objects. Clear general statements of the kind of view I have in mind were made by Paul Bernays in 1950 and by W. V. Quine somewhat later, in connection with his doctrine of 'ontological relativity'.[2] I want to exhibit some difficulties that arise in stating the view more fully. The purpose of this essay is to explore some directions in which we are led in dealing with the difficulties. I have indicated elsewhere that I think that something close to the

---

First published in *Synthese*, 843 (1990): 303–46. Reproduced here by permission of Kluwer Academic Publishers.

[1] M. Resnik, 'Mathematics as a Science of Patterns: Ontology and Reference', *Noûs*, 15 (1981): 530.

[2] P. Bernays, 'Mathematische Existenz und Widerspruchsfreiheit' (1950), in *Abhandlungen zur Philosophie der Mathematik* (Darmstadt: Wissenschaftliche Buchgesellschaft, 1976), 92–106. Quine is generally most explicit when speaking of the natural numbers. For a very explicit general statement, however, see *Ontological Relativity and Other Essays* (New York: Columbia University Press, 1969), 43–5.

structuralist view is true.³ It will turn out, however, that it does need some significant qualifications; in particular, some mathematical objects for which structuralism is not the whole truth must still have their place.

In spite of the problems I see in its precise statement, the structuralist view is familiar. It is not my purpose to go deeply into the arguments offered for it, but before proceeding I should remind the reader of the main considerations. Where it is most immediately persuasive is in the case of *pure* mathematical objects such as sets and numbers (in the broad sense, including the various number systems). In these cases, we look in vain for anything else to identify them beyond basic relations of the structure to which they belong: for the natural numbers 0, S (successor), and perhaps arithmetic operations, for sets' membership relations and perhaps whatever individuals might enter into their composition. A symptom of this is the problem of 'multiple reductions', much, probably excessively, discussed in the literature. The context in which it entered the philosophical literature is that of logicist or set-theoretic treatments of the natural numbers: if one identification of the natural number sequence with a sequence of sets or 'logical objects' is available, there are others such that there are no principled grounds on which to choose one.⁴ The existence of multiple reductions was, however, long familiar from the 'arithmetization of analysis' which gave rise to alternative constructions, on the basis of the natural numbers and set theory, of positive and negative integers, rational, real, and complex numbers.⁵ The structuralist point of view is intended to dissolve this problem; I shall remind the reader how in Section 2.

Pure mathematical objects are to be contrasted not only with concrete objects, but also with certain abstract objects that I call quasi-concrete, because they are directly 'represented' or 'instantiated' in the concrete. Examples might be geometric figures (as traditionally conceived), symbols whose tokens are physical utterances or inscriptions, and perhaps sets or sequences of concrete objects. The 'concrete objects' that David Hilbert talked about in his accounts of intuitive, finitist mathematics are in my terminology quasi-concrete. A purely structuralist account does not seem appropriate for quasi-concrete objects, because the representation relation is something additional to intrastructural relations. Because they have a claim to be the most elementary mathematical objects, and also for other reasons, quasi-concrete objects

---

[3] C. Parsons, *Mathematics in Philosophy* (Ithaca, NY: Cornell University Press, 1983), 189–90, also 20–2. The 'eliminative structuralism' briefly discussed in the latter place is the main theme of Sects. 3–8 below.

[4] From Paul Benacerraf, 'What Numbers Could Not Be' (1965), in P. Benacerraf and Hilary Putnam (eds.), *Philosophy of Mathematics: Selected Readings*, 2nd edn. (Cambridge: Cambridge University Press, 1983), 272–94, and my 'Frege's Theory of Number' (1965), Essay 6 of *Mathematics in Philosophy*, esp. 154–5.

[5] Philip Kitcher points out that multiple reductions also arise for the set-theoretic devices used in all these constructions. See 'The Plight of the Platonist', *Noûs*, 12 (1978): 119–36, esp. 126.

are important in the foundations of mathematics. It is their role, as we will see in Section 8, that leads to the main qualifications I wish to make on the structuralist view.[6]

1. THE CONCEPT OF STRUCTURE

To begin with, we should make clear that the view we are discussing is not yet expressed by the statement that mathematics is about, or primarily about, structures. That statement seems to single out structures as a kind of mathematical object of particular importance. If we then ask what a structure is, we seem to be led back to familiar kinds of mathematical objects.

What is meant by a structure is usually a domain of objects together with certain functions and relations on the domain, satisfying certain given conditions. Paradigm examples of structures are the elementary structures considered in abstract algebra. For example, a *group* $\mathfrak{G}$ consists of a (non-empty) domain G, together with a two-argument function on G, which we will write $\circ$, such that $\circ$ is associative, there is a (unique) element e which is an identity element for $\circ$ (i.e., for all $a \in G$, $a \circ e = e \circ a = a$), and every element a of G has a (unique) inverse with respect to $\circ$; that is, for every a there is a b such that $a \circ b = b \circ a = e$. One might regard the identity e along with $\circ$ as part of the structure. (A distinguished element can be taken to be a function of 0 arguments.) The language in which we have set forth this simple example already refers to familiar kinds of mathematical objects, 'domains', which one readily thinks of as sets, functions, and relations.

If we follow out this reading, then we arrive at the set-theoretic concept of structure, familiar, for example, from model theory, which provides a means of talking of structures as themselves mathematical objects, and offers the resources of set theory for reasoning about them. We take the domain of the structure as a set, and the functions and relations in the usual set-theoretic way. The structure itself is then a tuple, a set-theoretic object. For example, our group $\mathfrak{G}$ will be the pair $\langle G, \circ \rangle$ or the triple $\langle G, \circ, e \rangle$. This way of talking about structures incorporates the dictum that mathematics is about structures into the familiar conception of set theory as the canonical language for all of mathematics, so that all mathematical objects can be construed as sets.

Resnik's initial statement of his view seems to require cashing in on terms of the set-theoretic conception of structure or some variant of it. He talks freely of structures or 'patterns' (his preferred term). It thus appears that

---

[6] See also my 'Mathematical Intuition', ch. V above. The notions of pure and quasi-concrete apply to abstract objects generally, and not only in mathematics. They do not offer an exhaustive classification of abstract or even of mathematical objects.

patterns, or structures, are primary objects in his ontology, and one will naturally ask what these are. His statement of what a pattern is (p. 532) is very close to the informal statement above of what a structure is. It is clear, however, that Resnik regards sets as on the same footing as other mathematical objects. It would take us somewhat afield to consider how Resnik reconciles the fact that in order to state his view, he has to talk of patterns as if they were a fundamental kind of object, while the view itself would make the patterns themselves 'positions in patterns'. We have, however, come up against the first difficulty in stating the structuralist view: it seems to require *structures*, and this concept seems to involve familiar kinds of mathematical object, or perhaps to call for explication in 'structuralist' terms that would threaten circularity. One answer to this difficulty is simply to use the set-theoretic conception of structure. This answer has an obvious difficulty: if we use it in order to give a structuralist account of some kind of mathematical object, then we will be assuming sets. The question will then arise how we are to give such an account of sets themselves. I shall, however, put that inadequacy aside for the moment and explore a version of the structuralist view that uses the set-theoretic conception of structure.

2. DEDEKIND ON THE NATURAL NUMBERS

The set-theoretic conception of structure provides a very natural framework for giving accounts of particular kinds of mathematical object. We can illustrate it by the natural numbers, where an account of this kind was already given a century ago by Richard Dedekind in *Was sind und was sollen die Zahlen?*[7] In this work Dedekind presents a development of arithmetic which is remarkably close to modern developments of arithmetic in set theory, although Dedekind's concept of *system* (set) is not axiomatized. His analysis of number, however, is presented as a characterization of the *structure* of the natural numbers.[8] This is given by the definition in paragraph 71 of a *simply infinite system*. A simply infinite system is a system (set) N such that there is a distinguished element 0 of N, and a mapping $S: N \to N - \{0\}$ which is one–one and on to, such that induction holds, that is:

---

[7] R. Dedekind, *Was sind und was sollen die Zahlen?* (Brunswick: Vieweg, 1888; 3rd edn. 1911).

[8] Thus the aim of Dedekind's analysis of number is quite different from that of Frege's *Grundlagen der Arithmetik* (Breslau, 1884; repr. in English trans. as *The Foundations of Arithmetic*, 2nd edn. (Oxford: Blackwell, 1953), where the main work is done by an analysis of the concept of cardinal number. The works are none the less often classified together, for good reasons: Frege's primitive notions of concept and extension are related to, though different from, Dedekind's primitive notion of system, and they rely on essentially the same analysis of mathematical induction, as Dedekind remarks in his preface to the 2nd edn. (p. x).

(1) $\forall M\{[0 \in M \wedge \forall x(x \in M \to Sx \in M)] \to N \subseteq M\}.$

Let us abbreviate these conditions by $\Omega(N, 0, S)$. I will use Dedekind's term 'simply infinite system' for the structure $\langle N, 0, S \rangle$.[9] Dedekind has given an explicit definition of the kind of structure instanced by the natural numbers, although the presence of the quantifier $\forall M$ means that it is not first-order in N, 0, and S.

I want now to consider how Dedekind interprets talk of *the* natural numbers. So far, what we have been given is only a definition of an ω-sequence or progression. Although his intent is not entirely clear, his treatment of the natural numbers is meant to rest on two substantive claims: the claim that simply infinite systems exist (par. 72, resting on the argument of par. 66 that infinite sets exist) and the famous categoricity theorem (par. 132). What he actually proves in the latter section is, in our terminology, that any two simply infinite systems are isomorphic.

It is clear that Dedekind is not following the procedure common in contemporary books developing the number systems in set theory, and which was followed by Zermelo, of presenting a simply infinite system and then identifying the natural numbers with that structure. The explanation he gives of talk about the natural numbers is somewhat awkward:

If, in considering a simply infinite system N, ordered by a mapping ϕ, one abstracts from the specific nature of the elements, maintains only their distinguishability, and takes note only of the relations into which they are placed by the ordering mapping ϕ, then these elements are called *natural numbers* or *ordinal numbers* or simply *numbers*, and the initial element 1 is called the initial number (*Grundzahl*) of the *number series N*. (par. 73)

A reading of Dedekind that seems to me to accord reasonably well with this passage takes him as holding that statements about natural numbers are implicitly general, about *any* simply infinite system. A statement in the usual language of arithmetic will be expandable to one in which the arithmetical primitives are N, 0, and S, so that we can write it as $A(N, 0, S)$. Let $\Omega(N, 0, S)$ be as above. When $A(N, 0, S)$ is taken as a statement about *the* natural numbers, we are to understand N, 0, and S as *variables*, and the statement is elliptical for:

(2) For any N, 0, and S, if $\Omega(N, 0, S)$, then $A(N, 0, S)$.

---

[9] Dedekind treats the natural numbers as beginning with 1; I follow contemporary usage in beginning with 0.

Note that Dedekind uses the notion of mapping, i.e., function. He characterizes a mapping as a kind of law (par. 21). Once a reduction of functions to sets is available, however, multiple reductions arise again, as Kitcher pointed out ('Plight of the Platonist'). Thus we have an additional reason for looking for a version of structuralism more thoroughgoing than the set-theoretic.

A STRUCTURALIST VIEW

The categoricity theorem implies that (2) holds if $A(N, 0, S)$ holds for a single simply infinite system $N, 0, S$.

I will call this interpretation the *eliminative reading* of Dedekind. It clearly avoids singling out any one simply infinite system as the natural numbers and expresses the general conception I have in mind in speaking of the structuralist view. We shall soon see that it is probably not what Dedekind intended. But it is worth pursuing because it exemplifies a very natural response to the considerations on which a structuralist view is based, to see statements about a kind of mathematical objects as general statements about structures of a certain type and to look for a way of eliminating reference to mathematical objects of the kind in question by means of this idea. Such a programme I will call *eliminative structuralism*. The response, although not yet the explicit programme, is expressed by Benacerraf when he concludes from structuralist intuitions that numbers are not objects and writes, 'Number theory is the elaboration of the properties of all structures of the order type of the numbers'.[10] The eliminative structuralist programme has attracted a number of philosophers, and its tendency to be revived after attempts run into serious difficulties shows that the ontological intuition behind it exerts a powerful attraction.

But to return to Dedekind: his statement in the next sentence after the above quotation that one can call the natural numbers 'a free creation of the human mind' raises doubts as to whether the eliminative reading is the best reconstruction of his intention. He explains this idea somewhat further in another text, a letter to H. Weber of 24 January 1888 of which an extract was published in Dedekind's collected works. Weber had evidently suggested that the basic concept of number is that of cardinal; Dedekind argues for the priority of the ordinal concept, by which he means, I think, the structure of initial element and successor function. He then writes:

If one wants to take your way . . . then I would advise not taking the number as the class (the system of all finite systems similar to one another) but rather as something *new*, corresponding to this class, which the mind *creates*. We are of a divine kind and possess, without any doubt, creative power not only in the material realm (railroads, telegraph) but most especially in the mental. This is just the same question of which you speak at the end of your letter in relation to my theory of irrationals, where you say that the irrational number is nothing at all other than the cut itself, while I prefer to create

---

[10] Benacerraf, 'What Numbers Could Not Be', 291. This and some other remarks of Benacerraf intimate an eliminative structuralist programme, but he does not commit himself to it or indicate any way other than the set-theoretic of carrying it out. Note that the statement quoted quantifies over structures. On the same page he talks of 'systems of objects' and 'relations'. On the other hand, it appears that he does want to avoid Dedekind's position of exempting sets from his structuralism; note his reference (p. 290) to Takeuti's reduction of Gödel–Bernays set theory to a theory of ordinals.

something new, distinct from the cut, which corresponds to the cut and of which I say that *it* brings forth the cut, generates it.[11]

What Dedekind proposes in the cases of cardinals and real numbers is well captured by W. W. Tait's notion of 'Dedekind abstraction'. Suppose one has described a structure of a certain similarity type, such as the finite von Neumann ordinals $\langle \omega, \phi, \lambda x(x \cup \{x\}) \rangle$, as a simply infinite system. Then one can introduce a new structure $\langle N, 0, S \rangle$ of that type, together with an isomorphism from the given one.[12] Clearly this does not square with the eliminative reading, although it is still structuralist in the general sense with which we began. But I shall continue to explore the eliminative reading.

The eliminative reading avoids a difficulty faced by one crude attempt to deal with the 'multiple reduction' problem, namely, saying that *any* simply infinite system can be, or can be taken to be, the natural numbers. Some such choices seem absurd, outside very special contexts; for example, suppose one has the natural numbers N and then takes as one's 'natural numbers' the even numbers of N, with the appropriate successor function.[13] Again, suppose we have a certain simply infinite system N', 0, S', and suppose we take as 'the' natural numbers the system N, 0, S obtained by replacing the 17th element of N' by Richard Nixon, and adjusting S' accordingly.[14] On this choice, the sentences 'Richard Nixon is a natural number' and 'Richard Nixon = 17' become true, which seems absurd. It is clear that neither is even possibly true on Dedekind's reading.[15]

A structuralist view of the natural numbers faces another well-known objection. This is that the application of arithmetic requires relations of numbers

---

[11] P. Dedekind, *Gesammelte mathematische Werke*, ed. by Robert Fricke, Emmy Noether, and Öystein Ore (Brunswick: Vieweg, 1930–2), iii. 489. I was led to look for this passage by W. W. Tait, who rightly questioned my earlier acceptance of the eliminative reading. In his published comment on the matter, Tait claims that 'a new scandal will be created by taking Dedekind to be a structuralist' ('Critical Notice: Charles Parsons' *Mathematics in Philosophy*', *Philosophy of Science*, 53 (1986): 590–1). He is right if he means (as I did in the place he comments on) the eliminative reading of Dedekind or some related eliminative structuralist interpretation. But the interpretation of Dedekind he himself recommends makes Dedekind's view of the number systems structuralist in the more general sense of the Introduction. There is of course no question of attributing to Dedekind a structuralist view of sets and functions and thus a structuralist view of mathematical objects in general. [12] Ch. VII, n. 12.

[13] Cf. Quine, *Ontological Relativity and Other Essays*, 45.

[14] i.e., for m, n, in N, $n = Sm$ iff $m = S'(16)0'$ and $n =$ Richard Nixon, $m =$ Richard Nixon, and $n = S'(18)0'$, or otherwise and $n = S'm$.

[15] One might be worried because the sentence 'Richard Nixon = 17 ∨ Richard Nixon ≠ 17' is true, and therefore one or the other disjunct must be true. This is more harmless than it seems, since we are to think of '17' as containing implicit parameters for an unspecified simply infinite system. For each value of these parameters, one disjunct is true, but we can only assert what is true generally of simply infinite systems. The word 'true' is actually ambiguous in this context, since it might either presuppose particular values for the parameters, or mean 'generally true', as it does in number-theoretic assertions and in the remark in the text.

that are not internal to the structure of the numbers themselves or even to some larger universe of mathematical objects. For example, according to a widely accepted analysis of counting, in counting a group of objects a one-to-one correspondence is established between the numbers from 1 to a certain number n and the objects of the group. Such relations in which the elements of a structure may stand will be called *external* relations. How can Dedekind's view interpret such relations?

Clearly, those relations between elements of a simply infinite system and other objects will only count as relations of *numbers* if they are in a certain sense invariant under choices of realization of the structure. Let $\langle N, 0, S\rangle$ and $\langle M, 0, R\rangle$ be simply infinite systems. By Dedekind's theorem, they are isomorphic; let h be the isomorphic mapping of N on to M. If, now, for $n \in N$, f is a one–one correspondence of some objects, say the F's, and $\{m: m \in N \wedge m \leq_S n\}$, then of course, if we set $g(x) = h[f(x)]$, g is such a correspondence between the F's and $\{m; m \in M \wedge m \leq_R h(n)\}$.[16] Thus if, using N as 'realization' of the numbers, one concludes on the basis of f that there are n F's, using M, one would conclude on the basis of g that there are $h(n)$ F's, which is the right result.

This treatment of external relations is generally applicable; suppose $S = \langle S, \ldots\rangle$ and $T = \langle T, \ldots\rangle$ are structures, and h is an isomorphism of S on to T. Then if R is a relation of elements of S to elements of another set U, and we set $xR'y$ iff $\exists z \in S[x = h(z) \wedge zRy]$, then for any x in S, y in U, xRy obtains if and only if $h(x)R'y$, and R' can do the same work as R. This statement is of course vague and would have to be filled out in particular cases, as we have done for the example of counting. It should also be clear that it does not depend specifically on the set-theoretic conception of structure, and this treatment of external relations will be applicable to other structuralist views of such objects as numbers.

I want to consider another difficulty Dedekind's analysis faces. Dedekind thought it necessary to prove that simply infinite systems *exist* (pars. 66, 72). This seems intuitively necessary; otherwise it seems that one is not entitled to say (informally) that the natural numbers exist. More troubling, on the eliminative reading, if there are no simply infinite systems, then for any N, 0, S the statement (2) giving the 'canonical form' of an arithmetic statement A is vacuously true. But then both A and $\neg A$ have true canonical forms, which amounts to the inconsistency of arithmetic. A version of this difficulty is faced by eliminative structuralist accounts of natural number generally.

Dedekind sought to meet it by his famous, or notorious, argument for the existence of infinite systems (sets) in paragraph 66. This appeals to a kind of transcendental psychology. He argues that 'the totality of things, which can

[16] $\leq_S$ is the order relation induced by S as successor; similarly for $\leq_R$.

be objects of my thought' is infinite; for given such an object s, we can let $S(s)$ be the thought that s can be an object of my thought, and this will be a new object of my thought. S is then a one–one mapping of the potential objects of my thought into themselves, and thus this totality is infinite.

With our more critical attitude toward set theory, we would question the treatment of the totality of objects of my thought as a set,[17] and the intrusion of non-mathematical concepts into Dedekind's argument would also give rise to objections. To translate Dedekind's development of arithmetic into a modern set theory like ZF, we would need the axiom of infinity; that is, the existence of an infinite set would simply be assumed. The question how much better it is possible to do than Dedekind will then arise when we consider intuitive justifications of set-theoretic axioms.

Dedekind's analysis deals neatly with the problem of multiple reductions, since no one simply infinite system is identified with the natural numbers. It appears to do so at the price of set-theoretic economy: the conventional development of arithmetic in set theory, for example, by taking as the natural numbers the finite von Neumann ordinals, does not require the axiom of infinity.[18] The difference is not quite so great as it appears: Dedekind did not make the distinction between sets and classes, and in fact this argument does not require that there be a simply infinite system whose domain is a set. Thus what is needed is just quantification over classes; however, in the absence of the axiom of infinity, common applications of induction and recursion require impredicatively defined classes.

The comparison of Dedekind's with the now conventional approach makes clearer the significance of set-theoretic constructions of number systems. These constructions should be viewed, not as *identifying* a certain system of objects as the numbers in question, but as proving the 'objective reality' of the concept of that kind of structure, in something close to Kant's sense. If a set theory is assumed, this may not have much significance when the structure in question is the natural numbers; it is more significant when we are dealing with the positive and negative integers and the rationals,

---

[17] Cantor, in his famous letter to Dedekind of 28 July 1899 (*Gesammelte Abhandlungen*, ed. Ernst Zermelo (Berlin: Springer, 1932), 443–7), already mentions the *Inbegriff alles Denkbaren* as an example of an inconsistent multiplicity (therefore not a set) (p. 443). But he does not explicitly relate this remark to Dedekind's argument.

Zermelo is quite explicit in criticizing Dedekind on this point, observing that the *Inbegriff alles Denkbaren* cannot be a set according to his own axioms, in 'Untersuchungen über die Grundlagen der Mengelehre 1', *Mathematische Annalen*, 65 (1908): 266 n., or 204 n. in the translation in Jean van Heijenoort (ed.), *From Frege to Gödel: A Source Book in Mathematical Logic, 1879–1931* (Cambridge, Mass.: Harvard University Press, 1967).

[18] On the history of the development of arithmetic in set theory without infinity, see my 'Developing Arithmetic in Set Theory without Infinity: Some Historical Remarks', *History and Philosophy of Logic*, 8 (1987): 201–13.

assuming only the natural numbers, or with the complex numbers, assuming only the reals (or natural numbers and sets thereof). Then if we talk of numbers of these kinds as if they were objects *sui generis*, the construction ensures that we avoid the problem of vacuity that threatened Dedekind's account of natural numbers.

A problem with which we began our exploration of structuralism was that the statement of the view seemed to require reference to structures, and we were then faced with the problem what kind of objects these are. One lesson we might draw is that we should distinguish what is required for a structuralist account of a particular kind of mathematical object, such as the natural numbers, and what is required to give a general statement of the structuralist view. It is the latter which at the outset required reference to structures. That the former requires such reference is not evident. Our Dedekindian account might be found wanting on the grounds that it builds general reference to structures into its canonical form for arithmetical statements. In our further discussion of eliminative structuralism, we will consider whether it is possible to avoid this result.

3. ELIMINATIVE STRUCTURALISM AND LOGICISM

We have derived from our discussion of Dedekind an eliminative structuralist analysis of number which, however, still assumes sets and functions. More recent developments of the eliminative structuralist idea have sought to avoid such assumptions. But as we have seen, they provide a ready way of talking about structures. Moreover, if one is going to accept set theory anyway, it falls naturally into the role of ultimate court of appeal for questions of the existence of structures, or in our Kantian terms, of the objective reality of concepts of structures.

Because of the assumption of sets, however, we have so far not even stated a method of eliminating mathematical objects generally, whether or not we adopt a structuralist view of sets (and related objects such as classes and functions). Taken simply as an account of arithmetic, however, our Dedekindian analysis can be readily reformulated so that the role of set theory is taken over by *second-order logic*. We noted that the domains of simply infinite systems, although they need to be within the range of the variables, do not need to be sets. The same is true of other higher-order entities quantified over in Dedekind's account. His definition of a simply infinite system can be translated directly into second-order logic; this is simply a consequence of the fact that the structure of the natural numbers is second-order-definable. The induction clause (1) is replaced by

(3) $\quad \forall M\{[M0 \wedge \forall x(Mx \to M(Sx))] \to \forall x(Nx \to Mx)\}.$

We write the new definition as $\Omega'(N, 0, S)$. Now suppose that A is provable in second-order arithmetic. In second-order arithmetic, functions definable by the usual kinds of recursion (in particular, primitive recursion) are explicitly definable. We will suppose that in A expressions for such functions have been eliminated, so that A contains only 0, S, and logical expressions. If $A(N, 0, S)$ is the result of relativizing the quantifiers of A to N, then clearly the obvious reformulation of (2), which I will call (2'), is provable in pure second-order logic with identity (including comprehension). Moreover, if A is merely a truth in the language of second-order arithmetic, $A(N, 0, S)$ will hold in all standard models of $\Omega'(N, 0, S)$.

This simple translation of the language of arithmetic into that of second-order logic has been offered as a basis for a defence of the view that arithmetic is a part of logic. This form of logicism is a variety known as 'if-thenism' or 'deductivism'. We will consider shortly a well-known defence of this view offered some years ago by Hilary Putnam,[19] which, however, used first-order formulations. Harold Hodes has recently vigorously defended the second-order form presented above, following Frege in interpreting second-order variables as ranging over functions and concepts.[20] Such a use of Frege's theory of functions gives this sort of if-thenism a claim to eliminate mathematical objects. A virtue of the second-order language, however, is the variety of interpretations to which it is susceptible; hence there may be other ways of using the above translation, or modifications of it that are still second-order, in an eliminative programme. But as Putnam's example illustrates, first-order versions have been pursued. Since they have tended to proceed by trying to approximate the second-order version, we will concentrate on the latter.

This version of logicism does not escape all traditional objections, for example, to handling induction by means of a definition.[21] None the less, it is worth some further exploration, if for no other reason because of the general applicability of the strategy. For example, two- or three-dimensional Euclidean geometry also has a categorical second-order axiomatization; we can then interpret Euclidean geometry along the same lines. We can also, it seems, meet the demand for a treatment of set theory, by formulating the axioms of separation and replacement as single statements in a second-order language, so that ZF becomes finitely axiomatizable. This case illustrates that the

---

[19] H. Putnam, 'The Thesis that Mathematics is Logic' (1967), in *Mathematics, Matter, and Method: Philosophical Papers*, vol. 1 (Cambridge: Cambridge University Press, 1975), 12–42.

[20] H. Hodes, 'Logicism and the Ontological Commitments of Arithmetic', *Journal of Philosophy*, 81 (1984): 123–49.

[21] I discuss this in 'Frege's Theory of Number', 167–70, 173–5.

second-order variables are doing the work of the set-theoretic notion of class rather than that of set.

This logicist eliminative programme faces the same problem of the possible emptiness of the generalizations (2') as Dedekind's original formulation: if, no matter how N, 0, and S are interpreted, $\Omega'(N, 0, S)$ is false, then (2') is vacuously true, and is equally true if we replace A by $\neg A$. We can, as Putnam does, view the matter in terms of provability: if we can prove both (2') and its counterpart for $\neg A$, then $\Omega'(N, 0, S)$ will have proved inconsistent, and vice versa. The problem also arises on versions of the strategy that replace $\Omega'(N, 0, S)$ by a first-order statement.

What reason have we, on if-thenist grounds, to believe that things will not collapse in this way? We can in some cases appeal to another theory, within which it is possible to construct a model of the theory at hand or otherwise prove its consistency. Our proposal bears some resemblance to Frege's re-interpretation of Hilbert's idea that the axioms of geometry 'define' the concepts of geometry.[22] Hilbert had made copious use of arithmetical and analytical models to establish consistency and independence.[23] But for the most elementary mathematical structures, something more than reference to another *mathematical* theory is needed to convince ourselves of their 'objective reality' as types of structures, or that their theories are consistent. This is particularly true of the natural numbers. Recall that this is just the point at which Dedekind found it necessary to appeal to transcendental psychology. We may not be able to avoid going outside the strictly mathematical, at least as understood by eliminative structuralism, in order to convince ourselves of something so basic as the objective reality of the natural numbers.

At the time he defended if-thenism, Putnam was quite aware of this difficulty.[24] He replied that we do not need to assume the actual existence of structures satisfying the conditions in which we are interested, but only that they could exist (pp. 32–3). This would presumably preclude the logical validity, in any reasonable sense, of the canonical versions, in the manner of (2'), of a statement and its negation. The formulation suggests the use of modality in explicating mathematical existence that Putnam made shortly afterwards,[25]

[22] See Frege's correspondence with Hilbert and Korselt, in Gottfried Gabriel *et al.* (eds.), *Wissenschaftlicher Briefwechsel* (Hamburg: Meiner, 1976), and his two series of articles 'Über die Grundlagen der Geometrie', in Ignacio Angelelli (ed.), *Kleine Schriften* (Hildesheim: Olms, 1967). This material is collected in translation in Eike-Henner W. Kluge (ed.), *Frege on the Foundations of Geometry and Formal Theories in Arithmetic* (New Haven: Yale University Press, 1971). It is instructively discussed by Resnik in *Frege and the Philosophy of Mathematics* (Ithaca, NY: Cornell University Press, 1980). ch. 3.

[23] D. Hilbert, *Grundlagen der Geometrie* (Leipzig: Teubner, 1899: many later edns. with extensive appendices).

[24] Putnam, 'The Thesis that Mathematics is Logic', 26–33. For reasons which do not concern us, he saw the problem as arising in connection with the application of mathematics.

[25] Ch. VIII.

in which he has been followed by many others. What would be presupposed (in the case of arithmetic) is that it is possible that there are N, 0, S such that $\Omega'(N, 0, S)$ holds. This statement, however, is not interpreted in if-thenist terms, although it does not necessarily go outside the framework of eliminative structuralism more broadly conceived.

But in the earlier paper, Putnam seems to take the standpoint of provability and to interpret 'possible existence' as syntactic consistency. One might question the adequacy of this in view of the deductive incompleteness of whatever axiomatization is involved.[26] To this it can be replied that we do not have better assurance of the 'existence' of such a structure as the natural numbers than such consistency statements would give us. But for the if-thenist view, a problem arises about the interpretation of the consistency statement. A theory T (in this context, we can take Ω' or another antecedent as a theory) is consistent if there does not exist a logical proof of a contradiction from its axioms. Ordinarily we take this as a mathematical statement; a proof is an array of symbols satisfying certain conditions; these symbols are types and therefore abstract objects. This reading is at first sight outside the if-thenist framework; but note that symbols and proofs are not obviously pure mathematical objects and are plausibly quasi-concrete. It would, of course, be possible to regard the syntactical objects as themselves merely a structure, but then the appeal to consistency would be simply another form of reduction to another mathematical theory, or the consistency statement is actually being used in an applied context. What counts is what we can expect when we actually construct proofs. One way of taking the consistency statement,

---

[26] On the second-order version of if-thenism that we have emphasized, this incompleteness resides in the logic; on Putnam's own first-order version, it resides in the axioms whose schematized versions form the antecedent of the conditional that is the canonical form of a mathematical statement.

Positions closely related to what we call if-thenism are instructively discussed in Rose-Marie Rheinwald, *Der Formalismus und seine Grenzen* (Konigstein/Ts.: Hain, 1984), ch. 11, under the name *Implikationismus*. The distinction she makes between the syntactic and semantic version then recalls ours between the 'standpoint of provability' and that of semantic validity. As her name for the position suggests, however, there is an important difference between her interpretation and ours. She interprets a mathematical statement, made in the context of an axiomatic theory, as expressing logical implication of the 'naïve' statement by the axioms; it is for that reason that she says that if the mathematical statement is taken as an if–then statement, 'if . . . then . . .' cannot be the material conditional (p. 49). Then, of course, the question arises whether by 'implication' is meant logical derivability, semantic consequence, or something else; hence her distinction of syntactic and semantic implicationism.

In my view, Rheinwald's 'implicationism' introduces an element of reflection into the interpretation of mathematical statements which is not essential to the idea of if-thenism or of eliminative structuralism. In the canonical forms we consider, such as (1), 'if . . . then . . .' is the material conditional; the thesis is then (e.g.), that the truth of the arithmetical statement A consists in the logical validity of (1). Then, of course, the syntactic/semantic distinction arises for the interpretation of logical validity. But the reflection belongs not to the content of individual mathematical statements but to the account of that content.

which bypasses both the structuralist understanding of syntax and the interpretation of it as about quasi-concrete but abstract symbol-types, is to take it as a statement of nominalistic syntax, where the objects are physical inscriptions. This offers a way of reconciling usual ways of understanding mathematical theories with nominalism: we interpret the theories in the if-thenist way, but deal with the problem of objective reality by appealing to consistency, nominalistically interpreted. This suggestion offers us the opportunity to discuss some issues concerning nominalism.

4. NOMINALISM

I understand the term 'nominalism' in the sense most usual in contemporary philosophy of logic and mathematics, as the rejection (perhaps programmatic) of all abstract entities. It is this demand that would lead to the use of an interpretation of syntactical statements as referring to physical inscriptions. I shall take nominalism, moreover, to involve extensionalism, so that we are not allowed to use a modal language in talking about such physical objects as inscriptions.

It appears that in the eliminative structuralist programme, the problem of the objective reality of the natural numbers would have a simpler solution than Putnam's appeal to consistency: one would just describe (by a predicate) a domain of physical objects, a zero object, and a successor relation that would satisfy Ω'. The refusal of modality means, however, that the quantifiers range over *actual* objects. If we can give such a physical model of the natural numbers, then the physical universe is in some way infinite. Although recent nominalism has tended to be physicalist, the issue is the same if one seeks the model in some domain of mental objects such as sensations, ideas, or thoughts.[27] Now, that the physical world is in some respect infinite is no doubt true. I am not sure whether according to current physical theory there are infinitely many physical particles or other objects that can claim to be 'physical objects'; however, physical theories are based on a geometry in which space (or space-time) is infinitely divisible (whether or not it is infinite in extent). But should it be taken as a presupposition of elementary mathematics that the real world instantiates a mathematical conception of the infinite? This would have the consequence that mathematics is hostage to the possible future development of physics. Can we rule out the possibility that physics will abandon infinitely divisible space-time and replace it by some

[27] Note that Dedekind's model in par. 66 is not exactly of this kind, since he speaks of objects of his thought; moreover, the thoughts themselves seem to be not mental occurrences but rather thoughts in something like Frege's sense.

'quantized' conception? If not, and if this were to happen, then even accepting Hartry Field's claim that points and regions of space-time are physical and acceptable to the nominalist would not save the infinite in the physical world.[28] Would we be obliged to abandon the mathematics of the infinite, even the infinity of the natural numbers?[29] A great deal of the historically given mathematics would have to be jettisoned in this case.

The problem with the appeal to a physical model to deal with the problem of objective reality is that it makes mathematics presuppose a hypothesis that is stronger and more specific than needed, as I have argued elsewhere.[30] Any particular such hypothesis would be vulnerable to a refutation which would not upset the mathematics. The nominalist might call attention to the fact that the world as we now understand it contains many different models of infinity, and then take the holistic position that no one of them is required for the infinity of the natural numbers.[31] In view of the dependence of all infinities in the real world on infinities in space and time, it is not clear that this position is really an advance over the Fieldian one just proposed; at any rate, it seems conceivable that all the hypotheses about infinities in the real world might be abandoned.

Let us now return to Putnam's suggestion, mentioned at the end of Section 3, that the possible truth of the antecedent of the if-thenist's conditionals need amount to no more than formal consistency, interpreted as a proposition of nominalistic syntax. Such a syntax will have to be carefully formulated in order to avoid assuming that actual inscriptions are closed under the usual operations of logical syntax. But no matter how this is done, the statement of consistency says no more than that there is not an *actual* inscription that is a proof of a contradiction from whatever axioms are in question. This is of course not quite so weak as it seems, since future inscriptions are allowed: it says that no proof of a contradiction will ever be written down. None the less, it cannot be stronger than the statement that no proof of a contradiction is *physically possible*, and such possibilities are constrained, for example by the physical structure of space-time, in such a way as to make the consistency statement weaker than on its usual mathematical understanding. It is also hard to see what grounds other than inductive the nominalist can have for

---

[28] H. Field, *Science without Numbers* (Princeton: Princeton University Press, 1980), 31. Although Field is not an eliminative structuralist, in this work and subsequent papers (which importantly modify his position), he makes a number of moves that an eliminative structuralist might make, and some of the issues raised here arise for him. An adequate discussion of his views, however, would take us too far afield, and there is already a considerable critical literature on his work by others. We will discuss below only one aspect of his position, his conception of second-order logic.

[29] Observe that the simple (first-order) statement that S maps N one–one into N-$\{0\}$ already has only infinite models.   [30] Parsons, *Mathematics in Philosophy*, 184–6.

[31] This possibility was suggested to me by Isaac Levi.

believing consistency statements for theories having only infinite models to be true. The most elementary mathematics used in proof theory, such as primitive recursive arithmetic, will not be known to be sound on any nominalistic interpretation,[32] and the intuitive idea that the axioms of such theories describe coherent possibilities is tainted with modality. That the consistency-based if-thenist form of nominalism is on stronger ground than other forms of strict nominalism is not evident. The other strategy suggested by Putnam's response to the problem of objective reality, taking possible existence seriously, seems more promising.

Before we take leave of nominalism, however, I want to consider the possibility, more congenial to nominalism, that the logic of if-thenism should be first-order. In such cases as arithmetic and Euclidean geometry, it appears that the if-thenist has to jettison the intuition that these branches of mathematics describe a unique structure. In the if-thenist translation of an arithmetical statement, for example, $\Omega'(N, 0, S)$ will have to be replaced by a list of axioms that is no longer fixed once and for all but which must depend on the context of the statement. Consider, for example, the manner of doing this that disrupts the second-order formulation the least: return to the set-theoretic formulation $\Omega(N, 0, S)$ (leaving S as a function symbol); add to it comprehension axioms corresponding to those of second-order logic, which must now count as non-logical axioms. They will correspond closely to those for classes in NB; it is thus best to think of the objects involved as classes rather than sets.[33] In this language, we can describe a finitely axiomatizable conservative extension of the usual first-order arithmetic Z, related to it roughly as NB is to ZF; call the conjunction of its axioms $\Omega_1(N, 0, S)$.[34]

Of course, if we use Ω_1 instead of Ω in our canonical form (2), the truth of the statement A amounts to its provability in this conservative extension of Z; because of the completeness theorem, it makes no difference whether we

---

[32] If the inscriptions of nominalistic syntax do not satisfy the closure conditions of usual syntax, then the following kind of anomaly can arise: suppose that for a given formal theory T we can give instructions for constructing a proof in T of a contradiction, so that we can give a strictly constructive proof (say, in primitive recursive arithmetic) that T is inconsistent. We might, however, be able to give a lower bound for the length of a proof of a contradiction in T, so large that on physical grounds it will be impossible actually to write down such a proof. Then 'T is consistent' interpreted nominalistically will be true. If T contains primitive recursive arithmetic, we will be able to write down a proof in T of the (arithmetized) statement that T is inconsistent.

[33] It is simplest to have two sorts of variables, for individuals and classes; but of course a one-sorted formulation is readily available; then one needs the predicate 'is a class'.

[34] A little delicacy is needed to obtain finite axiomatizability. Comparison with the axiomatization of NB shows that one must have pairing and projection functions. These can either be primitive or defined in terms of simple arithmetic functions, but in the latter case a few of these must be primitive, with recursion equations as axioms.

Some issues about the interpretation of theories of this kind and their relation to more conventional axiomatizations by schemata are discussed in essays 1, 3, and 8 of my *Mathematics in Philosophy*.

understand logical validity as semantic consequence or in terms of provability. Thus arithmetic truth becomes relative to the axioms assumed. Putnam was willing to accept this consequence.[35] The formulation we have chosen has the advantage that it makes the relativism arise in a natural place: extensions of the axioms will be obtained by stronger comprehension axioms about classes or whatever other second-order entities one appeals to.

The relativistic position is quite counter-intuitive, not only because it contradicts the idea that the natural numbers are a unique structure (as are other number systems, and Euclidean space of a fixed number of dimensions), but because it makes the *meaning* of statements in these parts of mathematics depend on what axioms one is assuming, not only about the structure with which one is immediately concerned, but about classes and even about other objects, where reference to them may give rise to relevant comprehension axioms. But it would be difficult to refute categorically without more analysis of the intuition of the uniqueness of such structures as the natural numbers, and it would take us too far from our main theme to undertake that here. We will consider later whether the nominalist can after all use second-order logic.

What we have considered so far are versions of eliminative structuralism based on an extensional logic. Two main difficulties have arisen for it: the interpretation of second-order logic and the problem of avoiding vacuity in the conditionals used to interpret mathematical statements. No solution of the second problem has been found so far that preserves the purity of structuralism, but Putnam's observation that the consistency of the antecedents is sufficient, if interpreted conventionally rather than nominalistically, compromises structuralism only by admitting quasi-concrete mathematical objects, the expressions of formal syntax. (We have, however, not considered the problem of knowing such consistency presuppositions to be true.) The first problem is of course avoided by the use of first-order logic just considered, at the price of relativism. What it does, in effect, is simply to incorporate the second-order entities into the structure that is being considered only hypothetically.

At this point, to deal with the second problem, *modality* has been appealed to by a number of writers, in particular Putnam himself (see n. 25). It offers the most direct solution to the problem of non-vacuity, by taking literally the statement that it is sufficient if the existence of an instance of the structure is possible. Moreover, modal nominalism is a more promising account of syntax than strict, extensionalist nominalism. Though such an interpretation still leaves a dilution of the structuralist position, it suggests a general strategy for eliminating mathematical objects: pure mathematical objects are dealt

[35] Putnam, 'The Thesis that Mathematics is Logic', esp. 22–5.

A STRUCTURALIST VIEW

with in an eliminative structuralist way, while quasi-concrete objects are interpreted in some modal nominalist manner. We now turn to this strategy.

5. MODALISM

By 'modalism' I mean the programme of eliminating mathematical objects in favour of modalities, or the thesis that mathematical objects can be eliminated in this manner.[36] The simplest version of modalism would simply expand the resources of if-thenism by claiming the possible existence of the structure in the framework of which mathematical statements are interpreted; for example, statements about natural numbers could still have the canonical form (2'). But a presupposition for number-theoretic statements in general would be the possibility of a simply infinite system, that is,

(4) $\lozenge \exists M \exists x \exists R \Omega'(M, x, R).$

Once we have introduced modality, however, it would be more in keeping with the idea to hold that the mathematical truth of A consists not in the logical truth of (2'), but rather in its necessary truth. This can readily be incorporated into the canonical form by replacing (2') by its necessitation. There is then a simplification, because then the mathematical truth of A will consist simply in the truth of its canonical form, which in the case of a number-theoretic statement will be:

(5) $\square \forall M \forall x \forall R [\Omega'(M, x, R) \to A(M, x, R)].$

It is also possible to take the necessity operator as obviating the quantifiers; that is, we can think of the canonical form as simply

(6) $\square [\Omega'(N, 0, S) \to A(N, 0, S)].$[37]

At this point the question arises how the modal operators are to be understood. Four candidate interpretations naturally arise (in what one would take to be order of strictness of necessity):

1. strictly logical, in a sense connected with formal logic;
2. logical in the sense more usual in discussions of modality, which takes

---

[36] The term has been used in roughly this sense by other writers, e.g., Rheinwald, *Der Formalismus*, ch. 3. The inventor of 'modalism' was perhaps Putnam, who, however, seems to repudiate it at the outset (ch. VIII). But what he is denying is not that mathematical objects can be eliminated in favour of modalities, but that this account of them is in all respects clearer or more fundamental than taking mathematical language at face value as referring to mathematical objects (the 'mathematical-objects picture').

[37] (6) is obviously equivalent to the result of replacing '$\to$' by the strict conditional in the result of dropping the quantifiers in (2'). Whether the difference of (5) and (6) is significant depends on the interpretation of the modal operators and on refinements of modal logic.

account of the constraints of non-logical concepts, in my opinion best called 'metaphysical' (after Kripke);
3. mathematical;[38]
4. physical, or, more generally, 'physical or natural'.[39]

Although the issue has relevance outside the context of the present discussion, I shall have to reserve for another occasion the discussion of which of these interpretations is clearest and most appropriate. To state my view briefly: strictly logical modality seems to me an awkward conception and probably not clear independently of mathematical modality.[40] Physical or natural possibility is too restrictive, as we indicated above in our remarks on nominalism: it demands too much to ask that the structures considered in mathematics be physically possible; indeed, in the case of higher set theory, there is every reason to believe that they are *not* physically possible.[41] It may seem that in an eliminative structuralist account of mathematical objects we are dealing with statements that contain only logical expressions essentially and therefore that metaphysical and mathematical modalities should not differ. This is indeed true on certain understandings of mathematical modalities. But if one takes the language of mathematics at face value, then the usual intuition that mathematics is necessary has the consequence that any pure mathematical objects that there are exist necessarily. This does not comport with eliminative structuralism or even with the more general idea of thinking of mathematical existence as potential. For this reason, I interpret the necessity in statements (5) and (6) as mathematical necessity, where this is to be distinguished from metaphysical necessity.[42] But many of the remarks below will be independent of this choice.

Modalism seems to give a satisfactory solution to the problem of non-vacuity. Statement (4) demands no more than that a simply infinite system

[38] Ch. VIII; also Parsons, *Mathematics in Philosophy*, 21–2, 182–4, 327–9.
[39] The phrase is Quine's: 'Necessary Truth', in *The Ways of Paradox and Other Essays*, 2nd edn. (Cambridge, Mass.: Harvard University Press, 1976), 76.
[40] Parsons, *Mathematics in Philosophy*, 179–83. These remarks, however, took formal logic to be first-order logic and would generalize readily only to a logic sharing its basic properties, in particular recursive axiomatizability. With regard to standard second-order logic, in particular, the issues are not quite the same. Indeed, well-known problems about the line between logic and mathematics might imply that the distinction between 'strictly logical' and mathematical modality is not a sharp one.   [41] Ibid. 191–3.
[42] Cf. Ibid. 327–8. It should be clear to the reader of that essay that I allow mathematical modalities to be *de re*. I mention this here because Timothy McCarthy points out some technical difficulties for Putnam's modalist interpretation of set theory unless his modality is allowed to be *de re* ('Platonism and Possibility', *Journal of Philosophy*, 83 (1986): 288–9). McCarthy, however, seems to suppose that if that is conceded, then the modality is metaphysical in the above Kripke-derived sense. I do not agree. Moreover, in spite of remarks that give his interpretation a basis in the text. I think McCarthy is probably wrong in attributing to Putnam an intended interpretation of mathematical modalities close to what above is called strictly logical. It is this that gives rise to the difficulty he points out.

should be possible in a sense that may require some further explanation, but which is at any rate weaker than physical possibility. An analogous claim would hold in the case of other structures. There may be an epistemological problem about how we know a statement like (4) to be true. But its truth seems to follow from the supposition that theories in physics describe coherent possibilities, and perhaps it can be seen in more direct and intuitive ways.

One might, to be sure, object to it on constructivist grounds; in the context of classical logic, which we have not questioned, statement (4) seems to state the possibility of an 'actual' infinity. Indeed, a more thoroughgoing modalism might replace (4) by a weaker statement, to the effect that any structure of the type of an initial segment of the natural numbers can be extended.[43] I will not pursue or reply to this sort of objection here, since even the weaker statement allows various means of interpreting classical arithmetic, and constructivism is a large subject in itself.

There are two types of objections in the literature that I do wish to consider. The first questions the faithfulness to mathematical discourse of the canonical forms the modalist offers. The second questions whether, given the apparatus that the modalist's translations must use, there really is a reduction of reference to mathematical objects.

Thus our first question is whether it is at all reasonable to suppose that the modalist's canonical forms represent what the mathematician actually means by a mathematical statement of the relevant sort. Turning in particular to arithmetic and thus to statement (5) or (6), one obvious objection can, I think, be discounted. When the number theorist talks of numbers, surely he does not refer to objects of a merely hypothetical domain, although one ought perhaps not to make too much of what is involved in the idea that number theory involves genuine reference to objects called numbers; genuine reference there is. But the emphasis we have placed on the problem of nonvacuity should make clear that the views we have considered do not regard the reference involved as to purely hypothetical objects. The canonical form (5) or (6) arises in a context in which there is the *presupposition* (4); (4) plays a role analogous to that played by existential presuppositions in the use of singular terms.

None the less, there is a related discomfort, which perhaps will be felt more strongly about (5) (as well as its ancestor (2′)) than about (6): do the intuitions behind structuralism really support the interpretation of arithmetic statements as of this explicitly general second-order form? The generality also appears in the presupposition (4), since that whose possibility is presupposed is an existential generalization. The vagueness of Dedekind's talk of abstraction

---

[43] In the logical context we are assuming, this would follow from the sort of statement claimed to be intuitively evident in my 'Mathematical Intuition', Ch. V, pp. 95–113.

perhaps had its point. Even if we cannot single out one simply infinite system that is the numbers *par excellence*, perhaps it is falsifying the sense of discourse about natural numbers to take the step we took in interpreting Dedekind, of taking arithmetical statements to be really about *every* simply infinite system.

It should be observed that an objection of this kind can be made to any of the canonical forms proposed by different versions of eliminative structuralism. What force is granted to it depends on the importance of eliminating mathematical objects relative to other objectives. Given a domain of mathematical objects that, like the natural numbers, forms a second-order definable structure, statements about them will implicitly have the same kind of generality, by the counterpart of Dedekind's isomorphism theorem. Putting this generality into the explicit content of the statements will for some be a small price to pay for avoiding an ontology of mathematical objects. A non-eliminative version of structuralism should, however, be able to avoid it.[44]

6. DIFFICULTIES OF MODALISM: SECOND-ORDER LOGIC

Let us turn now to the second type of objection, whether the modalist's apparatus really does offer an elimination of mathematical objects. A first question would touch any reductive programme making use of modality: does not the modal operator itself involve us in an ontology of possible worlds? And although possible worlds are not paradigmatic mathematical objects, they are in one way or another problematic enough, so that eliminating mathematical objects in favour of possible worlds is a dubious gain.

There is no need to dwell on this objection, since the ontology of possible worlds arises not directly from modal discourse but from a semantical treatment of it that was expressly designed to use conventional (extensional) mathematics. If we simply take modalities at face value, there is no reason to take a statement, say, of possibility as involving the existence of anything. Commitment to possible worlds only becomes a problem in a framework in which the modalities are primitive if the modal logic is powerful enough to simulate quantification over possible worlds. Such a logic may be needed for some modalist treatments of set theory,[45] but in this case there is another more intuitive objection to the eliminative programme (see below).

We now turn to the matter of second-order logic, which the reader may well

---

[44] As does the position of Field, who insists on taking the language of mathematics at face value, and then claims that statements involving reference to pure mathematical objects are not true. This view is reiterated in Ch. XII, where Field enlarges the logic he admits to include modalities and adopts a position having some kinship with what I am calling modalism.

[45] Parsons, *Mathematics in Philosophy*, 324–5.

feel we have swept under the rug so far. In our discussion of arithmetic, the only alternative to using second-order logic in eliminative structuralist reductions was a relativism that we found unacceptable. Putnam still takes that position in his modalist treatment of arithmetic in 'Mathematics without Foundations' (pp. 168–84 above). When he turns to set theory, however, he claims to do better. He seeks to capture the notion of a *standard* model of set theory by means of first-order modal logic; he asserts that this notion 'can be expressed using no "non-nominalistic" notions except the "□" ' (p. 182). The models are to be certain graphs; Putnam writes: 'The model will be called standard if (1) there are no infinite-descending "arrow" paths; and (2) it is not possible to extend the model by adding more "sets" without adding to the number of "ranks" in the model (p. 182). This definition is not on its face first-order. But it appears that Putnam intends the graphs in question to be thought of as individuals, so that the modalized quantification over graphs in (2) is then first-order. Moreover, it seems that both the 'paths' of (1) and the 'ranks' of (2) are to be internal to the model. How the definition is supposed to work, without assumptions about graphs that would defeat the purpose, is still not clear to me; it is not clear to me how any definition of the sort he proposes can work given the non-finite axiomatizability of Zermelo set theory. Neither Putnam nor anyone else has published a working out of the idea.[46]

There remains the question whether second-order logic can be interpreted in a way that coheres with the idea of eliminating mathematical objects. It will be observed that in many of these cases, number theory included, impredicative second-order logic is needed for the deductive development of the theory translated in one of the ways proposed. Thus whatever entities comprise the range of the second-order variables will have to satisfy impredicative comprehension. An obvious candidate interpretation, advocated recently by Hodes (see n. 20), is by Fregean concepts. Frege effectively embodied a comprehension principle for concepts and functions in his logic.[47] Although the question of predicativity had not been raised at the time (nor did Frege take note of it when it became an issue in the Poincaré–Russell debate), Frege's procedure is no doubt an expression of realism about the reference of predicates, a view that would have been shared both before and since. What would be demanded of a use of second-order logic in an eliminative structuralist programme is a

---

[46] In understanding Putnam's definition, I have been much helped by McCarthy's exposition (see n. 42), in spite of the disagreements noted above. But I do not see how his own version would work without ordinals external to the model to serve as ranks.

[47] It is worth noting how comprehension arises in Frege's logic. It is embodied in his axiom IIb of universal instantiation for function quantifiers and in his rules of substitution for Latin letters (free variables). See *Grundgesetze der Arithmetik*, vol. 1 (Jena: Pohle, 1893), sects. 25, 48 (rule 9). Behind these is of course Frege's conception of a function name; he took function names to be closed under quantification over functions.

justification of such realism *independently of mathematical objects* and compatible with the eliminative structuralist's reasons for wanting to eliminate them. Frege, whose conception of object was certainly meant to accommodate mathematical objects, did not take on any such burden himself. Moreover, the most visible internal difficulty of Frege's theory of concepts is that he is not in practice able to do without nominalizing predicates in talking about concepts. But this nominalization has to be rejected by the programme of eliminating mathematical objects.[48] The history of the debate about Frege's theory does not encourage the view that concepts according to his conception are less problematic than mathematical objects.

Two further proposed interpretations of second-order logic in the literature may be relevant to our concerns. Hartry Field interprets Hilbert's geometry, a second-order theory, by taking the second-order variables to range over regions, which in turn can be taken to be sums of points in the calculus of individuals.[49] This can be generalized to theories in monadic second-order logic individual variables range over points. Field, as we have noted, claims that points and regions are physical and therefore nominalistically acceptable. In fact, he appears to go further and accept as a logic 'what might be called *the complete logic of the part/whole relation* or *the complete logic of Goodmanian sums*'.[50] One can characterize this logic *platonistically* as postulating a domain of individuals such that for any non-empty *set* of individuals in the domain, there is an individual that is the sum of all of them.[51] No comparable nominalistic characterization of this logic is available, but it is at least clear that a form of full comprehension holds: given a predicate of individuals that is true of at least one individual, there is a sum of just the individuals of which the predicate is true,[52] and moreover, the admissible predicates will be closed under quantification over all individuals, including these very sums. In the point-region case, one has for any predicate of points a region containing just the points of which the predicate is true (provided there are any), and these predicates will be closed under quantification over regions. The complete logic is, as Field is well aware, analogous to standard second-order logic, for example in not being recursively axiomatizable.

Field introduces this logic for somewhat different purposes from those of

---

[48] One can, of course, interpret this nominalization or the second-order language itself by means of semantic ascent. But then the theory one obtains is predicative relative to the underlying domain of objects, and does not serve the purpose at hand. (Cf. Parsons, *Mathematics in Philosophy*, essays 3 and 8.) The interpretation by semantic ascent is still highly relevant to structuralism, but only when we have given up the eliminative version, as we will see in Sect. 9.

[49] Field, *Science without Numbers*, 37–8. [50] Ibid. 38.

[51] This amounts to saying that the individuals form a complete Boolean algebra, minus the zero element.

[52] This is the principle (C_s) of H. Field, 'On Conservativeness and Incompleteness', *Journal of Philosophy*, 82 (1985): 250.

the eliminative structuralist programme we have been discussing, but still in the service of a programme for eliminating mathematical objects. Its adaptation to our setting poses some problems, because it offers an interpretation of the usual second-order language only when the objects the individual variables range over are atoms in the sense of the calculus of individuals, that is, without proper parts. That would restrict the interpretation of the kind of translation of mathematical statements that we have been considering. But it is still worthwhile to consider Field's approach to second-order logic in its own terms.

We can understand Field's conception better without the help of set theory by adapting an understanding of second-order language that will concern us later. This is that the comprehension principle will hold for *any* predicate true or false of individuals. The point is that what is a predicate well-defined in this sense is entirely open-ended; in particular, it is not limited by the resources of any particular formalized language. It is most reasonably regarded as indeterminate at a given time, since it depends on what we will come to be able to express and understand. Since predicates can contain parameters (both first- and second-order) this allows the possibility that even a single instance of comprehension will, by generalization, yield more second-order entities than there are predicates in a particular (say, formalized) language.

So far, this understanding corresponds to the understanding of second-order principles in mathematical practice, and also in the design of new formal systems, where one expects principles such as mathematical induction or separation and replacement to be valid for a new formalized language. Impredicativity enters only when we regard our stock of predicates as closed under quantification over second-order entities, in Field's case regions.

One version of this conception would regard the second-order entities as constituted in some way by the predicates themselves. Then the closure assumption just cited is very dubious; if it is indeed undetermined what predicates will or can be added to our language, then it is doubtful that a predicate quantifying over all such predicates will be definitely true or false of each individual. Appearances to the contrary arise from an implicit reducibility assumption; for example, that to each such predicate corresponds an extension a; then in an extensional context we can replace the predicate 'F' by the predicate '( ) $\in a$', and thus replace the quantification over second-order entities by quantification over sets.

Although this conception will interest us shortly, it is of course not Field's; for him regions are to be physical entities. Then the assumption that predicates are closed under quantification over regions is natural enough given a realistic attitude toward the physical world. If one were to question it, it would not be on the ground that it does not fit with the idea of regions as physical. What is more questionable is the transfer to this situation of the undoubted intuitive plausibility of comprehension principles. Field, as a nominalist, should reject the idea of an object associated with a predicate as something like its reference, as well as the 'quasi-combinatorial' picture of 'selecting' the elements of a given set that satisfy a given predicate that is often appealed to in motivating the axiom of separation.[53] It is not clear what direct argument for his principle Field can offer that would not trade on intuitions that support non-nominalist principles, in particular the existence of sets of points.[54]

The best position for Field would probably be to say that the comprehension principle is a hypothesis justified by its consequences, in systematizing the geometrical basis of physics that constitutes, according to Field, the central part of true (as opposed to fictive) mathematics. The most direct use made of it, however, is logical, in Field's arguments for the conservativeness of platonistic mathematics over nominalistic theories.[55] Field's view on this reading puts him in a position in which we have found other formulations of nominalism: making the justification of mathematics turn on a hypothesis about the physical world, which is more vulnerable to refutation than the mathematics. In any case, the interpretation of second-order logic that Field suggests would be of use to modalist treatments only of mathematics involving lower cardinalities.[56]

A bolder proposal concerning second-order logic is made by George Boolos, although he did not advance it in the service of eliminative structuralism or any other programme for eliminating mathematical objects.[57] Boolos

---

[53] e.g., H. Wang, *From Mathematics to Philosophy* (London: Routledge and Kegan Paul, 1974), 184.

[54] If regions are construed as arbitrary non-empty sets of points, then of course it can be proved set-theoretically that regions satisfy Field's logic; i.e., that they are a complete Boolean algebra without the zero element.
In 'On Conservativeness and Incompleteness', 251 n. 12, Field responds to the objection that 'the intuitive basis of the complete theory of the *part of* relation is derivative from the idea of a set'. He claims that 'set theory has the theory of the *part of* relation as its main intuitive basis'. It is certainly true that ideas of whole and part enter into the intuitive understanding of set-theoretic concepts and entered into the historical origin of the mathematical concept of set. Equally central to the origin of the set-theoretic approach to mathematics, however, were ideas about functions. But the crucial question for Field is what independent motivation he can give for logically strong principles of mereology such as his comprehension principle.

[55] Field, *Science without Numbers*, ch. 1, and id., 'On Conservativeness and Incompleteness'.

[56] Although Field thinks the structures arising in higher set theory logically possible, he does not argue for this on the basis of any claim that there is, or even that there possibly is, a physical model of them based on points and regions. See e.g. the remarks about the possible truth of the conjunction of the axioms of von Neumann–Bernays–Gödel set theory in Ch. XII. Here he relies on the fact that NB is essentially a first-order theory.

[57] G. Boolos, 'Nominalist Platonism', *Philosophical Review*, 94 (1985): 327–44. See also id., 'To Be is To Be the Value of a Variable (or To Be Some Values of Some Variables)', *Journal of Philosophy*, 81 (1984): 430–49.

reads second-order formulae by paraphrasing monadic second-order quantifiers by plural quantifiers of natural language. If the first-order quantifiers of a second-order language range over, say, G's, then the second-order quantifier $\exists F$ is to be read 'there are some G's' and an occurrence within its scope of, say, 'Fx', as 'x is one of them' (assuming that what 'them' cross-refers to will be unambiguous, but elaboration can take care of the problem if it is not). Thus consider, for example, in the second-order language of set theory, the comprehension axiom

(7) $\qquad\qquad\qquad \exists F \forall x [Fx \leftrightarrow \neg (x \in x)]$.

This will be paraphrased as

(8) There are some sets such that each set is one of them if and only if it is not an element of itself.

Boolos appears to hold that the paraphrase of a monadic second-order sentence[58] will not involve any ontological commitment to entities other than those that would be values of the individual variables. But at first glance, it appears that in a context of this kind a quantifier like 'there are some sets' is saying that there is a plurality of some kind. Cantor's notion of 'multiplicity' and Russell's of 'class as many' were more explicit versions of this intuitive notion, both attempting to allow that pluralities might fail to constitute sets.

With respect to a number of non-first-order-izable examples, Boolos argues that they do not involve commitment to classes.[59] In part this rests on the absence of explicit mention of classes in the sentences in question (such as (8)). Now the same could be said of pluralities (a possible commitment Boolos does not explicitly address). The difficulty is that speaking of pluralities is already a kind of reduction of plural quantification to singular; to put (8) in terms of pluralities, one would replace 'there are some sets' by 'there is a plurality of sets' and then 'each set is one of them' by something like 'each set

---

The application of Boolos's method of paraphrase in the eliminative structuralist programme faces the difficulty that it applies directly to monadic quantifiers. Thus a pairing function for individuals is needed to extend it to full second-order logic. That the individuals countenanced by the nominalist, or non-mathematical objects in general, come with a pairing function is very doubtful. Could one get by with only monadic second-order logic anyway? Let us consider a straightforward example, the natural numbers. The second-order characterization of the structure (i.e., $\Omega'(N, 0, S)$) is monadic. But in the deductive development of second-order arithmetic, polyadic quantifiers appear at an early and crucial point: the introduction of primitive recursive functions, to begin with addition and multiplication. But the eliminative structuralist can deal with this by assuming as part of the structure of the natural numbers either pairing or addition and multiplication (from which pairing can be defined by first-order means). The treatment of the natural numbers does lose something of its elegance by this change.

[58] By this, I mean a sentence in which the quantifiable second-order variables are monadic; in other respects, quantification (first- or second-order) may be polyadic. The same applies to talk of monadic second-order logic or languages.

[59] Boolos, 'Nominalist Platonism', 328–33.

belongs to it'. If the result is taken as the canonical form, Boolos's intention seems to be violated: on his view, it is plural quantification that is canonical.

Still, there is in sentences making essential use of such plural quantification a form of generalization and cross-reference that one does not find in straightforwardly first-order-izable sentences. Thus in (8) and other examples obtained by paraphrasing second-order formulae one has the pronoun 'them' referring back to 'some sets'; in effect, what follows 'there are some sets' has to say something about *some sets*: we would like to say, a certain (indefinitely indicated) plurality of sets; this is distinguished from saying something about *a set*, which would be indicated by an individual variable. The same is true in the example from Geach and Kaplan, more natural in English:

(9) Some critics admire only one another.

Here the cross-reference is carried by 'one another'; again, that they admire only one another is said about *some critics*, not about an indefinitely indicated critic.

These observations express a discomfort not unlike one felt by Boolos himself, which he undertakes to deal with. He quotes the following statement by Harold Hodes: 'Unless we posit such further entities [as Fregean concepts], second-order variables are without values, and quantificational expressions binding such variables can't be interpreted referentially.'[60] In reply to this, Boolos offers an inductive definition of satisfaction for a monadic second-order language which, he argues, obviates the notion of values of the second-order variables.[61] It is assumed that the language can express a pairing function for individuals (thus enabling a reduction of polyadic second-order logic to monadic) and indeed sequences of individuals, as well as the usual syntactical notions.[62] The inductively defined predicate is 'R and s satisfy F', where R is a second-order variable, s ranges over sequences of individuals, and F over formulae. All the clauses of the definition are the typical ones, except that for second-order quantification, which reads:

(10) If F is $\exists VG$, then R and s satisfy F iff $\exists X \exists T\{(\forall x(Xx \leftrightarrow T\langle V, x\rangle)) \wedge \forall U[(U$ is a second-order variable $\wedge U \neq V) \rightarrow \forall x(T\langle U, x\rangle \leftrightarrow R\langle U, x\rangle)] \wedge T$ and s satisfy $G\}$.

Looking at the matter 'platonistically', R codes an assignment of second-order entities to the variables of the language; to the variable V is assigned $\lambda x R\langle V, x\rangle$. Then what (10) says is that there is a (second-order entity) X and

[60] Hodes, 'Logicism and the Ontological Commitments of Arithmetic', 130.
[61] Boolos, 'Nominalist Platonism', 335–7.
[62] The role of sequences is the usual in satisfaction definitions; thus finite sequences would be sufficient. Boolos is concerned with the case where the individuals are sets and the non-logical apparatus is that of set theory, so that the above assumption holds.

A STRUCTURALIST VIEW 299

an assignment T such that T assigns X to V and agrees with R in what it assigns to other variables, such that T and s satisfy G.[63]

It is hard to agree with Boolos in finding that the treatment of second-order variables in this definition does not offer scope for the notion of a value comparable to what it offers to the notion of a value of an individual variable. The values of the individual variables, Boolos says, are just the terms of the sequences. Why should we not say, similarly, that X is a value of a second-order variable if it is $\lambda x R\langle V, x\rangle$ for some R and V, that is if $\exists R \exists V \forall x (Xx \leftrightarrow R\langle V, x\rangle)$? The difference between the treatment of first- and second-order variables seems to lie just in the facts that in the second-order case functions are coded by predicates and that a function from individuals to n-argument second-order entities can be coded by an $n + 1$-argument second-order entity.[64]

These considerations reinforce, it seems to me, an ontological intuition a little different from but complementing Quine's, according to which ontological commitment is carried by the expressions that play the role of subjects in language and thus indicate what one is talking about. In a primitive second-level predicate, second-order variables or expressions that can be substituted for them play that role. The same is evidently true of plural expressions in Boolos's paraphrases of second-order formulae, as I have indicated above. Boolos has not, in my view, made a convincing case for the claim that his interpretation of second-order logic is ontologically non-committal. The great interest of his reading, in my view, is that it breathes new life into the older conception of pluralities or multiplicities. As a source of second-order logical forms, the plural and plural quantification are rightly distinguished from what was so much emphasized by Frege, predication and, more generally, expressions with argument places. In particular, if it is the idea of

---

[63] This way of putting things assumes the second-order entities are extension-like; but if the language is not extensional, Boolos's definition will have to be modified in any case.

[64] It might also be pointed out that Boolos inductively defines a predicate with a second-order argument (the 'assignment' R), to express satisfaction for a language in which there are no such primitive predicates, that is, in which second-order variables occur only predicatively (in atomic contexts of the form Vv). The straightforward way of eliminating this predicate would be by third-order logic; thus whether one undertakes to eliminate it or not, one is stuck with atomic predication with second-order arguments. It is not clear to me how Boolos would use the English plural to handle such predication. Thus to try to read (10), let us call a *variable-pair* an ordered pair whose first term is a variable. Then (10) might be read: there are some objects and some variable-pairs such that each object is among the former if and only if it is the second term of one of the latter pairs whose first term is V, and every other second-order variable U is such that for each object, the pair whose first term is U and whose second term is it is one of the latter pairs if and only if it is one of the R's, and the latter pairs and s satisfy G. This has the difficulty that nothing marks the final 'the latter pairs' as a second-order argument; 'the latter pairs and s satisfy G' could say that all or most of the latter pairs are x's such that x and s satisfy G.

There are, however, cases where plural noun phrases unambiguously express second-order arguments; 'the latter pairs are infinite in number' would be an example. But what is the principle by which such cases are distinguished?

generalization of predicate places that we appeal to in making sense of second-order logic, then the most natural interpretations will be relative substitutional or by semantic ascent, and these will not license impredicative comprehension, and it is hard to see how that will be justified.[65] But if one views examples such as Boolos's as involving 'pluralities', they are more like sets as understood in set theory in that no definition by a predicate is indicated, so that one need not expect them to be definable at all. Thus one obstacle to the acceptance of impredicative comprehension is removed.

An advocate of Boolos's interpretation in an eliminative structuralist setting could grant my claims about ontological commitment, but then take a position analogous to the Fregean: second-order variables indeed have pluralities as their values, but these are not objects. Like Frege's claim about functions and concepts, this position would base an ontological difference of objects and pluralities on the grammatical difference between the singular and plural terms that refer to each. It does not seem to me to have the same intuitive force as Frege's position, since there is no analogue to the regress argument that can be made if one views the reference of a predicate as an object. There will still be, just as with Frege's concepts, the irresistible temptation to talk of pluralities as if they were objects, as we have already noted above. The only gain this interpretation offers over the Fregean is a more convincing motivation of impredicativity.[66]

We can sum up our discussion of second-order logic as follows: if the eliminative structuralist uses it, he will not be able to avoid ontological commitments more uncomfortable on balance than that to mathematical objects, either to Fregean concepts or to multiplicities that are not 'unities'. If he does not, he is faced with the relativistic consequences that have made if-thenism an unpersuasive view for some time.

7. REJECTION OF ELIMINATIVE STRUCTURALISM

The problems concerning second-order logic may not seem fatal to everyone. We might also point out that one serious problem we raised, that of

---

[65] Cf. Parsons, *Mathematics in Philosophy*, essay 8 and elsewhere.

[66] In reflecting on set-theoretic paradoxes, both Cantor and Russell entertain the idea of pluralities that are not 'unities': Cantor, in the letter cited in n. 17 above, uses the term 'inconsistent multiplicity'; Russell speaks of objects that form a class as many but do not form a class as one, such as 'the classes which as one are not members of themselves' (*Principles of Mathematics* (1903), 2nd edn. (London: Allen and Unwin, 1937), 102). In explaining the distinction of class as many and class as one, Russell appeals to the difference between plural and singular reference (ibid. 68–9); his use of the plural in deploying the former notion resembles that in Boolos's examples and in his reading of second-order formulae. It is of course an inference from certain pluralities failing to be unities to their failing to be objects. On this issue with reference to Cantor, see Parsons, *Mathematics in Philosophy*, 280–6.

A STRUCTURALIST VIEW

non-vacuity, seems to be solvable on a modal nominalist basis at least for arithmetic, and it will also be solvable for classical analysis if (like Field) one is willing to grant that the physical world instantiates in some way or other one of the standard Euclidean or non-Euclidean geometries. Even without this hypothesis, a lot of mathematics can be rescued: since, as we pointed out above, many basic results of analysis can be proved in rather weak theories, the modal nominalist solution to the problem of non-vacuity is available for a lot of everyday analysis. Moreover, the eliminative structuralist who admits second-order logic of course has second-order arithmetic at his disposal; that is, classical analysis with the real numbers interpreted as second-order entities rather than as objects.

The question remains how from this sort of point of view one might convince oneself of the possibility of more elaborate structures, either defined in higher than second-order terms or involving domains of objects of higher cardinalities. Any structuralist approach to set theory involves admitting the possibility of such structures; conversely, set theory allows rich possibilities of constructing models of theories and therefore of showing the possibility of structures where no more nominalist way of doing so is available.

Modalist eliminations of set existence assumptions have been proposed, such as that of Putnam in 'Mathematics without Foundations' (Ch. VIII) discussed briefly in Section 6. At least with the help of second-order modal logic, the technical part of such a programme can be carried out. But then, it seems to me, any such elimination faces a very simple and fatal difficulty. The non-vacuity assumption, (4) or perhaps some weaker alternative, will still have to be made. But then the concept of object in terms of which objects satisfying the required conditions are possible is a very general one; there is no reason to believe that structures of the required kinds are possible where the objects involved have any of the characteristic marks of concreteness. In cardinality, they will outstrip anything that can be represented in the physical world. One may, like Putnam, begin with a spatial conception that models some aspects of the concept of set, and then conclude that a *space* is possible whose cardinality is high enough so that it can contain a standard model of some set theory.[67] Here, I think, there is a slide between the notion of physical space and the notion of a space as a structure of the general sort considered in geometry. Putnam is, to be sure, not claiming that such a space is physically possible. What is in question is the mathematical possibility of a physical space satisfying certain conditions. If one goes beyond conditions of a geometric character, and the cardinality, what is to be added to make this possibility that of a *physical* space? And, more to the point, of what relevance would it be to the acceptability of set theory as mathematics?

[67] Ch. VIII; cf. Parsons, *Mathematics in Philosophy*, 192 n. 32.

It might be thought that matters are different if we suppose that there is possibly some structure of the *mind* that satisfies whatever conditions are required, or that there is some plausible weaker modal condition on operations of the mind whose truth is sufficient for an eliminative modalist interpretation of set theory. Many writers on the concept of set, who could appeal to remarks of Cantor himself, have given a central place to a mental operation of 'collecting' in their account of the concept. If the operation gives rise, by a kind of synthesis, to the set as a new object, it does not help in the present connection.[68] But the role of the object can perhaps be played by the operation itself.[69] Then, the idea goes, quantification over sets can be interpreted in terms of possibilities of collecting. Such a view, however, meets the same kind of difficulties. It is not only that we are asked to accept as mathematically possible mental operations which, because of their transfinite iteration, go well beyond the actual capabilities of the human mind. Such an extension of human capability, whether or not it is thought of mentalistically, is involved in the much more entrenched idea of computability or constructibility 'in principle'. Computations and at least the more elementary constructions, however, can at least be thought of spatio-temporally. But with collectings of a large transfinite number of objects, iterated through a large transfinite sequence, we need the possibility of at least a 'time' that would be much richer than actual space-time or any possible space-time that physical considerations would lead us to consider.[70] The concept of mental operation is thus extended by rather extravagant analogies. We may see the concept of object that allows such objects as large transfinite sets as an analogical extension of a concept of object that doubtless begins with the physical objects, organisms, and events of everyday experience. But particularly if it is understood so as to be purged of some of the pictures traditionally associated with platonism, this extension is limited to what the mathematics actually requires. It is hard to see what problem with mathematical objects is really helped by assuming the possibility of a physical space, or a system of mental operations, with the

---

[68] This was the proposal of Husserl in *Philosophie der Arithmetik* (1891), ed. Lothar Eley in *Husserliana*, vol. 12 (The Hague: Nijhoff, 1970), which, in spite of its psychologistic guise, is of particular historical interest because Husserl was Cantor's colleague at the time. Husserl evidently thought later that the basic idea was not essentially bound up with the psychologism that he had rejected, since it reappears in the late *Erfahrung und Urteil*, ed. Ludwig Landgrebe (Hamburg: Claassen, 1948; orig. Prague, 1939), sect. 61. The most immediate difficulty with Husserl's proposal seems to me to be not psychologism but how to make it apply to infinite sets.

[69] As is proposed by Philip Kitcher in *The Nature of Mathematical Knowledge* (New York: Oxford University Press, 1983), ch. 6, for whom, however, the operation of 'collecting' is not mental but an overt action.

[70] Compare the remarks on Wang in Parsons, *Mathematics in Philosophy*, essay 10, pp. 272–80, and on Kitcher in my review of *The Nature of Mathematical Knowledge*, *Philosophical Review*, 95 (1986): 133–4.

kind of structure higher set theory requires.[71] It is probably only the perhaps inchoate sense that there is some scandal to human reason in supposing that there are, or possibly there are, objects that stand in relations that the conception of a particular mathematical structure, such as a model of set theory, calls for, with perhaps nothing more to them than that (or only something more of a very abstract character) that leads philosophers to entertain such ideas. In my view, such an attitude should be turned on its head. We are able to understand higher set theory and to have enough of an intuition concerning the principles of set theory to create a highly developed theory, which shows no sign of being inconsistent or incoherent, and about which there is too much agreement for it to be *ad hoc*. If a 'formal' concept of object is the only one that can interpret set theory without assuming extravagant possibilities, that is a strong argument in favour of such a concept of object.

8. A NON-ELIMINATIVE STRUCTURALISM AND ITS LIMITS

Eliminative structuralism begins with the observation that there is in the end nothing to the pure abstract objects of mathematics than their being related in certain structures, and infers that talk of objects with so little of an intrinsic nature must be only a *façon de parler*. The conclusion of our discussion is that at least when we come to higher set theory, we have no reason to suppose even the possibility of objects that are more 'concrete' than pure mathematical objects. A structuralist view of higher set theory will then oblige us to accept the idea of a system of objects that is really no more than a structure. But then there is no convincing reason not to accept it in other domains of mathematics, in particular in the case of the natural numbers. It would be highly paradoxical to accept Benacerraf's conclusion that numbers are not objects and yet accept as such the sets of higher set theory.

Let us consider, however, whether sets should be an exception to a general structuralism about pure mathematical objects. The claim of the last section was only that the possibility of the kind of structure higher set theory commits us to could only be made out by domains of pure abstract objects. Does this imply a structuralist view of sets? Does the universe of sets, in other words, consist of more than a domain of 'objects' related in a relation called 'membership' satisfying conditions that can then be stated in the language

[71] In recent years, a favourite reason for seeking to eliminate mathematical objects has been the dilemma posed by Benacerraf in 'Mathematical Truth' (Ch. I above). It is easy to see that the kind of assumption considered here is of no help with Benacerraf's problem. First, only what is actual stands in causal relations to our knowledge; second, once we consider the sort of space or mental operation that is required, we no longer have any idea of what a causal theory of knowledge involving them would be like.

of set theory? Where individuals are involved, to be sure, the answer can be 'yes' for a relatively trivial reason, that the individuals might be identifiable independently and not be preserved in an isomorphic structure. That may, however, simply express an element of generality in the notion of the 'universe of sets': it is not a unique structure because set theory does not determine what is an individual, how many there are, or what structure they have other than belonging to the sets they do.

The more serious reason for thinking a structuralist view of set theory to be false comes from intuitions about sets of a more general ontological character. For example, there is the conception of a set as a totality 'constituted' by its elements, so that it stands in some kind of ontological dependence on its elements, but not vice versa. This would give to the membership relation some additional content, still very abstract but recognizably more than a pure structuralism would admit. The 'iterative conception of set' (a conception, not mainly of what a set is, but of what the 'universe of sets' is, and therefore of what sets there are) is usually explained with appeal to such general ontological conceptions, although also with appeal to ideas that cannot in the end be taken literally.

Whether these considerations show that an 'ontological' conception of the universe of sets should be substituted for the structuralist one is a serious issue which, however, I will have to leave for another occasion. It would still share many consequences with the structuralist view; sets would be 'incomplete' in the sense discussed below, and the distinction between possibility and actuality would not apply to their existence in the usual way. I wish here to focus on a more elementary qualification of the structuralist view that will emerge once we have finished the discussion of the formulation of the latter view.

If we abandon eliminative structuralism, we need no longer seek a canonical form for the languages of various mathematical theories that reinterprets them after the manner of the proposals we have been discussing. We can take the language of mathematics more at face value. If so, how is our position still structuralist? The main point is that taking the language of mathematics at face value does not require us to take more as objectively determined about the objects spoken of than that language itself specifies. This statement, however, is ambiguous, as we can illustrate by the case of arithmetic.

We might take as the language of arithmetic a language that has no reference to objects other than natural numbers. This will be, of course, a first-order arithmetic; it will contain '0' 'S', '=', and probably additional predicates or functors, such as '+' or '×'. Then its intended interpretation (or even an unintended one) transfers readily to any isomorphic copy of the natural numbers. Actual arithmetic, however, will generally not use such a restricted language. If it is applied, the numbers will number objects other than natural numbers. If it is pure, sets of numbers or functions with either domain or range contained in N are likely to be appealed to, or perhaps rational or real numbers, systems that themselves contain non-negative integers.

Our remarks in Section 2 about 'external relations' indicate why the existence of this more comprehensive linguistic usage does not force us to say that the natural numbers are more than structurally determined. When, however, the structuralist thesis says that talk about mathematical objects presupposes a background structure, what this structure is is context-dependent. 'Interstructural' relations in pure mathematics are generally better understood if we take the background structure to be sufficiently comprehensive to incorporate all the structures involved. It would certainly be counter-intuitive, for example, to interpret the relations of natural numbers to sets of numbers, of which an individual number may or may not be an element, as external relations in precisely the sense of Section 2, since that treatment takes the identity of the 'external' objects, in this case sets, to be fixed when one goes to an alternative realization of the structure of the numbers. This conflicts with an entrenched principle about sets, that they are determined by their elements. The more natural way of looking at things, obviously, is to suppose that once we begin talking of sets of numbers at all, our 'background structure' is one involving numbers *and* sets.[72]

Attribution of a 'background structure' to a mathematical discourse is not only context-dependent; it is also a matter of interpretation. For this reason, one should be cautious in making such assertions as that identity statements involving objects of different structures are meaningless or indeterminate. There is an obvious sense in which identity of natural numbers and sets is indeterminate, in that different interpretations of number theory and set theory are possible which give different answers about the truth of identities of numbers and sets. In a lot of ordinary mathematical discourse, where different structures are involved, the question of identity or non-identity of elements of one with elements of another just does not arise (even to be rejected). But of course some discourse about numbers and sets makes identity statements between them meaningful, and some of that (such as the standard development of arithmetic in axiomatic set theory) makes commitments as to the truth value of such identities. Thus it would be quite out of order to say (without reference to context) that identities of numbers and sets are meaningless or that they lack truth values.

---

[72] Such a structure is not uniquely determined, even given the underlying principles about sets. Informal mathematics would not settle whether to treat the natural numbers as individuals or to identify them with one or another sequence of sets. Thus, although '$2 = \{\phi, \{\phi\}\} \vee 2 \neq \{\phi, \{\phi\}\}$' will be forced to be true (unless one thinks of the matter type-theoretically), there is room for convention as to which disjunct is true. For most mathematical purposes, moreover, one does not need to make a convention.

The conception of mathematical objects expressed here has sometimes been expressed by saying that mathematical objects are 'incomplete', meaning by this that there is only a certain specific range of predicates such that there is a fact of the matter as to whether they are true of the object in question. As a description of the situation, this is itself incomplete, for it neglects the fact that the *relations* of a structure are themselves given only by formal conditions, and in different realizations of a type of structure, not only will there be different domains of objects, but these relations will have different realizations. The 'incompleteness' of the pure abstract objects of mathematics is for this reason more radical than that of another kind of object often said to be incomplete: fictional or non-existent objects (assuming, for the sake of argument, that there are such objects). Fictional objects are taken to be undetermined with respect to properties and relations whose holding or not cannot be reasonably inferred from the story; a more drastic incompleteness obtains for Meinongian objects like the golden mountain, since what we have to go on about it is only that it is a mountain and that it is golden. In these situations, however, we do not envisage any reinterpretation of the predicates applied to the objects; Sherlock Holmes is a detective in a sense that we can take to be fixed, also when we consider other detectives (real or fictitious). We have, independently of the story, an understanding of notions such as that of a detective, of a murder, of London, of Baker Street (since these are real places). There is at least some level of understanding of this kind of simple mathematical notions like addition, multiplication, or set membership, and of more complex ones such as curve or surface or computation. On a purely structuralist view, however, none of these notions is fixed in a way in which the non-fictional vocabulary used to describe a fictional situation is. Their role in mathematics, rather, is in the genesis and motivation of mathematical conceptions, and in the application of mathematics.

The absence of notions whose non-formal properties really matter, even more than incompleteness or the absence of a non-relational characterization of the nature of any object, makes mathematical objects on the structuralist view continue to seem elusive, and encourages the belief that there is some scandal to human reason in the idea that there are such objects. My claim is that something close to the conception of objects of this kind, already encouraged by the modern development of arithmetic, geometry, and algebra, is forced on us by higher set theory. The reason why I do not outrightly say that such objects are forced on us is the question of the status of the general ontological conceptions mentioned above in connection with set theory.

Let us now turn to the concept of structure itself. I have resisted the interpretation of structuralism that would make it an interpretation of mathematical statements as *about* structures, thus giving rise to the question what manner of objects these are. On the contrary, a structuralist account of a particular kind of mathematical object does not view statements about that kind of object as about structures at all (except in the special case where structures are themselves mathematical objects, as in model theory). Still, a concept of structure is needed to state the view itself. The most fundamental notion of structure for this purpose is metalinguistic: the 'domain' is given by a predicate, and the relations and functions by further predicates and functors.[73] The most concrete way of giving a structure would be by predicates and functors that are antecedently understood, so that there is some independent verification of fundamental propositions about the structure (say, axioms for the theory of this kind of structure). Brouwer and Hilbert could be taken to have been attempting something like this in their descriptions of the intuitive basis of arithmetic. A less problematic, but also less philosophically interesting, case is where a type of structure is defined in the abstract and then examples are given from different branches of mathematics, but in this case, the 'realizations' of one type of structure are simply found in another structure.

We can now see the difference between arithmetic and set theory in a different light: in the case of arithmetic, a realization of the structure can be described in the more concrete way by giving an intuitive model of which the most developed example is Hilbert's strings of strokes. In the latter case, the predicate 'x is a string of strokes' defines the domain, and the zero element and successor functor can also be directly explained.[74] Induction can be stated, at the cost of vagueness, in the metalinguistic way mentioned in the discussion of second-order logic in Section 6 above.[75] Tait's idea of Dedekind abstraction (Section 2 above) can be adapted to give a convincing description of discourse about *the* natural numbers.[76]

Many will not agree with this picture of number theory according to which the objective reality of the natural numbers can be made out by an example in which the objects are quasi-concrete. However that may be, the situation in

---

[73] By a functor I simply mean a singular term with one or more argument places; hence the use of a functor does not of itself commit one to the existence of any function.

[74] Cf. Ch. V, where also the claim is defended that the elementary Peano axioms are intuitively known to be true in this model.

[75] Cf. my 'The Impredicativity of Induction', in Leigh S. Cauman, Isaac Levi, Charles Parsons, and Robert Schwartz (eds.), *How Many Questions: Essays in Honor of Sidney Morgenbesser* (Indianapolis: Hackett, 1983), 132–53, where some difficulties about knowing induction to be true are also mentioned.

[76] Since first-order arithmetic is not categorical, it might seem that impredicative second-order logic is needed for the proof of Dedekind's categoricity theorem. That is not true. To show that two predicates N_1 and N_2 define isomorphic domains, it is sufficient that for each, we can apply induction to first-order predicates containing the other.

Tait, using an idea of F. W. Lawvere, points out that the categoricity theorem also holds in an intuitionistic setting. See 'Against Intuitionism: Constructive Mathematics is Part of Classical Mathematics', *Journal of Philosophical Logic*, 12 (1983): 173–95, esp. 177 and n. 12.

set theory is different, and our understanding of set theory has to proceed by pulling ourselves up by our bootstraps. In its developed form in set theory, the concept of set draws on two elementary notions, neither of which is a purely structural one: that of a set as a 'collection' or 'totality' of its elements, and that of the extension of a predicate, that is, of an object associated with a predicate, with extensional identity conditions. Although both these ideas would allow the construction of in some sense intuitive models, the first one seems to me to depart from concrete intuition at least when it admits infinite sets, the second when it admits impredicatively defined sets. The result of these extensions, however, is that the elements of the original ideas that are unquestionably preserved in the theory have a purely formal character. For example, the priority of the elements of a set to the set, which is usually motivated by appealing to the first of these two informal conceptions, is reflected in the theory itself by the fact that membership is a well-founded relation. It is important to our understanding of set theory and to the objective reality of sets, however, that for the bottom of the set-theoretic hierarchy we do have more intuitive models.

In this situation, what does the metalinguistic notion of structure mean? We can define 'structures' to interpret the language of set theory by using our mathematical vocabulary, including the predicates '( ) is a set' and '( ) is an element of [ ]'. Without such vocabulary, or other mathematical vocabulary or similar abstraction, we will not be able to describe a structure satisfying the axioms of set theory; that is the 'bootstrapping' aspect of the understanding of set theory. But otherwise, the metalinguistic conception of structure works in the same way as it does in the more elementary cases.

Why should I regard the metalinguistic conception of structure as more appropriate for my purpose than the set-theoretic? I have just denied the most obvious reason, that it would enable one to describe structures for set theory independent of the concept of set. Indeed, the 'bootstrapping' involved in the understanding of the concept of set weakens resistance to the use of the set-theoretic concept of structure here. What decides in favour of the metalinguistic conception in the present context is a feature of set theory itself: we want to talk of structures for set theory without supposing that their domains are sets. Also, the set-theoretic conception rests on a commitment as to the ontology of set theory itself (say, in favour of ZF rather than a von Neumann-style formulation), which the most fundamental conception of structure should not make.

Although I have presented what I consider to be a defensible version of the structuralist view, an outcome of our investigation is that structuralism is not the whole truth about mathematical objects. In the statement of the view using the metalinguistic conception of structure, we appeal to linguistic objects such as predicates and functors. These are quasi-concrete objects, and so long as they are viewed in this way, the structuralist view will not hold for them. The relation of linguistic types to their tokens (and in general of quasi-concrete objects to their concrete 'representations') is not an external relation in the sense of Section 2.

It will be objected that any mathematical theory, that of linguistic objects included, can be interpreted as talking about objects for which the structuralist view holds. With regard to mathematical structuralism based on the metalinguistic conception of structure, the proposal is that syntax be viewed in this way, with the notions of string and concatenation, and perhaps others, as basic relations. There will be vagueness about this, but not necessarily more so than on the construal I have proposed; with regard to the domain of objects over which predicates are interpreted, such vagueness is unavoidable.[77] I also don't think it can be objected to on the ground of circularity.

I am not prepared to argue that unifying in this way the concept of structure used in stating the structuralist view with other reference to objects in mathematics is wrong. The philosophical gain it achieves, however, is only apparent. It applies in this case as well the transition in the development of mathematics from dealing with domains of a more concrete nature to speaking of objects only in a purely structural way. But this transition leaves a residue. The more concrete domains, often of quasi-concrete objects, still play an ineliminable role in the explanation and motivation of mathematical concepts and theories. In particular, this is true of any mathematical treatment of formalized or natural languages. Thus, if the structuralist view of mathematical objects is taken to mean that all mathematical objects are only structurally determined, it has to rest on legislation about what counts as a mathematical object. The explanatory and justificatory role of more concrete models implies, in my view, that it is not the right legislation even for the interpretation of modern mathematics.[78]

---

[77] In response to the difficulty I state in Section 2 above, Resnik considers a theory in which structures themselves are considered directly as 'only structurally determined' objects ('Mathematics as a Science of Patterns', 538–9). The interpretation of this theory will give rise to the same systematic ambiguities. My preference for the metalinguistic conception rests in part on the fact that on it such systematic ambiguity arises at a place, with the notion of truth, where it will arise anyway.

[78] This essay is based on lectures given to meetings of the Society for Exact Philosophy in Toronto, 17 May 1985, and of the Association for Symbolic Logic in Washington, DC, 29 Dec. 1985. Some of the material had been presented in lectures at Dartmouth College in 1981 and at the University of Padua in 1983. I am indebted to comments from all four audiences, and also to correspondence with Michael Resnik and especially W. W. Tait. This essay was written while I was at Columbia University, to which I and it owe much. Support of the John Simon Guggenheim Memorial Foundation is gratefully acknowledged.

NOTES ON THE CONTRIBUTORS

PAUL BENACERRAF is Professor of Philosophy at Princeton University.
GEORGE BOOLOS is Professor of Philosophy at the Massachusetts Institute of Technology.
MICHAEL DUMMETT was the Wykeham Professor of Logic at Oxford until his retirement.
HARTRY FIELD is Professor of Philosophy at the Graduate School and University Center of the City University of New York.
W. D. HART is Professor of Philosophy at the University of Illinois at Chicago.
DANIEL ISAACSON is a Fellow of Wolfson College, Oxford.
PENELOPE MADDY is Professor of Philosophy at the University of California at Irvine.
CHARLES PARSONS is Professor of Philosophy at Harvard University.
HILARY PUTNAM is Professor of Philosophy at Harvard University.
W. V. QUINE is Professor Emeritus of Philosophy at Harvard University.
STEWART SHAPIRO is Professor of Philosophy at Ohio State University, Newark branch.
W. W. TAIT is Professor of Philosophy at the University of Chicago.

SUGGESTIONS FOR FURTHER READING

BOOKS

Barker, Stephen F., *Philosophy of Mathematics* (Englewood Cliffs, NJ: Prentice-Hall, 1964).

Barrett, R., and Gibson, R. (eds.), *Perspectives on Quine* (Oxford: Blackwell, 1990).

Benacerraf, P., and Putnam, H. (eds.), *Philosophy of Mathematics: Selected Readings*, 2nd. edn. (Cambridge: Cambridge University Press, 1983).

Beth, Evert W., *Mathematical Thought* (Dordrecht: Reidel, 1965).

Bishop, Errett, *Foundations of Constructive Analysis* (New York: McGraw-Hill, 1967), ch. 1 and appendix B.

Black, Max, *The Nature of Mathematics* (New York: Littlefield, Adams and Co., 1959).

Bostock, David, *Logic and Arithmetic: Natural Numbers* (Oxford: Clarendon Press, 1974).

——, *Logic and Arithmetic: Rational and Irrational Numbers* (Oxford: Clarendon Press, 1979).

Carnap, Rudolf, *Foundations of Logic and Mathematics* (Chicago: University of Chicago Press, 1939).

——, *Meaning and Necessity*, enlarged edn. (Chicago: University of Chicago Press, 1956), supplement.

Castonguay, Charles, *Meaning and Existence in Mathematics* (Berlin: Springer, 1972).

Chihara, Charles S., *Constructibility and Mathematical Existence* (Oxford: Clarendon Press, 1990).

——, *Ontology and the Vicious Circle Principle* (Ithaca, NY: Cornell University Press, 1973).

Curry, Haskell B., *Outlines of a Formalist Philosophy of Mathematics* (Amsterdam: North Holland, 1958).

Dedekind, Richard, *Essays on the Theory of Numbers* (New York: Dover, 1963).

Dummett, Michael, *Elements of Intuitionism* (Oxford: Clarendon Press, 1977).

——, *Frege and Other Philosophers* (Oxford: Clarendon Press, 1991).

——, *Frege: Philosophy of Language* (London: Duckworth, 1973).

——, *Frege: Philosophy of Mathematics* (London: Duckworth, 1991).

——, *The Interpretation of Frege's Philosophy* (London: Duckworth, 1981).

——, *Truth and Other Enigmas* (London: Duckworth, 1978).

Field, Hartry, *Realism, Mathematics and Modality* (Oxford: Blackwell, 1989).

——, *Science without Numbers* (Oxford: Blackwell, 1980) (reviewed in articles by Friedman, Malament, Manders, Resnik, and Shapiro).

Frege, Gottlob, *The Foundations of Arithmetic*, trans. J. L. Austin (Oxford: Blackwell, 1959).

Goodstein, R. L., *Essays in the Philosophy of Mathematics* (Leicester: Leicester University Press, 1965).

Gottlieb, Dale, *Ontological Economy: Substitutional Quantification and Mathematics* (Oxford, 1980).

Hahn, L. E., and Schilpp, P. A. (eds.), *The Philosophy of W. V. Quine* (La Salle, Ill.: Open Court, 1986).

Hale, Bob, *Abstract Objects* (Oxford: Blackwell, 1987).
Hallett, Michael, *Cantorian Set Theory and Limitation of Size* (Oxford: Clarendon Press, 1984).
Hellman, Geoffrey, *Mathematics without Numbers: Towards a Modal-Structural Interpretation* (Oxford: Clarendon Press, 1989).
Heyting, Arendt, *Intuitionism: An Introduction*, 3rd edn. (Amsterdam: North Holland, 1971).
Kielkopf, Charles F., *Strict Finitism* (The Hague: Mouton, 1970).
Kitcher, Philip, *The Nature of Mathematical Knowledge* (New York: Oxford University Press, 1984).
Körner, Stephan, *The Philosophy of Mathematics* (New York: Harper, 1960).
Lakatos, Imre, *Mathematics, Science and Epistemology*, ed. G. Currie and J. Worrall (Cambridge: Cambridge University Press, 1978).
——, *Proofs and Refutations*, ed. J. Worrall and E. Zahar (Cambridge: Cambridge University Press, 1976).
Lakatos, Imre (ed.), *Problems in the Philosophy of Mathematics* (Amsterdam: North Holland, 1967).
Lehman, Hugh, *Introduction to the Philosophy of Mathematics* (Oxford: Blackwell, 1979).
Maddy, Penelope, *Realism in Mathematics* (Oxford: Clarendon Press, 1990).
Moore, A. W. *The Infinite* (London: Routledge, 1990).
Parsons, Charles, *Mathematics in Philosophy: Selected Essays* (Ithaca, NY: Cornell University Press, 1983).
Poincaré, Henri, *Mathematics and Science: Last Essays* (New York: Dover, 1963).
——, *Science and Hypothesis* (New York: Dover, 1952).
——, *The Value of Science* (New York: Dover, 1958).
Putnam, Hilary, *Mathematics, Matter and Method, Philosophical Papers*, vol. 1 (Cambridge: Cambridge University Press, 1975).
——, *Philosophy of Logic* (New York: Harper, 1971).
——, *Realism and Reason*, in *Philosophical Papers*, vol. 3 (Cambridge: Cambridge University Press, 1983).
Quine, W. V., *From a Logical Point of View*, 2nd edn. (Cambridge, Mass.: Harvard University Press, 1961).
——, *Ontological Relativity and Other Essays* (New York: Columbia University Press, 1969).
——, *Philosophy of Logic* (Cambridge, Mass.: Harvard University Press, 1986).
——, *Pursuit of Truth* (Cambridge, Mass.: Harvard University Press, 1990).
——, *The Roots of Reference* (La Salle, Ill.: Open Court, 1973).
——, *Set Theory and its Logic*, rev. edn. (Cambridge, Mass.: Harvard University Press, 1969).
——, *Theories and Things* (Cambridge, Mass.: Harvard University Press, 1981).
——, *The Ways of Paradox and Other Essays*, rev. and enlarged edn. (Cambridge, Mass.: Harvard University Press, 1976).
——, *Word and Object* (Cambridge, Mass.: MIT Press, 1960).
Ramsey, Frank, *Philosophical Papers*, ed. D. H. Mellor (Cambridge: Cambridge University Press, 1990).
Resnik, Michael D., *Frege and the Philosophy of Mathematics* (Ithaca, NY: Cornell University Press, 1980).
Russell, Bertrand, *Essays in Analysis*, ed. D. Lackey (New York: George Braziller, 1973).

Russell, Bertrand, *Introduction to Mathematical Philosophy* (London: George Allen and Unwin, 1919).
——, *Logic and Knowledge*, ed. R. Marsh (London: George Allen and Unwin, 1956).
——, *The Principles of Mathematics*, 2nd edn. (London: George Allen and Unwin, 1937).
Shapiro, Stewart, *Foundations without Foundationalism* (New York: Oxford University Press, 1991).
Steiner, Mark, *Mathematical Knowledge* (Ithaca, NY: Cornell University Press, 1975).
Troelstra, A. S., *Principles of Intuitionism* (Berlin: Springer, 1969).
Wang, Hao, *From Mathematics to Philosophy* (London: Routledge and Kegan Paul, 1974).
Weyl, Hermann, *Philosophy of Mathematics and Natural Science* (London: Atheneum, 1963).
Wittgenstein, Ludwig, *Lectures on the Foundations of Mathematics*, ed. Cora Diamond (Brighton: Harvester, 1976).
——, *Remarks on the Foundations of Mathematics*, trans. E. Anscombe, ed. G. von Wright, R. Rhees, and E. Anscombe, 3rd edn. (Oxford: Blackwell, 1978).
Wright, Crispin, *Frege's Conception of Numbers as Objects* (Aberdeen: Aberdeen University Press, 1983).
——, *Wittgenstein on the Foundations of Mathematics* (London: Duckworth, 1980).

ARTICLES

Boolos, George, 'Nominalist Platonism', *Philosophical Review*, 94 (1985): 327–44.
——, 'To Be is To Be a Value of a Variable (or To Be Some Values of Some Variables)', *Journal of Philosophy*, 81 (1984): 430–49.
Burgess, John P., 'Dummett's Case for Intuitionism', *History and Philosophy of Logic*, 5 (1984): 177–94.
——, 'Synthetic Mechanics', *Journal of Philosophical Logic*, 13 (1984): 379–95.
——, 'Why I Am Not a Nominalist', *Notre Dame Journal of Formal Logic*, 24 (1983): 93–105.
Castañeda, Hector-Neri, 'On Mathematical Proofs and Meaning', *Mind*, 70 (1961): 385–90.
Ewing, A. C., 'The Linguistic Theory of *A Priori* Propositions', in H. D. Lewis (ed.), *Clarity is Not Enough* (London: George Allen and Unwin, 1963), 147–69.
Friedman, Michael, review of Field's *Science without Numbers*, *Philosophy of Science*, 48 (1981): 505–6.
Gasking, Douglas, 'Mathematics and the World', repr. in A. Flew (ed.), *Logic and Language*, 2nd ser. (Oxford: Blackwell, 1961), 204–21.
Goldfarb, Warren D., 'Logic in the Twenties: The Nature of the Quantifier', *Journal of Symbolic Logic*, 44 (1979): 351–68.
Hanson, Norwood Russell, 'Number Theory and Physical Theory', in P. R. Cohen and M. Wartofsky (eds.), *Boston Studies in the Philosophy of Science* (Atlantic Heights, NJ: Humanities Press, 1965), 93–120.
Hardy, G. H., 'Mathematical Proof', *Mind*, 38 (1929): 1–25.
Hart, W. D., 'On an Argument for Formalism', *Journal of Philosophy*, 71 (1974): 29–46.
Hodes, Harold, 'Logicism and the Ontological Commitments of Arithmetic', *Journal of Philosophy*, 81 (1984): 123–49.

Jubien, Michael, 'Ontology and Mathematical Truth', *Noûs*, 11 (1977): 133–50.
Kitcher, Philip, 'The Plight of the Platonist', *Noûs*, 12 (1978): 119–36.
Kneale, William, 'Are Necessary Truths True by Convention?', in H. D. Lewis (ed.), *Clarity is Not Enough* (London: George Allen and Unwin, 1963), 133–46.
Kneebone, G. T., and Mackie, J. L., symposium on proof, *Proceedings of the Aristotelian Society*, suppl. vol. 40 (1966): 23–46.
Lorenzen, Paul, 'Constructive and Axiomatic Mathematics', *Synthese*, 12 (1960): 114–19.
Maddy, Penelope, 'Believing the Axioms', *Journal of Symbolic Logic*, 53 (1988): 481–511, 736–64.
——, 'Proper Classes', *Journal of Symbolic Logic*, 48 (1983): 113–39.
Malament, David, review of Field's *Science without Numbers*, *Journal of Philosophy*, 79 (1982): 523–34.
Manders, Kenneth L., review of Field's *Science without Numbers*, *Journal of Symbolic Logic*, 49 (1984): 303–6.
Mehlberg, Henryk, 'The Present Situation in the Philosophy of Mathematics', *Synthese*, 12 (1960): 380–414.
Myhill, John, 'Some Philosophical Implications of Mathematical Logic', *Review of Metaphysics*, 6 (1952): 165–98.
——, 'Some Remarks on the Notion of Proof', *Journal of Philosophy*, 57 (1960): 461–71.
Parsons, Charles, 'The Impredicativity of Induction', in L. S. Cauman, I. Levi, C. Parsons, and R. Schwartz (eds.), *How Many Questions* (Indianapolis: Hackett, 1983), 132–53.
——, 'The Uniqueness of the Natural Numbers', *Iyyun*, 39 (1990): 13–44.
Pollock, John L., 'Mathematical Proof', *American Philosophical Quarterly*, 4 (1967): 238–44.
Resnik, Michael, 'Mathematics as a Science of Patterns', *Noûs*, 15 (1981): 529–50.
——, review of Field's *Science without Numbers*, *Noûs*, 27 (1983): 514–19.
Shapiro, Stewart, review of Field's *Science without Numbers*, *Philosophia*, 14 (1984): 437–44.
Skolem, Thoralf, 'The Logical Nature of Arithmetic', *Synthese*, 9 (1955): 375–84.
Tait, W. W., 'Against Intuitionism', *Journal of Philosophical Logic*, 12 (1983): 173–95.
——, 'Finitism', *Journal of Philosophy*, 78 (1981): 524–46.
Tymoczko, Thomas, 'The Four–Color Problem and its Philosophical Significance', *Journal of Philosophy*, 76 (1979): 57–83.
White, N. P., 'What Numbers Are', *Synthese*, 27 (1974): 111–24.

INDEX OF NAMES

Not including Suggestions for Further Reading

Ackermann, Wilhelm 212
Anselm 235
Aristotle 35, 164 n.
Armstrong, David 122

Belnap, Nuel 256
Benacerraf, Paul 1, 3, 5, 7, 10, 11, 13, 99 n., 110, 115, 118, 144, 147, 148–9, 166, 201 n., 271 n., 275 n., 301
Berkeley, George 168, 169
Bernays, Paul 181, 270
Boole, George 67, 68
Boolos, George 12, 294–8
Brouwer, L. E. J. 9, 63, 88, 106–7, 305
Burgess, John 118 n., 188–90

Camp, Joseph 256
Cantor, Georg 278 n., 295, 300
Carnap, Rudolf 6, 31 n., 34, 35, 38 n., 40 n., 41, 45, 46, 50 n., 51, 151 n., 181, 268
Chihara, Charles 6 n., 108 n., 135–41, 150, 156, 252 n.
Church, Alonzo 31 n.
Cocciarella, Nino 268–9 n.
Cohen, Paul 139

Davidson, Donald 7, 31 n.
Dedekind, Richard 49, 93, 156 n., 165 n., 205–11, 273–9
Descartes, René 96, 235
Detlefsen, Michael 252 n.
Duhem, Pierre 6, 52, 53
Dummett, Michael 7, 8, 9, 10, 54, 57, 58, 59, 113 n., 143, 147–66, 211

Edgington, Dorothy 55 n.
Einstein, Albert 12
Euclid 2, 12, 60

Feferman, Solomon 209
Field, Hartry 11, 12, 29 n., 108 n., 152 n., 225–34, 284, 290 n., 292–4
Fitch, F. B. 55 n.
Fraenkel, Adolf 134
Frege, Gottlob 7, 8, 12, 15 n., 32, 77, 127, 157, 185–202, 205 n., 237 n., 244–5, 273 n., 280, 291–2, 298
Friedman, Harvey 215, 219–20

Gettier, Edmund 4, 134
Gödel, Kurt 10, 25, 26, 56, 60, 95–6, 113, 114–26, 133, 135–41, 146, 147, 174, 177, 203, 212–15
Goldman, Alvin I. 23 n., 118, 122, 134–5
Goodman, Nelson 31 n., 56 n., 104 n.
Goodstein, R. L. 215–17
Gottlieb, Dale 253 n.
Grice, Paul 4, 61, 121
Grover, Dorothy 256

Hacking, Ian 6 n.
Harrington, Leo 217–19
Harman, Gilbert 6, 23 n., 56 n., 116
Hart, W. D. 4 n., 10 n.
Hellman, Geoffrey 13 n.
Hempel, Carl 54
Heyting, Arend 63
Hilbert, David 15, 16, 17, 21, 27, 67, 68, 69, 97 n., 103, 179, 210, 212–15, 227, 271, 281 n., 305
Hintikka, Jaako 97 n., 181, 269 n.
Hodes, Harald 188, 280, 296
Hume, David 4, 12, 31, 44, 45, 47, 60, 188–96
Husserl, Edmund 97–9, 105, 300 n.

Isaacson, Daniel 10

INDEX

Kant, Immanuel 9, 12, 13, 31–2, 97, 99, 102, 107, 164 n., 201, 235–7
Kaplan, David 62
Kirby, Laurie 216–17
Kitcher, Philip 13 n., 165 n., 271 n., 300 n.
Kleene, S. C. 245 n.
Kline, Morris 174
Kreisel, Georg 76, 81, 139 n.
Kripke, Saul 15 n., 115, 117, 181, 241–2, 258, 266–8, 288

Lear, Jonathan 115, 117
Leibniz, Gottfried 31, 34, 99, 164 n.
Levi, Isaac 284 n.
Lewis, C. I. 38 n., 44 n., 51
Loar, Brian 62
Locke, John 44, 45, 47, 106

McCarthy, Timothy 288 n., 291 n.
McGinn, Colin 57, 59
Malament, David 226 n., 231 n.
Maddy, Penelope 3 n., 7, 147
Mill, John Stuart 174
Montague, Richard 269 n.
Mostowski, Andrzej 139

Paris, Jeff 215–19
Parsons, Charles 9, 13, 147, 193 n.
Peirce, C. S. 5, 44
Pitcher, George 122
Plato 1, 26, 164 n.
Poincaré, Henri 291
Polya, George 87
Putnam, Hilary 6, 10, 11, 21 n., 101–2 n., 115, 153 n., 225, 231, 240 n., 280–3, 284, 286, 288 n., 291, 299

Quine, W. V. 5, 6, 7, 11, 27, 28, 29, 52–4, 56, 57, 60–2, 66, 101, 115, 150 n., 174, 181, 207, 225, 231 n., 270, 288 n.

Rabinowicz, Wlodzimierz 55 n.
Ramsey, Frank 217–19, 269 n.
Reichenbach, Hans 170
Resnik, Michael 252 n., 264 n., 270–307, 281 n.
Rheinwald, Rose-Marie 282 n., 287 n.
Russinoff, I. S. 185 n.
Russell, Bertrand 12, 30, 32, 45, 47, 49, 57, 62, 137, 169, 183, 185–6, 198–200, 291, 295, 298 n.
Ryle, Gilbert 61

Scott, Dana 245 n., 269
Shapiro, Stewart 11
Skyrms, Brian 23 n.
Steiner, Mark 4 n., 24 n., 96 n., 115 n., 158, 234 n.

Tait, W. W. 9, 10, 276 n., 305 n.
Tarski, Alfred 2, 6, 18, 19, 21, 28, 29, 31 n., 58, 80, 81, 85, 86, 144 n., 148–9, 258
Tooke, John Horne 44
Troelstra, A. S. 81, 84

von Neumann, John 100 n., 181, 197, 212

Wang, Hao 211, 294 n.
Weyl, Hermann 97
White, Morton 31 n.
Wilkie, Alex 208–9
Williamson, Timothy 55 n.
Wittgenstein, Ludwig 72, 145, 149, 150, 155, 158, 159
Wright, Crispin 3, 189, 193 n.

Zermelo, Ernst 100 n., 134, 181, 182, 210, 274, 278 n.